Population Genetics in Animal Breeding

Second Edition

Population Genetics in Animal Breeding

Second Edition

Franz Pirchner
Munich University of Technology
Freising–Weihenstephan, Federal Republic of Germany

Translated from German
with the assistance of D. L. Frape

PLENUM PRESS • NEW YORK AND LONDON

Library of Congress Cataloging in Publication Data

Pirchner, Franz.
    Population genetics in animal breeding.

    Translation of: Populationsgenetik in der Tierzucht.
    Bibliography: p.
    Includes index.
    1. Population genetics. 2. Livestock — Breeding. 3. Animal genetics. I. Title.
QH455.P5613 1983   636.08'21   83-2164
    ISBN 0-306-41201-2

4, 743

This volume is a revised and expanded translation of the second
edition of *Populationsgenetik in der Tierzucht: Eine Einführung
in die theoretischen Grundlagen* by Franz Pirchner, published in
1979 by Paul Parey Verlag, Hamburg and Berlin. German edition
© 1964 and © 1979 Paul Parey Verlag, Hamburg and Berlin.

© 1983 Plenum Press, New York
A Division of Plenum Publishing Corporation
233 Spring Street, New York, N.Y. 10013

Printed in the United States of America

Preface

This book attempts to outline population genetics and quantitative genetics as they pertain to animal breeding and to discuss the theoretical aspects of this field of agricultural activity. Therefore, it brings into focus the basic principles of animal breeding, which are illustrated with pertinent examples; however, it is not intended to give recommendations for particular situations.

Since the first edition, considerable development has occurred both in the basic and in the more applied fields. This has modified and in some cases even changed previously held conceptions, necessitating a thorough revision of the first edition.

The extent of work in this sphere has reached dimensions which preclude exhaustive discussion of all its aspects in a volume of this size. Nevertheless it is hoped that this introductory text will stimulate the reader to explore the subject in greater depth and inspire study of the original literature.

It is further hoped that my teaching experience has had some noticeable impact on style and presentation. I owe much to constructive critical comments on the first edition. I am grateful to Dr. D. L. Frape for his help in changing my own translation into readable English. M. Asbeck and E. Fuchshuber have completed an admirable job in typing my handwritten script and A. Pickal accomplished the careful drawing of fresh illustrations.

FRANZ PIRCHNER

Weihenstephan

Contents

1

Genetic Structure
of Populations

An objective of scientific breeding is the exploitation of inherited variability so that the performance of domestic animals and cultivated plants can be improved. More generally, it attempts to change the inherited characteristics of populations of domestic animals in a desirable direction. Population genetics describes the genetic structure of a population and the laws which govern changes in it. Therefore, an understanding of the processes which underlie breeding involves a knowledge of population genetics. Such a knowledge allows a rational approach to breeding, which should lead to greater success than an entirely empirical approach.

Inherited variation among animals is mediated largely by changes in genes. However, it should not be overlooked that chromosomal variation is responsible for no small proportion of inherited differences. Chromosomal variability is caused by differences in chromosome number, by deficiencies and duplications, where parts of chromosomes are lacking or are duplicated, and by changes in the arrangement of genes on chromosomes, i.e., by translocations and inversions. Chromosomal variability is recognized when changes are discernible, i.e., visible by light microscope. Smaller changes, i.e., those on a submicroscopic level, add to the genetic variability in the narrow sense. Therefore there is no fixed boundary between chromosomal and genetic variability. The extent of chromosomally caused variability in humans is impressive, but this has become apparent only recently, with the application of improved methods of preparation. It is estimated that between 0.15 and 0.20% of all newborn babies carry trisomy 21, which causes mongoloidy. About 0.5% of newborn children possess a detrimental aneuploidy and about 7% have chromosomal aberrations, i.e., aneuploidies and structural changes, without immediate consequences to health. The proportion of chromosomal aberrations in miscarriages and lost embryos is much larger and estimated to be about 40%. Although for domestic animals investigations comparable in thoroughness and extent to those for humans are not available, those that have been made indicate a similar situation (Fechheimer, 1977). Therefore the importance of chromosomal aberrations as a cause of apparent infertility in domestic animals should not be overlooked. The introduction of techniques such as chromosome banding will yield fresh insights, and chromosomal variability may assume even more importance as an explanation of inborn differences in fertility, health, and performance than it has now.

Cytoplasmic inheritance, via so-called plasmons, in contrast to that by means of the genome, which encompasses all genetic inheritance, causes matrocline inheritance, i.e., the progeny resembles the dam more than the sire. In general it cannot be differentiated from maternal influences where the maternal environment affects the fetus and the young animal. Maternal influences are important for young animals and in general recede with increasing age. These influences are discussed in more detail in Section 6.4.

Environmental effects are the subject of the field of animal nutrition and management and therefore will not be treated in any detail in this discussion. In animal breeding, environmental influences have a masking effect, in that the action of the genotype is camouflaged. Therefore, environmental influences are important, albeit in a somewhat negative sense, in the recognition of the effects of the genotype. Also, the manifestation of the genotype depends in many instances in a special way on the environment in which it performs. This phenomenon is denoted genotype–environment interaction and is discussed in Chapter 8. In the case of some threshold characteristics a particular environment is required for the manifestation of a trait. For example, disease resistance can only be expressed in the presence of disease (Chapter 13).

The potential for genetic variability is truly enormous: the presence of n loci with two alleles each permits 3^n possible genotypes, and multiple allelic loci where m alleles occur at a locus i lead to $\pi_i m_i(m_i + 1)/2$ genotypes. More than 20 years ago Stormont (1958) reported the blood-group alleles of cattle as given in Table 1.1. These figures permit more than 5400×10^9 different genotypes. At present, world cattle number some 1.2×10^9. If the presently known alleles of enzyme and protein loci

TABLE 1.1
Blood Groups in Cattle[a]

System	Number of phenogroups	Number of genotypes	Number of recognizable phenotypes
A–H	6	21	6
B	164	12,530	10,000
C	35	630	200
D	2	3	2
F–V	2	3	3
J–Oc	4	6	4
L	2	3	2
M	3	6	3
S–V	6	21	15
Z	2	3	3
Z'	2	3	2

[a] Source: Stormont (1958).

such as transferrins, hemoglobins, and caseins are added, the number of possible combinations is increased several tenthousandfold. Therefore, the potential variability is huge and the probability of identical animals is practically nil. The exceptions are products of one zygote, such as monozygous twins.

Only polymorphic loci were used for these calculations. Biochemical genetics has demonstrated in a rather direct way that between 30 and 50% of all structural loci are polymorphic in humans (Harris, 1966), *Drosophila* (Lewontin and Hubby, 1966), and mice (Selander, 1976). Zwiauer (1975) compiled data from the literature and it appears that in populations of domestic animals a similar proportion of loci are polymorphic. Nozawa *et al.* (1978) found nine of 27 loci to be polymorphic in Japanese island goats.

The development of molecular genetics led to a change in the conception of the gene. When population genetics was founded the gene was considered to be the unit of mutation, recombination, and function, which, of course, is no longer tenable. Nevertheless, the following presentation implies largely the classic conception of the gene. However, it has to be understood as the smallest unit of the chromosome that segregates in the course of an experiment or during the time span of a breeding project. In this sense, the whole casein complex, which involves the closely linked α_{S1}-, α_{S2}-, β-, and κ-casein loci, or the whole *B* blood-group complex, may be considered as one gene which influences performance. Of course, molecular genetics has had a large impact on theoretical population genetics and evolutionary genetics, where it has led to renewed discussion about the importance for evolution of natural selection and genetic drift.

Population genetics inquires into the genetic structure of populations, i.e., the frequencies of genes and genotypes, and the rules that govern their changes. It is a logical development of the Mendelian rules. A Mendelian population is understood as a group of individuals within which mating of any pair is equally probable, so that they share a common gene pool. Among domestic animals, the limit of a population, or its definition, is not always clear. In an investigation of Black-and-White cattle, the population may involve all living or breeding animals of Friesian origin, or all possible genotypes given the present gene frequencies of Friesians, the majority of which, however, will never be realized. In a practical situation, however, the population is usually clearly defined, e.g., the Holstein-Friesian population of the U.S.

A genetic or Mendelian population has a common gene pool and its members have, relative to other populations, an increased probability of mating. Therefore, population can be conceived of on several levels. On a species level, where *Bos taurus* is clearly delimited and matings with other species, such as *Bison bonasus,* are possible but subfertile, the concept is well defined. Population genetics conceives of a species as a population with integrated gene complexes, implying that matings between animals of different species are without success or at least have less success than intraspecific matings. On a different level a subspecies such as *Bos indicus* may be considered a population. Then population may be defined on a breed level, where a greater probability exists for matings among, for example, Friesian cattle than between them and cattle of other breeds. Finally, strains within breeds may be considered as populations—Holstein-Friesians, Canadian Friesians, British Friesians, Dutch Friesians, etc. In earlier times such strains were subdivided further and these substrains even had their own herd books, such as the Friesian Herdbook and the Holland Herdbook in the Netherlands. Also, lines of a poultry breeder and even herds may be construed as populations. However, temporary aggregations of animals such as are found at shows and auctions are not Mendelian populations.

A population is characterized by the frequency distribution of the genotypes and by the prevailing system of mating. Therefore, it is a statistical concept. For example, Guernseys are all spotted or piebald. Nevertheless, some Jerseys also carry white markings and when one inspects Jersey cattle before selection, one will see an occasional animal with spotting like a Guernsey. However, both breeds are clearly discernible by the frequency of spotting. Another example is provided by

B-hemoglobin, which is common in zebus and is not infrequent in European continental and Channel Island breeds, but is practically absent in lowland cattle. However, when very many Friesian cattle are blood-typed, an animal homozygous for B-hemoglobin will likely turn up.

Gregor Mendel discovered the particulate and duplicate nature of genes and their random distribution into gametes. He used inbred lines, their F_1 crosses, backcrosses, and F_2 and F_3 progeny. The gene frequencies in these are either 100%, nil, 50%, or, in the case of backcrosses, 25 or 75%. In populations of domestic animals it is exceptional for such frequencies to occur. The exceptions are crosses between breeds that are homozygous for complementary genes, such as crosses between Friesians and Herefords, or Friesians with Red-and-White breeds, horned with polled breeds, etc. Lauprecht (1930) found regular Mendelian ratios in progeny of such crosses.

The prediction of the outcome of matings in regular populations of domestic animals requires in addition to the satisfaction of Mendelian rules a knowledge of the frequency of mating types. It must be known how frequent matings are between homozygotes of one type (D), between them and heterozygotes (H), between them complementary homozygotes (R), etc. This presupposes a knowledge of the mating system. The simplest mating system is random mating, or panmixis, as it was called by Weinberg (1908). Random mating implies that the choice of the partner is independent of its phenotype and thus of its genotype. Therefore, the frequency of a mating combination results from the product of the frequencies of the participating phenotypes and genotypes.

Panmixis may appear unrealistic and improbable when it concerns genes that influence the performance or appearance of animals, if these are important to the breeder. However, it is a surprisingly good approximation for traits that are not of immediate concern to breeders. Consider a population with the following frequencies of an autosomal genotype: 50% AA, 20% AB, 30% BB (homozygous dominants, heterozygotes, and homozygous recessives will be denoted henceforth by D, H, and R, respectively): The two sexes have equal genotype frequencies. The frequencies of the mating combinations are derived from the product of the genotype (respectively, phenotype) frequencies:

$$(D + H + R)^2 = (0.5AA + 0.2AB + 0.3BB)^2$$

The argument can be turned around. Panmixis is present when the frequency of mating combinations is given by the square of the frequencies of the phenotypes (respectively, genotypes): f(mating combinations) = $[f(\text{genotypes})]^2$.

TABLE 1.2
Frequency of Progeny Genotypes after Random Mating

Mating type	Frequency	Offspring distribution			Offspring frequency		
		AA	Aa	AA	AA	Aa	aa
AA × AA	D^2 = 0.25	1			0.25		
AA × Aa	2DH = 0.20	1/2	1/2		0.10	0.10	
AA × aa	2DR = 0.30		1			0.30	
Aa × Aa	H^2 = 0.04	1/4	1/2	1/2	0.01	0.02	0.01
Aa × aa	2HR = 0.12		1/2	1/2		0.06	0.06
aa × aa	R^2 = 0.09			1			0.09
Sum	1.00				0.36	0.48	0.16

The frequencies of progeny with individual combinations are found by applying the Mendelian rules (Table 1.2, columns 2–5). The total probability of a progeny genotype in a population is the product of the frequency of a mating combination (column 2) and the probability of a genotype, given the mating combination (columns 3–5).

The composition of the progeny generation of our example differs widely from that of the parent generation and no direct connection is apparent. This is less startling when one recalls that a generation transmits not genotypes, but genes.

1.1. Gene Frequencies

Gene frequency or allele frequency denotes the proportion of loci occupied by the gene in question. On a diallelic locus, such as the *FV* blood-group locus in cattle or the hemoglobin locus, only three genotypes occur in the population, as shown in the examples of Table 1.3. When heterozygotes are discerned, genes can be counted directly and thus the gene frequency estimated in a simple manner. Homozygous *FF* animals have two, heterozygotes one *F* gene. Their number in the Pinzgau sample of Table 1.3 (genotype symbols also denote numbers of animals with the respective genotype) is found as follows:

$$2FF + FV = 2 \times 44 + 48 = 136$$

TABLE 1.3
Blood-Group and Transferrin Distribution in Central European Cattle[a]

Blood group	Genotype	Simmental O	E	χ^2	Brown Alpine O	E	χ^2	Pinzgau O	E	χ^2	Transferrin Simmental	O	E	χ^2
F	FF	594	592		366	365		44	45		AA	13	13.4	
FV	FV	351	355	0.10	220	221	0.02	48	46	0.14	AD	139	135.8	
V	VV	55	53		34	31		11	12		AE	6	8.3	
Sum		1000			620			103			DD	339	343.1	2.64
											DE	47	42.1	
p_F		0.770			0.768			0.660			EE	0	1.3	
q_V		0.230			0.232			0.340						

$p_A = 0.157$, $q_D = 0.794$, $r_E = 0.049$

[a] Source: Buschmann (1962), Kramser (1972). O, Observed number. E, Expected number.

The total number of genes at this locus in the sample equals twice the number of animals, $2 \times 103 = 206$. Therefore the gene frequency for F is $136/206 = 0.66$, or in symbols

$$p_F = \frac{2FF + FV}{2N}$$

Division of numerator and denominator by 2 yields

$$p_F = \frac{FF + FV/2}{N} = D + \frac{H}{2}$$

(usually D denotes dominants, which is not quite correct in this case, since F and V are codominant alleles). The frequency of the alternative allele can be computed in a similar manner:

$$q_V = \frac{FV + 2VV}{2N} = R + \frac{H}{2} = 0.107 + \frac{0.466}{2} = 0.34$$

Alternatively, since only two alleles are present and since $p + q = 1$, it follows that

$$q = 1 - p = 1 - 0.66 = 0.34$$

The frequencies of multiple alleles that are codominant, i.e., where heterozygotes can be recognized, may be estimated in an analogous fashion. The genotype frequencies of the multiple transferrin locus of a Simmental sample are given in Table 1.3 as an example.

1.2. Genetic Equilibrium

In large populations with random mating and in the absence of mutation, migration, and selection, the frequencies of the genotypes are given by the squares of the gene frequencies. This equilibrium law was discovered in 1908 independently by Weinberg and by Hardy.

Random mating implies random union of gametes. The probability that A_1-carrying sperm unite with A_1-carrying ova, where the proportion of each is p, is $f(A_1A_1) = pp = p^2$ (it is assumed that the gene frequency is equal in the two sexes). Similarly, the probability that an A_1 sperm fertilizes an A_2 ovum equals pq, and the probability of the reverse combination is qp. The probability of the union of two A_2 gametes equals q^2. In symbols this important relationship is

$$(pA_1 + qA_2)^2 = p^2A_1A_1 + 2pqA_1A_2 + q^2A_2A_2$$

which is illustrated further in Table 1.4 and in Fig. 1.1.

Under random mating the genotypic composition of a population remains constant over generations. This is intuitively understandable if one recalls that gene frequencies remain unchanged. Therefore panmixis must lead to the same proportion of genotypes.

When a disturbance of the genotypic proportions has occurred, one generation of panmixis will restore the equilibrium proportions. The Hardy–Weinberg law can be derived directly. In the progeny generation the gene frequencies are as follows:

$$A_1: \quad p^2 + pq = p(p+q) = p$$

$$A_2: \quad pq + q^2 = q(p+q) = q$$

This simple relationship can be demonstrated with the example of Table 1.2:

$$A_1: \quad 0.36 + 0.48/2 = 0.6$$

$$A_2: \quad 0.48/2 + 0.16 = 0.4$$

The rule can also be derived from the products of mating combinations and the Mendelian probabilities as demonstrated in Table 1.2.

Assume that the population has deviated from the equilibrium, i.e., its genotypic proportions no longer correspond to the square of the gene frequencies. The former are $uA_1A_1:2vA_1A_2:wA_2A_2$ and $u + 2v + w = 1$. The gene frequencies are $(u+v)A_1$ and $(v+w)A_2$. The distribution of genotypes after one generation of panmixis is

$$(u+v)^2A_1A_1, \qquad 2(u+v)(v+w)A_1A_2, \qquad (v+w)^2A_2A_2$$

TABLE 1.4
Hardy–Weinberg Rule

		♂ Gametes		
		A_1 (p)	A_2 (q)	
♀ Gametes	A_1 (p)	A_1A_1 (p^2)	A_2A_1 (pq) $\Big\}$	$= p^2A_1A_1 + 2pqA_1A_2 + q^2A_2A_2$
	A_2 (q)	A_1A_2 (pq)	A_2A_2 (q^2) $\Big\}$	

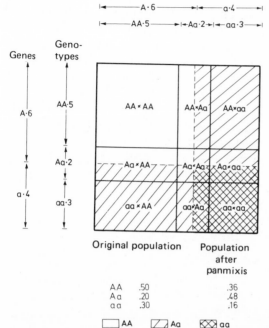

FIGURE 1.1. The Hardy–Weinberg law.

This generation in turn produces gametes in the following proportions:

A_1: $(u+v)^2 + (u+v)(v+w) = (u+v)(u+v+v+w) = u+v$

A_2: $v+w$

The gene frequencies are equal to those of the parental generation. Therefore, one generation of panmixis has sufficed to restore the genotypic equilibrium and the population will remain in this state until disturbed again.

In Table 1.3, the genotype frequencies expected in an equilibrium population are juxtaposed to the observed genotype frequencies. These served for the computation of the gene frequencies in the case of the Pinzgau breed, where $p_F = 0.66$, $q_V = 0.34$. The expected numbers of genotypes were then

$$FF: \quad p^2N \quad = 0.4356 \times 103 = 45$$

$$FV: \quad 2pqN \quad = 0.4498 \times 103 = 46$$

$$VV: \quad q^2N \quad = 0.1156 \times 103 = 12$$

The agreement between the expected and the observed numbers is excellent (Table 1.3). The occurrence of deviations of this size between the two kinds of numbers is to be expected by chance in more than 50% of samples of this size. In other words, the deviations are insignificant. The example demonstrates how well the simple mathematical rule serves to describe the situation and how useful it is.

The rule is also valid for multiple alleles, as Weinberg (1909) pointed out at a time when multiple alleles were not, or were hardly, known. Assume three alleles at a locus with frequencies p_A, q_D, and r_E, which, of course, sum to 1. The genotypic distribution is then

$$p^2AA + 2pqAD + 2prAE + q^2DD + 2qrDE + r^2EE$$

The rule may be applied to the transferrin data of Table 1.3. As the tf^e gene is very rare, only 0.24% of homozygous EE animals are to be expected, and so it is not very surprising that none of them appeared in the small sample.

The Hardy–Weinberg rule permits one to describe the genetic structure of a panmictic population solely by giving the gene frequencies. A knowledge of gene frequencies allows the estimation of the proportion of heterozygotes in the case of dominant inheritance. The immediate restoration of the genetic equilibrium after one generation of panmixis demonstrates the "lack of memory" of a population, but this pertains only to neutral alleles. It reflects the fact that the genes provide the connection between generations. In the absence of disturbing factors the gene frequencies characterize a population. These, together with the system of mating, determine in turn the genotypic frequencies.

The equilibrium rule is frequently used to test for panmixis and thus for the presence of disturbing factors such as selection. However, as pointed out by Lewontin and Cockerham (1959), such tests are not very sensitive and lack power. Figure 1.1 illustrates the restoration of the equilibrium proportions of the three genotypes of a biallelic locus.

Heterozygotes reach their maximal proportion at a gene frequency of 0.5. For multiallelic loci heterozygotes have the highest proportion when the gene frequencies are uniform, i.e., when $p = q = r$. While the maximal proportion of heterozygotes on a biallelic locus of a panmictic population cannot surpass 0.50, it increases for n multiple alleles to $(n-1)/n$, in case of three alleles to 2/3, of four alleles to 3/4, etc.

The ratio of heterozygotes to homozygotes at low gene frequencies is of some interest in connection with rare detrimental and lethal genes. The ratio of the frequency of a gene in heterozygotes to the frequency in homozygotes is $pq = q^2 = p{:}q$. This ratio is 1 when gene frequencies are 1/2, but it rises to 9 for $q = 0.1$ and to 99 for $q = 0.01$. A gene frequency of 1% is in the range of frequencies of some undesirable recessive genes. It implies a frequency of 1/10,000 for the rare homozygote. In such a case the phenotypically normal heterozygotes harbor 99 times more harmful genes than do the lethal or otherwise undesirable homozygotes.

The Hardy–Weinberg law explains the constancy of the variability in panmictic populations. This phenomenon is contrary to expectation under the assumption of blending inheritance, where the variability is halved in each generation. This fact caused considerable problems to the Darwinian theory in the last century. In Table 1.5 the behavior of the variance under blending inheritance is illustrated. Genotypes with frequencies of 1/4, 1/2, and 1/4 and phenotypic effects of 0, 1, and 2 are assumed. The variance in the parental generation is 1/2 and decreases under blending inheritance to 1/4 in the daughter generation, to 1/8 in the granddaughter generation, etc., while under Mendelian inheritance each generation has the same composition and therefore the same variance.

1.3. Multiple Loci

When several loci are to be considered which deviate from random alignment, the equilibrium will only gradually be restored by random mating, a fact which was pointed out by Weinberg (1909). This applies to independent loci as well as, *a fortiori*, to linked loci. These parental

TABLE 1.5
Decrease of Genetic Variability under "Blending Inheritance"[a]

Genotype	aa	Aa	AA
Frequency	$\frac{1}{4}$	$\frac{1}{2}$	$\frac{1}{4}$
Genotypic effect	0	1	2
	0 0	$\frac{1}{2}$	1
	1 $\frac{1}{2}$	1	$1\frac{1}{2}$
	2 1	$1\frac{1}{2}$	2

$M = \Sigma fx = 1$

$V(\text{parents}) = \Sigma fx^2 - (\Sigma fx)^2 = \frac{1}{4} \times 0^2 + \frac{1}{2} \times 1^2 + \frac{1}{4} \times 2^2 - 1^2 = \frac{1}{2}$

$V(\text{progeny}) = \frac{1}{16} \times 0^2 + \frac{1}{4} \times (\frac{1}{2})^2 + \frac{1}{8} \times 1^2 + \frac{1}{4} \times 1^2 + \frac{1}{4} \times (1\frac{1}{2})^2$
$\qquad\qquad + \frac{1}{16} \times 2^2 - 1^2 = \frac{1}{4}$

[a] M, Mean. V, Variance.

combinations tend to remain intact over generations. Assume the following gamete (or haplotype) frequencies:

$$f_1AB \quad f_2Ab \quad f_3aB \quad f_4ab; \quad f_1 + f_2 + f_3 + f_4 = 1$$

A, a and B, b are pairs of alleles at two loci. Then the gene frequencies at the individual loci are

$$p = f_1 + f_2, \quad q = f_3 + f_4, \quad r = f_1 + f_3, \quad s = f_2 + f_4$$

As an example, one may take the haplotypes encompassing the halothane *(Hal)* and the phosphohexose isomerase loci *(PHI)* of pigs. The haplotype frequencies for Danish Landrace pigs are given by Jørgensen (1977) as $f(Hal^SPHI^B) = 0.477$, $f(Hal^SPHI^B) = 0.313$, $f(Hal^SPHI^A) = 0.210$, and $f(Hal^SPHI^A) = 0$. The gene frequencies are as follows:

$$p = f(Hal^SPHI^A) + f(Hal^SPHI^B) = 0.687$$

$$q = f(Hal^SPHI^A) + f(Hal^SPHI^B) = 0.313$$

$$r = f(Hal^SPHI^A) + f(Hal^SPHI^A) = 0.210$$

$$s = f(Hal^SPHI^B) + f(Hal^SPHI^B) = 0.790$$

TABLE 1.6
Gamete Frequencies in Dihybrid Inheritance

Genotypes		Gametes			
		AB	Ab	aB	ab
AA BB	f_1^2	f_1^2			
AA Bb	$2f_1f_2$	f_1f_2	f_1f_2		
Aa BB	$2f_1f_3$	f_1f_3		f_1f_3	
Aa Bb	$2f_1f_4$	$(1-c)f_1f_4$	cf_1f_4	cf_1f_3	$(1-c)f_1f_4$
	$2f_2f_3$	cf_2f_3	$(1-c)f_2f_2$	$(1-c)f_2f_3$	cf_2f_3
AA bb	f_2^2		f_2^2		
Aa bb	$2f_2f_4$		f_2f_4		f_2f_4
aa BB	f_3^2			f_3^2	
aa Bb	$2f_3f_4$			f_3f_4	f_3f_4
aa bb	f_4^2				f_4^4

$$f_1^1 = f_1^2 + f_1f_2 + f_1f_3 + (1-c)f_1f_4 + cf_2f_3$$

$$= f_1 - c(f_1f_4 - f_2f_3) = f_1 - cd$$

$$f_2^1 = f_2 - c(f_2f_3 - f_1f_4) = f_2 + cd$$

$$d = (f_1f_4 - f_2f_3)$$

$$d^1 = (f_1 - cd)(f_4 - cd) - (f_2 + cd)(f_3 + cd) = d^0(1-c)$$

The frequencies of possible genotypes are given in Table 1.6. They result, under panmixis, from the multiplication of the gamete frequencies analogous to computation for genotypes at a single locus, the frequencies of which result from multiplication of the gene frequencies. The double heterozygote exists in two forms, as a coupling heterozygote AB/ab and as a repulsion heterozygote Ab/aB.

Each double heterozygote produces four types of gametes, those without crossing-over (parental gametes) and those with crossing-over. The crossing-over frequency is denoted by c. Then a double heterozygote in the coupling phase will generate the fraction $1 - c$ of AB-type gametes and the fraction c of Ab (respectively, c of aB) gametes with crossing-over. In the example mentioned above $c = 0.06$ and double heterozygotes will produce 94% parental gametes and 6% new combinations. If the cis and transconfigurations of genotypes are taken separately (i.e., AB/ab and Ab/aB are considered to be different), ten genotypes can be differentiated instead of nine when both types of heterozygote are identical. The frequency of AB-type gametes in generation 1 is given by

$$
\begin{aligned}
f_1^{(1)} &= f_1^2 + f_1 f_2 + f_1 f_3 + (1 - c)f_1 f_4 + c f_2 f_3 \\
&= f_1(f_1 + f_2 + f_3 + f_4) + c(f_2 f_3 - f_1 f_4) \\
&= f_1 - c(f_1 f_4 - f_2 f_3) \\
&= f_1 - cd
\end{aligned}
$$

The difference $d = f_1 f_4 - f_2 f_3$ is the linkage disequilibrium. The frequencies of gametes in the progeny generation are different from the parental gametic frequencies, but the gene frequencies remain equal:

$$
\begin{aligned}
p^{(1)} &= f_1^2 + 2f_1 f_2 + f_2^2 + f_1 f_3 + f_1 f_4 + f_2 f_3 + f_2 f_4 \\
&= f_1(f_1 + f_2 + f_3 + f_4) + f_2(f_1 + f_2 + f_3 + f_4) \\
&= f_1 + f_2 \\
&= p^{(0)}
\end{aligned}
$$

For independent loci $c = 1/2$. Therefore, a double heterozygote forms one-half coupling and one-half repulsion gametes. It follows that, even with independent loci, the approach to the equilibrium is gradual. The linkage disequilibrium is halved in each generation. However, the equilibrium is restored immediately when $p = r = 0.5$, i.e., the same gene frequencies at both loci, since here $d = f_1 f_4 - f_2 f_3 = 0$. Weinberg (1908) showed this to be the case in his classic paper.

The linkage disequilibrium diminishes in the following generations:

$$d^{(1)} = (f_1 - cd^{(0)}) (f_4 - cd^{(0)}) - (f_2 + cd^{(0)})(f_3 + cd^{0})$$

$$= (f_1 f_4 - f_2 f_3) - (f_1 + f_4 + f_2 + f_3)cd^{(0)}$$

$$= d^{(0)}(1 - c)$$

and after t generations

$$d^{(n)} = d^{(0)}(1 - c)^t$$

This relation makes it clear that with close linkage, i.e., small c, the linkage disequilibrium will decrease only very gradually. The number of generations necessary to decrease the linkage disequilibrium by a certain amount can be easily derived:

$$d^{(t)} = d^{(0)}(1 - c)^t$$

so that

$$t \log (1 - c) = \log \frac{d^{(t)}}{d^{(0)}}$$

and

$$t = \frac{\log(d^{(t)}/d^{(0)})}{\log(1 - c)}$$

For example, the number of generations of random mating necessary to halve the linkage disequilibrium between the *Hal* and the *PHI* loci in Danish Landrace pigs would be

$$t = \frac{\log 0.5}{\log 0.94} = 11.2$$

In Table 1.7 the numbers of generations necessary to reduce linkage disequilibria to certain fractions are shown for the case where individual loci are independent or where they have recombination frequencies of 1/10 and 1/100. It is obvious that close linkage requires many generations of panmixis to reduce the linkage disequilibrium by a large amount. Even independent loci would require seven generations of panmixis before d is decreased to 1% of its original value. Linkage disequilibria should be rare and unimportant in populations that have been panmictic for many generations. In contrast, they should be more frequent and

TABLE 1.7
**Number of Generations of Random Mating
to Reduce Linkage Equilibrium**[a]

c \ $\dfrac{d^{(n)}}{d^{(0)}}$	1/2	1/10	1/100
1/2	1	3.3	6.6
1/10	6.6	22	44
1/100	69	229	485

[a] c, Crossing-over frequency. $d^{(n)}$ and $d^{(0)}$, Linkage disequilibrium in generations n and 0.

important in populations derived from recent crosses, such as the Santa Gertrudis cattle. Another cause of disequilibrium is selection. Selection for intermediate optima should favor repulsion genotypes, while divergent selection should increase coupling genotypes.

In domestic animals, linkage exists, but is not readily demonstrated. The supergenes of the B and the C blood-group loci and the casein locus consist of several closely linked loci.

In cattle the hemoglobin and the J blood-group loci are known to be linked, and so are the loci for halothane susceptibility, PHI, blood group H, and the 6-glucose phosphate dehydrogenase (6-GPD) in pigs. New methods in cytogenetics, such as chromosome banding, and in somatic cell genetics should increase our knowledge in the near future.

By use of such methods, the linkage relations for several hundred loci of the human genome have been established (McKusick, 1980).

Förster *et al.* (1980), but using somatic cell hybrids between cattle and mice, and between swine and mice, were able to demonstrate that the loci for 6-GPD in cattle and those for in pigs occur on the X chromosome.

1.4. Sex Linkage

Sex chromosomes occur paired in the homogametic sex—XX in female mammals, WW in male birds—but they are unequal in the heterogametic sex—XY in male mammals, WZ in female birds. According to the Lyon hypothesis of dosage compensation, one of the X chromosomes is inactivated in the somatic cell of the homogametic sex. In general there is independence between cells in this inactivation.

Therefore, both X chromosomes influence the phenotype. In the heterogametic sex, only one X chromosome is present in each cell and it is assumed that the Y chromosome is largely devoid of structural loci, i.e., genetically it is in great measure inactive. In the following discussion the heterogametic sex is assumed to be the male (in birds the situation is the reverse). Assume that the two sexes have different gene frequencies, as is the case when crossbreds are produced from males of one and females of the other population. The next generation will have the average gene frequency $(2/3)\,p_F^{(0)} + (1/3)p_M^{(0)}$, where subscripts refer to the population from which the males and females, respectively, originated and superscripts to the generation number. Every daughter receives an X chromosome from each of her parents, but a son receives only one from his dam. Therefore, the gene frequencies in the two sexes will be

$$p_F^{(1)} = \tfrac{1}{2}(p_F^{(0)} + p_M^{(0)}), \qquad p_M^{(1)} = p_F^{(0)}$$

While the average gene frequency of the population remains unchanged under random mating, the sex-specific gene frequencies change:

$$p_F^{(2)} - p_F^{(1)} = \tfrac{1}{2}(p_M^{(1)} + p_F^{(1)}) - p_F^{(1)} = \tfrac{1}{2}(p_F^{(1)} - p_F^{(0)})$$

and

$$p_M^{(2)} - p_M^{(1)} = p_F^{(1)} - p_M^{(1)} = \tfrac{1}{2}(p_M^{(0)} + p_F^{(0)}) - p_M^{(1)} = -\tfrac{1}{2}(p_M^{(1)} - p_M^{(0)})$$

Thus, the difference in frequency of sex-linked genes between two successive generations is equal to half the difference between the preceding two generations. The difference between the sexes is

$$p_M^{(1)} - p_F^{(1)} = p_F^0 - \tfrac{1}{2}(p_F^0 + p_M^{(0)}) = -\tfrac{1}{2}(p_M^0 - p_F^0)$$

that is, half of the sex difference of the preceding generation with the sign reversed.

The development of sex-specific gene frequencies over generations is shown in Table 1.8. The successive differences between sexes diminish and change sign. The equilibrium frequency is approached asymptotically in each sex, but when averaged over sexes it is already present in the first generation. The sex-specific gene frequencies oscillate around the equilibrium frequency in ever-decreasing amplitudes.

The differences between successive generations are 0.3, 0.15, 0.075, etc. The difference between successive generations of males appears one generation later than the difference in females.

Introduction of genetic material generally occurs via males or via

TABLE 1.8
Attainment-of-Equilibrium
Frequencies with Sex-Linked Genes

Generation	P_M	P_F
0	0.2	0.8
1	0.8	0.5
2	0.5	0.65
3	0.65	0.575
.	.	.
.	.	.
.	.	.
∞	0.6	0.6

deep-frozen semen. Therefore, the fact that F_1 males lack the import X chromosome appears to be of some importance. This X chromosome could be lost if, for example, only F_1 males were to be used in successive generations.

1.5. Estimation of Gene Frequency

The gene frequency is the proportion of loci occupied by the respective gene. Its estimation is rather direct when the genes can be counted. This is possible in cases of intermediate inheritance or codominance in immunogenetics, where heterozygotes can be recognized. The variance of a gene frequency thus estimated equals the variance of a proportion $pq/2N$, the product of the gene frequencies divided by the number of loci in the population. The example of Simmentals given in Table 1.3 yields the following estimates for the genes at the FV bloodgroup locus:

$$p_F = \frac{2 \times 594 + 351}{2000} = 0.7695, \qquad q_V = 1 - 0.7695 = 0.2305$$

$$V_p = \frac{0.77 \times 0.23}{2000} = 88.55 \times 10^{-6}, \qquad s_p = 0.0094$$

The frequency of dominant genes can be estimated from the proportion of recessive homozygotes, provided that the population is in genetic equilibrium. In dominant inheritance only two phenotypes can be differentiated—recessives and dominants. While the recessives are all homozygotes, the dominants P are of two types, homozygote PP and heterozygote Pp. For example, the gene for polled is dominant in lowland cattle at least. Horned animals are recessive pp, polled cattle are either Pp or PP. This situation is common to many or most lethal genes, to blood-group genes, genes for coat color, etc. When the population is in genetic equilibrium the frequency of recessives R is q^2. Therefore, the frequency of the recessive gene can be estimated as

$$q = \sqrt{R} = \sqrt{q^2}$$

The frequency of the dominant gene results from the simple relation

$$p = 1 - q$$

For example, assume that a blood-group laboratory has available only an F-antiserum but not anti-V. Then only F and not-V animals can be differentiated. Thus F becomes an operational dominant. The example demonstrates that dominance is frequently a matter of diagnosis. In many instances refinement of the diagnosis, such as improved laboratory methods, may permit one to recognize heterozygotes. In our Simmental example there are 945 F-positives and 55 F-negatives. The latter are the homozygous recessives, from which gene frequencies can be estimated:

$$q = \sqrt{R} = \sqrt{0.055} = 0.235, \quad p = 1 - q = 0.765$$

In this case the result is very close to that obtained by direct gene counting. However, the variance of the estimate is considerably greater, an outcome which is intuitively obvious, as it is derived from the proportion of recessives only. The variance is derived from the variance of a function:

$$V_{f(R)} = V_R \left[\frac{df(R)}{d(R)} \right]^2$$

The gene frequency is a function of R: $q = f(R) = \sqrt{R}$, so that

$$V_R = \frac{(1-R)R}{N}$$

and

$$\frac{dq}{dR} = \frac{dR^{1/2}}{dR} = \frac{1}{2}R^{-1/2}$$

Therefore,

$$V_q = \frac{(1 - R)R}{N} \left(\frac{1}{2\sqrt{R}}\right)^2 = \frac{1 - R}{4N}$$

$$s_q = \left(\frac{1-R}{4N}\right)^{1/2} = 0.0154$$

Knowledge of the frequency of recessive genes permits the estimation of the proportion of heterozygotes, which is of considerable importance in cases of lethals or other rare genes. For example, Stegenga (1970) estimated the gene frequency for red coat in Friesian cattle to be 6–8%, so that only 1/3–2/3% of red calves are to be expected. Assuming q = 0.07, the frequency of heterozygotes will be 2 × 0.07 × 0.93 = 0.13. Therefore more than one-eighth of all animals are carriers, a fact which could be of significance to an AI (artificial insemination) station when test bulls have to be bought and homozygosis for black is desired.

The comparison of the numbers of observed genotypes with those expected permits a test for genetic equilibrium and indirectly for the neutrality of genes. The "goodness-of-fit" test is applied and the χ^2 values resulting in the examples are given in Table 1.3. This is not possible when dominance exists. However, consideration of mating types and data of two generations permits a test for panmixis. Mating-type frequencies that result from random mating are given in Table 1.9. For example, the first three lines encompass matings between dominants which cannot be subdivided phenotypically. In this group a recessive progeny is born only in families of the mating heterozygotes × heterozygote. Here, the probability is 1/4, and therefore the total probability of recessive progeny in the mating $D \times D$ equals p^2q^2. The ratios of the expected frequencies of recessives from the mating combinations $D \times$ D, $D \times R$, and $R \times R$ are $p^2q^2{:}2pq^3{:}q^4 = p^2{:}2p^2{:}q^2$, and they sum to q^2, as is expected. The ratios can be used for a test of the existence of the Hardy–Weinberg equilibrium, when all mating combinations are possible. Kramser (1972) found in Austrian Simmental the Z blood types given in Table 1.9.

The gene frequencies are $q = (61/142)^{1/2} = 0.655$ and $p = 0.345$. These permit the estimation of the expected numbers of recessives from the different mating types, which are compared in Table 1.9 with the actual numbers. The deviations barely miss the level of significance, which may be explained by the heterogeneous nature of the data (several breed societies).

TABLE 1.9
Test for Genetic Equilibrium under Dominant Inheritance[a]

Mating combination	Offspring			$f(z)$	$E(z)$	Δ	χ^2
	n	Z	z				
$Z \times Z \begin{cases} p^4 \\ 4p^3q \\ 4p^2q^2 \end{cases}$	62	51	11	$q^2p^2 = 0.051$	7.2	3.8	2.0
$Z \times z \begin{cases} 4pq^3 \\ 2p^2q^2 \end{cases}$	51	30	21	$2pq^3 = 0.194$	27.5	-6.5	1.54
$z \times z \quad q^4$	29	—	29	$q^4 = 0.43$	26.3	2.2	0.18
	142	81	61	$q^2 = 0.43$			3.726

[a] Source: Kramser (1972). $E(z)$, Expected number of z-type animals. Δ, Difference between observed and expected numbers of z-type animals.

Frequently information derives both from the parent and from the offspring generation, such as when blood-group information is obtained from dams and their offspring. An optimal way to estimate gene frequencies is to take account of both. The information from the individual generations is not independent. Therefore, a maximum-likelihood method should be employed, as is demonstrated for the data of Table 1.10, following a presentation given by Li (1976). Gene frequencies can be computed from the following formulas:

$$p = \frac{3a_1 + 2(a_2 + a_3) + a_4 + a_5 + a_6}{3N - a_4}$$

$$= \frac{3 \times 143 \times 2(22 + 35) + 48 + 6 + 13}{804 - 48} = 0.807$$

The formulas are comprehensible when it is realized that of the four genes common to parent and offspring, two are identical, i.e., the gene of the offspring that derives from the parent. This gene should not be counted twice. In the combination FF-dam, FV-offspring the F gene derives from the dam and it should be counted but once. The other two genes, one maternal F gene and the offspring V gene, are independent and supply information on the gene frequency in the population. An exception is provided by the combination FV-dam, FV-offspring. Here, both offspring genes may derive from the dam. Therefore, only two of the four genes are independent, one from the dam and the one from the offspring that derives from the sire. The total gene number in the

TABLE 1.10
Gene Frequency Estimation from Dam–Daughter Pairs[a]

Dams	Offspring			Σ Dams
	FF	FV	VV	
FF	$143a_1$	$22a_2$	—	$165\ a_1 + a_2$
FV	$35a_3$	$48a_4$	$6a_5$	$89\ a_3 + a_4 + a_5$
VV	—	$13a_6$	$1a_7$	$14\ a_6 + a_7$
Σ Offspring	178	83	7	268
	$a_1 + a_3$	$a_2 + a_4 + a_6$	$a_5 + a_7$	

[a] Source: Kramser (1972).

sample is then $3N - a_4$. This is three-fourths of all genes ($4N$) except for the combination heterozygous dam and heterozygous offspring, where only half of the genes contribute information. The gene frequency is the ratio of the number of independent F genes to the total number of independent genes. The standard deviation of the gene frequency is

$$s_p = \left(\frac{pq}{3N - a_4} \right)^{1/2} = 0.014$$

Gene frequency estimation from only the parental generation results in

$$p = 0.782, \qquad q = 0.218, \qquad s_p = 0.018$$

and from the offspring generation in

$$p = 0.819, \qquad q = 0.181, \qquad s_p = 0.017$$

The increase in precision is relatively modest in this example, but it is not negligible, and it demonstrates an increase in information that would be even more important with less numerous data.

Sex-linked loci permit direct gene counting, even in cases of dominance, in the heterogametic sex. Recessive genes are hemizygous in the heterogametic sex and gene frequency is identical with genotype frequency. Of course the quality of the estimate is improved when information from both sexes is combined. The orange-black color of cats can serve as an example. Homozygous O^+O^+ and hemizygous O^+ animals are black, homozygous OO and hemizygous O are orange, and heterozygotes OO^+ are tortoise shell and should be females (the very rare OO^+ males may have an XXY genotype and thus have the Klinefelter syndrome). When the numbers of phenotypes are denoted by O, B, and C (for tortoise shell), the gene frequencies can be estimated as follows:

$$p = \frac{2O_F + C_F + O_M}{2N_F + N_M}, \qquad q = \frac{2B_F + C_F + B_M}{2N_F + N_M}$$

N_F and N_M denote numbers of female and male cats. The estimation is illustrated in Table 1.11. With dominant inheritance, simultaneous gene frequency estimation from data of both sexes involves the solution of quadratic equations.

Multiple allelism presents no particular problems as long as genes can be counted. For example, as shown in Table 1.3, several transferrin alleles occur in European cattle which can be recognized phenotypically even though differentiation may be difficult in certain cases, e.g., D_1D_2. Only three alleles are considered in Table 1.3, tf^a, tf^d, tf^e. These lead to six genotypes and also six phenotypes: AA, AD, AE, DD, DE, EE. Let these symbols also denote the numbers of animals with the respective phenotypes and let N be the total; the gene frequencies are then found from the following type of expression:

$$p_A = \frac{2AA + AD + AE}{2N}$$

The variances are given by $p_i(1 - p_i)/2N$.

TABLE 1.11
Frequencies of Sex-Linked Genes[a]

Phenotype[b]	Genotype	Observed number	Expected number
♀ { O	OO	155	149.3
C	Oo^+	34	39.2
B	o^+o^+	2	2.5
♂ { O	O	172	179.6
B	o^+	30	23.4

$$p = \frac{2 \times 2 + 34 + 30}{382 + 202} = 0.116$$
$$q = 1 - p = 0.884$$
$$\chi^2 = 3.8, \quad 3 \text{ df}$$

[a] Source: Baxa (1973).
[b] O denotes an Orange phenotype, C, a calico phenotype, and B, a black phenotype.

When genes show dominance, the methods become considerably more complicated. Bernstein (1930) has elaborated a method for simple systems, such as the human *ABO* blood-group system, with two codominant and one recessive allele, where four phenotypes can be differentiated. Phenotypes are composed of genotypes as follows (phenotype symbols also denote frequencies):

$$O = r^2 \qquad\qquad B = q^2 + 2qr$$
$$A = p^2 + 2pr \qquad AB = 2pq$$

Bernstein (1930) used the following relationships:

$$r + p = (A + O)^{1/2}, \qquad r + q = (B + O)^{1/2}, \qquad r + p + q = 1$$

The first estimates are then

$$r' = O^{1/2}, \qquad p' = 1 - (B + O)^{1/2}, \qquad q' = 1 - (A + O)^{1/2}$$

The estimates are neither the most efficient nor are they unbiased, i.e., they do not sum to 1: $p' = q' + r' \neq 1$. Bernstein suggested the following remedy:

$$D = 1 - r' - p' - q'$$
$$p = p'(1 + \tfrac{1}{2}D), \qquad q = q'(1 + \tfrac{1}{2}D), \qquad r = (r' + \tfrac{1}{2}D)(1 + \tfrac{1}{2}D)$$

The improved estimates sum to $1 + D^2/4$. Sometimes subtypes occur, where one blood group is dominant over all others, the second over all but the first, etc., i.e., sequential dominance exists:

$$A: \qquad AA, AA_1, AA \qquad p^2 + 2pq + 2pr$$
$$A_1: \qquad A_1A_1, A_1a \qquad q^2 + 2qr$$
$$a: \qquad aa \qquad\qquad r^2$$

Similar to Bernstein's approach, the estimation can proceed as follows:

$$r = a^{1/2}$$
$$p = 1 - (A_1 + a)^{1/2}$$
$$q = 1 - r - p \quad \text{or} \quad q = (A_1 + a)^{1/2} - a^{1/2}$$

TABLE 1.12
Allocation of *SH'* Blood Types to Genotypes

Possible genotypes	Preliminary frequency	Probable distribution of genotypes
SH'/SH'	$p^2N = 0.0133 \times 17$	2.3
SH'/H'	$2pqN = 0.0384 \times 17$	6.6
SH'/s	$2prN = 0.0384 \times 17$	6.6
S/H'	$2qsN = 0.0085 \times 17$	1.5
Sum		17

$p = 0.115, r = 0.167, q = 0.167, s = 0.025$

[a] Source: Erlacher (1970).

Large allelic series demand more complicated methods. The allocation method (Neiman-Sørensen, 1958; Ceppelini *et al.*, 1955) proceeds as follows: First gene frequencies are estimated from frequencies of known genotypes, e.g., from parental frequencies where the genotype has been ascertained from progeny phenotypes. These first estimates serve to allocate the unknown phenotypes to genotypes. After the first round of allocation, an improved estimate of gene frequencies, based on known and on allocated genotypes, is possible. The procedure is iterated until the estimates converge. Table 1.12 gives an example where 17 animals with blood type *SH'* are to be allocated to four genotypes. The preliminary frequencies of the four alleles involved were derived from known genotypes in the same sample. They permit proportional allocation of the 17 animals, which is given in the last column. These figures are then combined with the originally known numbers and the resulting sums serve as bases for the next round of gene frequency estimates.

2

Changes in Gene
Frequency

2.1. Mutation

For simplicity we assume two alleles per locus, A_1 and A_2, where A_1 is the wild-type allele. A fraction u of the alleles A_1 mutate to A_2. This changes the frequency of A_2 genes by $+pu$. Therefore, the change in gene frequency is proportional to the product of mutation rate and gene frequency. The mutant allele A_2 mutates back to A_1 at a rate v which in general is much smaller than u. Rates of forward and back mutations for some color loci of mice are given in Table 2.1. The back mutation rate is about one-fifth of the rate of forward mutation.

The mutation rate is very small, of the order of 10^{-5}–10^{-7}. Therefore, mutation-caused changes in gene frequencies are also very small. Neglecting back mutations, we find the gene frequency in a generation t to be given by

$$p_t = (1 - u)p_{t-1} \quad \text{after two generations}$$

$$p_{t+1} = (1 - u)^2 p_{t-1} \quad \text{and so on}$$

Therefore, the ratio of gene frequency in generation t to the frequency in the base population is

$$p_t/p_0 = (1 - u)^t$$

TABLE 2.1
Spontaneous Mutation Rates in Mice[a]

Locus	$u \times 10^{-6}$	CI	$v \times 10^{-6}$	CI
a	71.1	1.8–39.6	4.7	2.3–8.7
b	0	0–13.3	0	0–4.5
c	9.7	0.2–54.1	0	0–4.3
d	19.2	6.2–44.8	3.4	0.4–12.4
en	15.1	0.4–84.0	0	0–39.
Average	11.1	4.8–21.8	2.7	1.4–4.7

[a] Source: Schlager and Dickie (1966). CI, Confidence interval (95%).

The number of generations necessary for a given change can easily be computed (Li, 1976). For example, to change the frequency from 0.1% to 1% by accumulation of mutations would require 9050 generations ($u = 10^{-6}$), a time span that exceeds by several orders of magnitude time intervals relevant to animal breeders. Note also that back mutations, which would slow the rate of change even further, were neglected.

Assume that mutation pressure only affects loci, with wild-type A_1 genes mutating to A_2 with rate u, and A_2 genes mutating back to A_1 genes with rate v. When the number of A_1 genes lost by mutation is balanced by the number gained by back mutation, an equilibrium ensues:

$$\Delta p = -up + v(1 - p) = 0$$

This results in an equilibrium gene frequency of

$$\tilde{p} = \frac{v}{v + u}$$

Since back mutations are much rarer than forward mutations, the gene frequency of the mutant must be much higher in order to maintain the equilibrium. In the data given in Table 2.1, $u{:}v \approx 5{:}1$, which would lead to a frequency of the wild-type allele of 1/6 [$= 1/5/(1/5 + 1)$], which obviously is not the case. Therefore, one must conclude that forces other than mutation rates are much more important in determining gene frequencies. However, this conclusion may not pertain to a large proportion of molecular variants where mutation rates in different directions may be similar. The widespread occurrence of genetic polymorphism in the case of enzymes and other proteins may be largely due to mutation equilibria of the gene frequencies.

Most newly arisen mutations are lost. The distribution of family size can be described by a Poisson series. The probability of k offspring is given by

$$P(n = k) = e^{-2} \frac{2^k}{k!}$$

where an average number of two offspring is assumed. A mutated gene (A_2) occurs in a heterozygote. The probability that it will be lost in families with 0, 1, 2, 3, etc., offspring from the mating $A_1A_1 \times A_1A_2$ equals 1, 1/2, $(1/2)^2$, $(1/2)^3$, etc. The probability of gene loss in the first generation sums to 37%, in the second to 53%, in the 15th to 89%. Therefore, only recurring mutants have a chance of becoming established in a population.

Even though mutation rates are very small, recurrent mutations are known in domestic animals. An example is polledness, which appeared independently in several breeds (Ayrshire, Hereford, Angus, Pinzgau). Most of these are due to a dominant gene. However, even though identical on the gene level, the mutations are probably different on the level of DNA. Therefore, from the point of view of molecular genetics, recurrent mutations at the nucleotide level may be nonexistent for all practical purposes.

Mutation rates were estimated for a number of human traits, mostly for hereditary diseases, and were found to lie in the range from 4×10^{-4} to 5×10^{-6}, similar to the rates for color loci in mice (Table 2.1). Since the probability of finding a mutant is proportional to its mutation rate, these figures are probably overestimates and the rate of gene mutation may be about 10^{-6}. We may assume that this value also pertains to domestic animals. Mutation rates are not constant. They can be influenced by various factors, such as irradiation and chemicals, and they vary among loci. Mutation creates the raw material for evolution and for breeding which is to be exploited by selection.

Even though they are rare for any single locus, mutations are not so infrequent when the whole genome is considered. Mukai (1964) showed in *Drosophila* that on chromosome II, polygenes that influence viability mutate about 20 times more frequently than do major genes that mutate to lethals. However, even these lethal mutations reach a frequency of 0.6% per chromosome II. Since this contains about 40% of the genetic material, it may be assumed that about 1½% of the gametes carry a new lethal mutation. Bailey (1959) showed that in inbred lines of mice, mutations induce skeletal variants that lead to fairly rapid differentiation of lines.

There have been speculations and experiments about the possibility of utilizing mutations for breeding. Most of these have been in connection with selection plateaus where the majority of the genes may be homozygous and genetic variability exhausted. It has been speculated that an increase of the mutation rate would create new genetic variability and thus permit more efficient selection. Most of the experiments have been performed with *Drosophila*, but some have also been done with poultry. The *Drosophila* experiments (Clayton and Robertson, 1964) showed that extremely high doses of X-rays are necessary to create the genetic variability available in unselected populations.

Mukai's experiments indicated, for example, that 800 generations of naturally accumulated mutants would be required to create the genetic variance for abdominal bristles extant in unselected populations. Experiments with Leghorns (Abplanalp *et al.*, 1964) were not unequivocal. The genetic variance was found to be increased in the irradiated population, but the errors of the estimates prevented a firm conclusion. Furthermore, the line used, which was thought to have plateaued and which therefore was assumed to have no or little genetic variability, responded to renewed selection and therefore made the inference from the genetic variance even more insecure. In summary, the rather few experiments indicate that mutation breeding as practiced with plants offers rather little prospect of success for animal breeding.

2.2. Migration

In the past migration probably played an important and possibly dominant role in the formation and development of populations of domestic animals. The introduction of breeding animals from outside the local populations and their use for grading up was, and is, one of the most effective means of changing a population. Formally, the effect of migration upon gene frequencies can be treated like mutation. Genes can come into the population via immigrants, and the population can lose genes via emigrants.

Let the gene frequency of the local population be p_L and that of the immigrants be p_I. A fraction m of the local population, or rather of its genes, is to be replaced by genes from immigrants. The gene frequency in the offspring population is then

$$p'_L = mp_I + (1 - m)p_L$$

and the gene frequency change due to immigration is

$$\Delta p = m(p_I - p_L)$$

Gene frequency change is proportional to the rate of immigration and to the difference in gene frequencies between the local population and the source population of the immigrants (provided that they constitute a random sample of their population of origin). Emigration has not been considered.

As an example, assume that half of a herd of Hereford cattle are crossed with Simmentals. The frequencies of the hemoglobin B gene in the two breeds are, respectively, 0 and 0.2. Therefore, the gene frequency change will be

$$\Delta p = \tfrac{1}{4}(0.2 - 0) = 0.05$$

and 10% of the offspring will be heterozygous at the Hb locus.

Immigration will cause populations to become more similar. Let σ_q^2 denote the variance between the gene frequencies of the populations before immigration; One generation of migration will reduce it to

$$\sigma_{q_1}^2 = (1 - m)^2 \sigma_{q_0}^2$$

Several models have been proposed to treat migration in population genetics. The simplest model assumes that immigrants originate from a large, more or less stable population whose gene frequencies remain constant over generations and are not affected by the loss of the migrants. Continuous migration will cause the gene frequencies of the receiving populations to become like those of the source population. The grading up of native breeds with breeds considered to be superior corresponds to such a model.

S. Wright introduced the island model, where several subpopulations exchange genes. A genetic equilibrium will be approached when the gene frequencies in the various subpopulations are equal. Nozawa *et al.* (1978) found that such an island model described satisfactorily the variation in marker gene frequencies of Japanese island goats. Further modifications involve models in which exchanges between neighboring populations are more frequent than between distant groups. Kimura and Weiss (1964) considered such models. Populations may be arranged in one dimension—along a valley (the Nile, for example) or a coast—or in two dimensions over an area. Subpopulations may be distributed discontinuously in clusters or the whole population may constitute a continuum without barriers.

As already mentioned, migration formally resembles mutation. Gene frequency change due to mutation is described by

$$\Delta p = -(u + v)p + v = -(u + v)(p - \bar{p})$$

[recall that $\bar{p} = v/(u + v)$]. Therefore, the combined gene frequency change due to mutation and migration is

$$\Delta p = -(m + u + v)(p_L - \bar{p})$$

where $\bar{p}$ is the average gene frequency of all populations that participate in the migration. However, mutation affects a single locus and migration the whole genome.

In general, immigrants come from neighboring populations, which tend to be similar, so that $p_I - p_L$ will be small. However, long-range migrations did occur in the past and they are increasingly common nowadays as transport costs less and is comparatively easy. This is true in particular when the introduction of genes is performed via frozen sperm or embryos.

In some herd books, animals are accepted when they resemble the required type, and after several generations of topcrossing with pure-bred sires. It is obvious that the difference in gene frequencies between such animals and the purebreds will be very small and that they will do little to alter the genetic makeup of a breed in particular since the rate of their admission into the herd book is small.

The intensity of genetic migration may vary between wide limits. It may consist of an isolated import of a few animals or it may result in nearly complete replacement of a local gene pool by repeated introductions of foreign breeding animals. Numbers of imported animals do not necessarily yield information on the rate of gene immigration since frequently imported animals are used more heavily than locally bred ones. For example, Robertson and Asker (1951) found that two generations after importation into Britain of Friesian cattle, bulls had on average 70 progeny and cows 20 (before advent of artificial insemination) and 11 years after importation one-fifth of the genes of registered heifers derived from import animals.

There may exist an exchange of animals between breeds which is unknown or forgotten, and which becomes evident when marker genes are studied. Farrel et al. (1971) reported a casein gene in Holstein and in Red Danes. The common occurrence of this rare gene may be explained by the neighborhood of original Holsteins and Red Danes in Southern Jutland. Kidd et al. (1974) found similar marker gene frequencies in South Devons and German Yellow cattle explainable by the use of some Devon sires in Germany in the 19th century.

Populations of domestic animals frequently form a hierarchy. The movement of genes from one stratum to a lower one can be considered as genetic migration, which may result in complete or partial replacement of the gene pool in lower strata. This topic will be considered in greater detail in Chapter 18.

2.3. Selection

The number of progeny varies among animals. This variation is frequently random, particularly in nature, i.e., the distribution of family size does not deviate from a random distribution such as a Poisson curve.

If the number of progeny is correlated with the genotype, this will have genetic consequences and it may lead to changes in the gene pool. Selection is present when progeny numbers of genotypes differ in a nonrandom manner (Lerner, 1958). Natural selection is present when the variability in family size between genotypes comes about without human intervention. Natural selection then is an *a posteriori* statement: When numbers of progeny differ between phenotypes and genotypes, natural selection is assumed to have acted. In contrast, in artificial selection the breeder prefers certain phenotypes and therefore genotypes, and gives to them preference to reproduce the population.

It is assumed that locus *A* influences the reproduction, or, more precisely, the selective or adaptive, value. These terms are defined as the proportional contribution of a genotype to the gene pool of the next generation. Fitness is frequently used to denote adaptive value. However, its colloquial meaning is quite different from the specific meaning it has in selection theory, so that the use of a notion which is more neutral may be advisable.

The selective value of the genotypes of an asexual organism is given by the number of progeny per parent individual. In the case of a sexually reproducing population the selective value could be half the number of offspring which a genotype produces, or it can be defined as the number of "successful gametes." The numbers of progeny or "successful gametes" must be counted at comparable ages or reproductive stages in the parent and in the progeny generation. Therefore, when the selective value of an adult individual is to be estimated, its progeny, or its "successful gametes," have to be counted when they reach the adult stage. The selective value embraces both viability and fertility of an individual.

The selective values of three genotypes of locus *A* are given in Table 2.2. The genotype A_1A_1 is taken as standard. The reduction in the selective value of the other genotypes relative to the standard is defined as the selection coefficient and it is denoted by s_1 and s_2, respectively. It is complementary to the selective value, i.e., $s_1 = (w_{11} - w_{12})/w_{11} = 1 - w_{12}/w_{11}$. It is not implied that the selective value or the selection

TABLE 2.2
Selection against Recessives

	Genotype		
	A_1A_1	A_1A_2	A_2A_2
Frequency of progeny before selection	p^2	$2pq$	q^2
Frequency of progeny after selection	p^2	$2pq$	$q^2(1-s)$
Relative frequencies	$\dfrac{p^2}{1-sq^2}$	$\dfrac{2pq}{1-sq^2}$	$\dfrac{q^2(1-s)}{1-sq^2}$
Gametes produced			
A_1	$\dfrac{p^2}{1-sq^2}$	$\dfrac{pq}{1-sq^2}$	
A_2		$\dfrac{pq}{1-sq^2}$	$\dfrac{q^2(1-s)}{1-sq^2}$

coefficient of a genotype is constant for all individuals with this genotype. Rather, it is the average selective value of this genotype in a particular population and in a particular environment. The reduction in the adaptive value of heterozygotes is frequently measured relative to the reduction in the undesirable homozygotes: $s_1 = hs_2 = hs$, where h is defined as the degree of dominance. Its use permits neglect of subscripts and s refers to the undesirable homozygote.

The various degrees of dominance are illustrated in Fig. 2.1. Complete recessives have a zero degree of dominance ($h = 0$) and the heterozygote has an adaptive value equal to the desirable homozygote. In heterozygotes intermediate between the two alternative homozygotes, the degree of dominance is 1/2, but it is 1 if the undesirable gene is completely dominant. In partial dominance the degree of dominance varies between the two extremes, i.e., $0 < h < 1$. Overdominance implies a negative h ($s_1 = -hs_2$). When the superior heterozygote is taken as

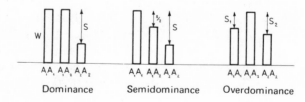

FIGURE 2.1. Selective values w and selection coefficients s.

the standard, the selection coefficients give the impairment of the selective values by both homozygotes, as is illustrated in Fig. 2.1.

The concept of selective value is illustrated with data from ranch mink published by Johansson (1965) (Table 2.3). Selective value is considered to be equal to the number of progeny per female mated. Of course this is not equal to the real selective value, since it does not embrace viability nor even male fertility. Also, these selective values pertain to two-locus genotypes, while in the further treatment it is assumed that single loci cause the difference in reproduction, etc.

In Johansson's material, standard mink have the largest number of offspring, 4.41 pups per mated female, and therefore also the highest selective value. Relative to the standard, the two mutants "blue-eyed pastel" and "sapphire" have selective values of 89 and 71%, respectively, which result in selection coefficients of 0.11 and 0.29, respectively. In the paper, fertility and litter size of heterozygotes are not given. For further discussion it is assumed that the standard is completely dominant.

The gene frequency change due to natural selection may be derived as follows: The gene frequency of the desirable gene, after selection, will be $(p^2w_{11} + pqw_{12})/\bar{w}$, where $\bar{w}$ denotes the average selective value of the population. This is given by

$$\bar{w} = p^2w_{11} + 2pqw_{12} + q^2w_{22}$$

and

$$\Delta p = \frac{p^2w_{11} + pqw_{12}}{\bar{w}} - p = \frac{pq}{\bar{w}}[p(w_{11} - w_{12}) + q(w_{11} - w_{22})] \quad (2.1)$$

This can be expressed in terms of selection coefficients:

$$\Delta p = \frac{pqs[q + h(p - q)]}{\bar{w}}$$

TABLE 2.3
Selective Values w and Selection Coefficients s of Mink Genotypes[a]

	n	Genotype	Percent infertile	Litter size	Offspring/♀ mated	w	s
Standard	3,681	$++/++$	17.5	5.29	4.41	1	0
Brown-eyed pastel	12,731	bb/tt	21.3	4.91	3.91	0.89	0.11
Sapphire	4,736	aa/pp	37.0	4.49	3.13	0.71	0.29

[a] Source: Johansson (1965).

which in the case of complete recessives simplifies to

$$\Delta p = \frac{spq^2}{1 - sq^2}$$

For illustration it is assumed that the "standard" of Table 2.3 is completely dominant, that the population mates randomly after selection, and that the gene frequency for "pastel" is

$$q_0 = \left(\frac{12,731}{16,412}\right)^{1/2} = 0.776^{1/2} = 0.881$$

and for standard is $p_0 = 0.129$ (subscripts denote generation number). thus,

$$\Delta p = \frac{0.11 \times 0.129 \times 0.776}{1 - (0.11 \times 0.776)} = \frac{0.011}{0.915} = 0.012$$

$$p_1 = 0.129 + 0.012 = 0.141$$

S. Wright derived the selective value of genes as the weighted average of the values of the genotypes that contain the gene:

$$w_i = \frac{p^2 w_{11} + pq w_{12}}{p}$$

Gene frequency change can be derived simply as

$$\Delta p = \frac{pw_1}{w} - p = \frac{p(w_1 - \overline{w})}{\overline{w}} = pq \frac{w_1 - w_2}{\overline{w}}$$

Differentiating the average selective value for gene frequency yields

$$\frac{\delta \overline{w}}{\delta p} = 2(w_1 - w_2)$$

The average selective value of the population should change by this quantity when p increases. Substituting it in expression (2.1) yields

$$\Delta p = \frac{pq}{2\overline{w}} \frac{\delta \overline{w}}{\delta p}$$

Gene frequency change is proportional to the variance of gene frequency and to the relative increase in the average selective value of the population caused by increasing the gene frequency.

The formulas reveal that gene frequency change depends both on s (or on w) and on the gene frequencies. The numerators of the expressions contain quantities such as pq or pq^2, which has the consequence that changes will become very small when gene frequencies are either low or high. The dependence of Δp on gene frequency can be easily demonstrated with the example in Table 2.3. Assume $p_0 = 0.3$, $q_0 = 0.7$. With $s = 0.11$, $\Delta p = 0.017$. When the frequencies are reversed ($p = 0.7$, $q = 0.3$), $\Delta p = 0.007$. Figure 2.2 illustrates the gene frequency changes in relation to gene action and to gene frequency.

The influence of the frequencies is understandable when it is realized that at low frequency for the recessive gene most of them will be in heterozygotes and therefore be hidden against the influence of selection. Further reduction of the already rather few homozygotes will accomplish very little in terms of reduction of the frequency of undesirable genes (in our example undesirable from the point of view of the selective value, not in terms of their value to fur production).

In general s will be small for most loci, or, if s has a larger magnitude, q will be very low. Therefore, the reduction of the mean selective value of a population compared with its wild type will be small and can be neglected, so that the denominator in the formulas for gene frequency change can be set to unity. The approximate formulas are:

selection against recessive genes $\qquad \Delta p \approx spq^2$

genes with intermediate effects $\qquad \Delta p \approx spq/2$

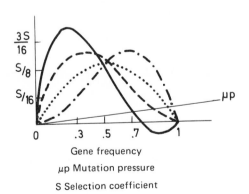

FIGURE 2.2. Change of gene frequency by selection (Wright, 1921). (—) Overdominant, (– –) dominant, (—·—) recessive, (···) intermediate.

genes with dominant effects $\Delta p \approx sp^2q$

genes with partial dominance $\Delta p \approx shpq$

Natural selection will have moved the frequency of a recessive lethal gene very close to 0, so that the frequency of the normal gene will be close to 1. Gene frequency then changes by about q^2; if $q = 0.01$, $\Delta q \approx 0.0001$. Partial dominance of such a lethal gene, even with a small degree of dominance, will permit a much greater decrease in the gene frequency, about hq when s and p are near 1. So when we assume $h = 0.02$, as was found for *Drosophila* lethals, $\Delta q \approx -0.0002$, which is twice as big as with complete dominance. Therefore, even a small degree of dominance makes selection much more efficient.

It is to be expected that undesirable genes which adversely affect heterozygotes will be very rare. This may help to explain the rarity of lethals in inbred offspring of domestic animals.

The formulas and Fig. 2.2 illustrate that selection is most efficient at intermediate gene frequencies. In the case of recessives the optimum is shifted toward higher frequencies where homozygotes are abundant, while selection against dominants is more effective around gene frequencies of about one-third. Dominants can be selected against in heterozygotes. Dominant phenotypes are comparatively frequent at lower gene frequencies. When these frequencies are high, dominant homozygotes cannot be distinguished from heterozygotes and the efficiency of selection decreases.

The time required to change gene frequencies by a certain amount can be estimated. In general rather long periods of time are necessary for changes of any magnitude when the assumed selection coefficients are those thought to be common in nature. Also, changes take a very long time when frequencies are already low. Therefore, such computations have a largely theoretical value, since it must be assumed that environmental conditions change during selection and thus selection coefficients are altered. Occasionally, strong selection against a certain genotype may become possible and rapid gene frequency changes may ensue.

The time required to change a gene frequency can be demonstrated most simply with a recessive lethal, or a gene where the homozygote is sterile. When the undesirable gene is A_2 and the normal gene is A_1, the genotype frequencies after selection will be

$$A_1A_1: \quad p^2, \qquad A_1A_2: \quad 2pq, \qquad A_2A_2: \quad 0$$

The frequency of A_1 is then

$$p_1 = \frac{p}{p^2 + 2pq} = \frac{1}{1 + q}$$

and that of A_2 is $q_1 = q/(1 + q)$.

After one generation of random mating, and after the ensuing selection, genotype frequencies will be as follows:

$$A_1A_1: \frac{1}{(1 + q)^2}, \quad A_1A_2: \frac{2q}{(1 + q)^2}, \quad A_2A_2: \quad 0$$

and the gene frequencies are

$$p^2 = \frac{1 + q_0}{1 + 2q_0}, \quad q_2 = \frac{q_0}{1 + 2q_0}$$

This can be generalized:

$$q_n = \frac{q_0}{1 + nq_0}$$

The number of generations necessary to change q_0 to q_n can easily be derived:

$$q_n + nq_0q_n = q_0, \quad n = \frac{q_0 - q_n}{q_0q_n} = \frac{1}{q_n} - \frac{1}{q_0}$$

The comparatively low efficiency of selection against rare recessives becomes evident when the number of generations required to reduce the fraction of lethals is computed. For example, when the lethal frequency is 1%, one may wish to know the time required to lower it to 0.1%. The gene frequencies are $q_0 = 0.1$ and $q_n = 0.0316$, and

$$n = \frac{1}{0.0316} - \frac{1}{0.1} = 31 - 10 = 21 \text{ generations}$$

In cattle, where the generation interval is about 5 years, 21 generations span about a century. Mutations will slow this change down even further.

Changes in the intermediate range of gene frequencies are faster. For example, to lower the gene frequency from 0.5 to 0.1 will require eight generations. Such a situation may occur when a breeder decides to rid his population of a recessive which may have become economically undesirable (for example, to decrease the frequency of genes for red coat color in a cattle population heretofore polymorphic for red and black).

When selection is incomplete more complicated expressions for the time required to change gene frequencies become necessary (Table 2.4). Their derivation requires the assumption of a model with overlapping generations and very small changes in gene frequency. The number of generations is computed by integrating the approximate rates of change, such as $\Delta q = -spq^2$. For a recessive the resulting formula is

$$n = \frac{1}{s}\left(\frac{1}{p_0} - \frac{1}{p_n} + \ln\frac{q_0 p_n}{q_n p_0}\right)$$

The time required is inversely proportional to the selection coefficient. In the mink example the number of generations necessary to change the frequency of the "pastel" gene from 0.881 to 0.500 can be computed as

$$\frac{1}{0.11}\left(\ln\frac{0.50}{0.50}\frac{0.881}{0.119} + \frac{1}{0.5} - \frac{1}{0.881}\right) = 26.1 \text{ generations}$$

In other gene frequency ranges the time required will be different. For example, to change p by the same amount from 0.500 to 0.119 would require 41.1 generations.

TABLE 2.4
Length of Time to Change Gene Frequency

Mode of inheritance	Time, number of generations[a]
Semidominant	$\dfrac{2}{s}\ln\dfrac{p_n q_0}{q_n p_0}$
Dominant	$\dfrac{1}{s}\left(\ln\dfrac{p_n q_0}{q_n p_0} + \dfrac{1}{q_n} - \dfrac{1}{q_0}\right)$
Recessive	$\dfrac{1}{s}\left(\ln\dfrac{p_n q_0}{q_n p_0} + \dfrac{1}{p_0} - \dfrac{1}{p_n}\right)$

[a] p, Frequency of favored gene. s, Selection coefficient. Subscripts denote generations.

Multiple loci complicate the expressions a great deal. Only comparatively simple situations have been clarified. One approach consists of treating gametes as genes. In the case of two loci, the four gametes AB, Ab, aB, and ab are treated as multiple alleles, and the change in gamete frequency is given as $\Delta p_i = p_i(w_i - \overline{w})/\overline{w}$ or, when j identifies all gametes other than i, $\Delta p = pq\,(w_i - w_j)/\overline{w}$. With close linkage, and when only a short time span is considered, this should be a reasonable approximation.

When linkage is loose and/or the time span is long, recombination will take place. Kimura (Crow and Kimura, 1970) has described this with a continuous model. The selective value of a genotype, or more precisely, the Malthusian rate of population growth of a genotype (m_{ij}), equals the difference between birth and death rates. Recombination between loci affects genotypes of newborn individuals, while dead individuals still have the parental combination. If b, c, and D symbolize, respectively, birth rate, recombination frequency, and linkage disequilibrium, the change in gamete frequencies is (subscript denotes the 1 coupling gamete, 2 the repulsion gamete)

$$\Delta p_1 = p_1(m_1 - \overline{m}) - cbD$$

$$\Delta p_2 = p_2(m_2 - \overline{m}) + cbD$$

Linkage disequilibrium changes proportionately to c and also to b, the birth rate of double heterozygotes.

Epistasis can create disequilibrium without linkage. Kimura (Cross and Kimura, 1970) and Felsenstein (1965) have demonstrated such consequences from additive epistasis in haploids, but in diploids the situation is similar. Epistasis is considered, in analogy to linkage, as the superiority of double dominants, or recessives, over mixed haplotypes which have both dominants and recessives: $E = m_1 - m_2 - m_3 + m_4$.

When the difference between this measure of epistasis and the recombination rate is large, linkage disequilibrium may ensue. This may last for a long time, so that Kimura denotes it as quasiequilibrium. Both linkage disequilibrium D and epistasis measure E have the same sign. Positive epistasis leads to positive D, and *vice versa*. However, the linkage measure that best reflects the relationship with E is the ratio p_1p_4/p_2p_3 rather than the difference $p_1p_4 - p_2p_3$.

Kimura and Crow point out that positive epistasis will increase the frequency of cis gametes even when one of these is somewhat undesirable. This increases the variability in the population. In contrast, negative epistasis furthers trans gametes. Therefore, variability will diminish and intermediary phenotypes increase.

The effects of epistasis in diploids are qualitatively similar. Regular epistasis, i.e., superiority of double dominants or double recessives, causes positive disequilibrium, whereas superiority of mixed types leads to negative disequilibrium. Irregular epistasis has consequences which generally cannot be predicted but which depend on gene frequencies.

One would assume that natural selection can only increase the adaptive value of a population. This seems intuitively plausible. The formulas for gene frequency change by selection imply that changes occur only insofar as the adaptive value of the population is improved. Fisher (1930) formulated the "Theorem of Natural Selection," which states that the increase in fitness of any population at any time is equal to the genetic variance of fitness in the population at the time.

Kempthorne and Pollak (1970) point out that this law holds only for clonally propagated populations and not in a strict sense for sexually reproducing diploid organisms. In these, at least two components of fitness exist, viability, which is an attribute of an individual, and reproductive performance, which really is to be attributed to the two partners of a mating. Fisher's law can be shown to hold approximately with regard to the fitness of a population of young animals which has not been exposed to viability selection. Fitness would increase in proportion to the quotient of genic variance of fitness divided by one plus the population mean fitness.

No such comparatively simple relation exists for the fitness of an adult population which has been already exposed to viability selection. However, if there is only selection for viability and equal fecundity of all matings, the rule holds again. In contrast, when no viability differences exist and selection is only for fecundity, the change in population fitness is proportional to a quantity which is about double the aforementioned quotient. If fitness and viability are closely correlated, the change in fitness may even be negative. Kempthorne and Pollak emphasize that different concepts of fitness lead to different expressions of the change of population fitness relative to the genetic variance of the population. They are pessimistic about the possibility of proving Fisher's Fundamental Theorem, even for rather simple models.

Crow and Kimura emphasize that the validity of the theorem presupposes panmixis, a continuous generation model, and constancy of selective values for genotypes. They put forward an expression describing the change in mean fitness of a population which contains the genic variance of fitness, the average change of the population fitness due to change in the external situation (environment or competition with other populations), and a product which involves nonadditive genetic effects and the deviation from genetic equilibrium.

When epistasis exists its effect may cancel recombination loss. Loose linkage may lead, under selection, to a linkage disequilibrium whose effects may cancel the epistatic effects. Therefore, the Fundamental Theorem may hold under such presumably widely occurring conditions. The authors emphasize that Fisher's Theorem describes complex biological situations but that it, like many models, only approximates nature in a rather incomplete fashion.

2.3.1. Equilibria under Selection

Even though selection is weak at near extreme gene frequencies, eventually it should fix or remove a gene depending on whether its action is positive or negative with regard to the adaptive value of the genotype, and also depending of course on the existence of a large population. In small populations loss and fixation may occur by random drift.

In large populations a gene locus will remain polymorphic, i.e., it will be occupied by several alleles, when forces prevent fixation. Such forces may be mutation and/or migration pressure, or temporal and/or spatial differences in selection pressures.

Genetic polymorphism, an expression coined by E. B. Ford, is said to exist when several alleles occur at a locus and when the frequency of the rare allele surpasses the mutation–selection equilibrium frequency (Section 2.3.2).

Overdominance is much discussed and is a cause of genetic polymorphism. It is present when the adaptive value of the heterozygote is greater than that of the better homozygote. An extreme example for overdominance would be balanced lethals. In Table 2.5, frequencies and adaptive values of genotypes are given. Both homozygotes are at disadvantages relative to the heterozygote. This situation can be simply described by using selection coefficients (s_1, s_2) against the homozygotes.

TABLE 2.5
Selective Values and Selection Coefficients under Overdominance

	A_1A_1	A_1A_2	A_2A_2
Frequency before selection	p^2	$2pq$	q^2
Selective value	$w(1 - s_1)$	w	$w(1 - s_2)$
	w'	$w'(1 - hs)$	$w'(1 - s)$

Another notation gives a negative value to the degree of dominance h. The change in gene frequency is given by

$$\Delta p = \frac{-pq(s_1 p - s_2 q)}{\overline{w}} = pqs \frac{q + h(p - q)}{\overline{w}}$$

$$\overline{w} = w_{A_1 A_2}(1 - p^2 s_1 - q^2 s_2) = w_{A_1 A_2}[1 - sq(2ph + q)]$$

The gene frequency change may be positive or negative, depending on the sign of the parentheses in the numerator. The curve given in Fig. 2.2 is based on the following selection coefficients: $s_1 = 1/6$, $s_2 = 1/2$, or $s = 2/5$, $h = -1/2$. At frequencies of $p = 0.2$, $q = 0.8$,

$$\Delta p = 0.0704/0.808 = 0.08713$$

and at gene frequencies of $p = 0.9$, $q = 0.1$,

$$\Delta p = -0.0108/1.032 = -0.010465$$

The expression equals zero when the quantity in the parentheses in the numerator equals zero, i.e., $s_1 p = s_2 q$, or $q = -h(p - q)$. For the present example this happens at $q = 1/4$. This equilibrium gene frequency is found from

$$q_E = \frac{s_1}{s_1 + s_2} = \frac{-h}{1 - 2h}$$

The equilibrium is stable. When gene frequencies are above the equilibrium point, selection will decrease them, and it will increase them when the frequencies are below equilibrium. Overdominance prevents a large population from becoming homozygotic and will tend to anchor gene frequencies to the equilibrium value. In small populations overdominance may contribute to an even more rapid fixation of genes (Robertson, 1964).

In our example the $A_1 A_1$ genotype is superior to the alternative homozygote and it should have two-thirds more progeny than the latter. However, the heterozygote is superior by one-fifth to the $A_1 A_1$ genotype. In $A_1 A_2$ the A_2 gene, inferior in the homozygote, exerts a beneficial influence. An equilibrium is achieved when the number of A_2 genes contributed by the superior heterozygote cancels the number lost by the inferior $A_2 A_2$ genotype. The adaptive value of the population is at its peak when the gene frequencies are at the equilibrium (Fig. 2.3).

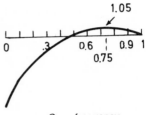

FIGURE 2.3. Population mean selective value w under overdominance.

Gene frequency

The occurrence and significance of overdominance, or single-locus heterosis, are much discussed. Wright suggested a model which can explain overdominance by pleiotropic effects of the alleles (Fig. 2.4). Each gene has several effects, some advantageous, some detrimental. One may assume that the advantageous effects are at least partially dominant. Under such conditions, the beneficial effects of the two alleles in a heterozygote outweigh their detrimental effects and the heterozygote will be superior to each homozygote.

Direct demonstration of heterozygote superiority is relatively rare. The best-documented case involves sickle cell anemia, which fits Wright's model very well. The normal gene (Hb^1) is dominant with respect to oxygen transport. The sickle cell gene (Hb^s) is dominant with respect to resistance to malaria. Therefore, the heterozygote combines, at least partially, the advantages of both genes. A similar situation exists for warfarin resistance of rodents, where one allele is dominant with regard to resistance to toxicity but recessive with regard to increasing vitamin K requirements (Greaves and Ayres, 1977). The heterozygote combines the advantages of resistance to the toxin with nearly normal vitamin K requirements. The gene for stress resistance in pigs also provides an example of overdominance (Section 17.2), where this is really due to the antagonism of natural and artificial selection: Natural selection favors the normal stress resistance allele, while artificial selection for muscularity favors the stress susceptibility allele. Other examples of heterozygote superiority due to the interaction of natural and artificial selection are provided from fur breeding, where some heterozygous phenotypes are favored by the market even though in some cases homozygotes are

FIGURE 2.4. Wright's model of overdominance. Gene A_1 affects component M (———) favorably, component S (— — —) unfavorably. Gene A_2 acts in the opposite direction. Under complete dominance, value of heterozygote equals 18 units.

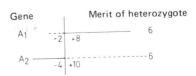

Gene | Merit of heterozygote

lethal, e.g., the gray karakul. In Wyandotte chickens the rose comb is breed standard. However, Crawford (1964) showed that cockerels homozygous for the rose comb have impaired fertility. Therefore, the heterozygous rose-comb cockerels, which are normally fertile, have superior adaptive values and the gene for single comb remains in the population.

Temporal and spatial differences in the adaptive value of genotypes will enhance genetic polymorphism. When the adaptive values change between different life stages of an individual, in many cases heterozygous superiority ensues. For example, Briles *et al.* (1957) found genetic polymorphism for *B* blood-group alleles in some highly inbred lines of chickens. It appeared that one allele improved pullet viability, while the other allele enhanced layer viability, so that the heterozygotes had advantages in both stages of the chicken's life. At that time Marek's disease was the principal cause of losses before housing. It has subsequently been controlled by vaccination, so that the allele conferring resistance to this disease should have lost much of its advantage and overdominance at this locus should disappear. This situation is similar to that of sickle cell anemia, where eradication of malaria changes the Hb^s allele to an ordinary detrimental recessive.

As for regional differences in selection pressure, Levene (1953) showed that where the harmonic mean of selective values in different niches displays overdominance, i.e., h is negative, genetic polymorphism will exist. Crow and Kimura emphasize that Levene's assumptions are restrictive. In reality it is to be expected that phenotypes prefer niches suited to them. If migration between the niches is restricted, polymorphism should be expected. Assume that breeders in different regions have different breeding goals. For example, in one region high milk fat-% may be emphasized (for example, in New Zealand), in another milk yield (in the U.S.). The blood-group gene BO_1Y_2D' (or an adjacent gene) improves fat-% but seems to lower milk yield (Andresen *et al.*, 1960; Conneally and Stone, 1964). Its frequency should be enhanced by selection for milk fat-% and possibly lowered by selection for yield. Occasional exchange of breeding animals will prevent fixation or loss of the BO_1Y_2D' gene.

Felsenstein (1976) has summarized theoretical work dealing with consequences of spatial and temporal variation of environments. Both should maintain genetic variability at a higher level than should uniform environments. Mackay (1980) tested such predictions with *Drosophila* and found that spatially heterogeneous nutritional environments, as well as temporally changing nutrition, will increase genetic variability, the latter more than the first, contrary to the theoretical expectation. At the same time varying environments improve fitness. Therefore the author concludes that in varying environments heterozygous animals seem to be preferred by natural selection.

Selection may be different in the two sexes and thus lead to genetic polymorphism. Zurkowski and Bouw (1966) report that gene frequencies in male and female progeny of heterozygous bulls differ in opposite directions from the expected 1:1 ratio. The gene frequency of registered daughters for the $I'H_4$ allele (B blood-group locus) was 0.45 and that of registered sons was 0.55, the difference being statistically significant. The authors point out that bulls were selected to conform to certain type standards and cows to production standards, respectively. Muscularity and milk yield are negatively correlated genetically, and possibly the $I'H_4$ and I_2 alleles influence these traits in opposite directions.

A stable equilibrium may arise with multiple alleles. Heterozygotes need not exceed every homozygote in adaptive value, but must have on average a higher adaptive value than the average of the homozygotes. Li (1967) discusses the conditions which must hold in order to maintain polymorphism under such conditions. Examples can be provided by the large series of blood-group alleles in domestic animals. Conneally *et al.* (1963) found indications that cattle embryos heterozygous for B alleles have higher chances of survival than do homozygous genotypes.

In Fig. 2.2 mutation pressure is given by the straight line. Equilibrium frequencies are at the intersections between this line and the curves that reflect the gene frequency changes caused by selection. Gene frequencies lower than these points will be increased by selection, while those greater will be decreased by mutation. In this region selection is too weak to balance the mutation pressure. The net change of gene frequencies is given by the difference on the same vertical axis between the point on the line and the point of the curve describing the selection influence.

When the mutation rate is 10^{-6}, selection coefficients of various magnitudes will give the following frequencies Q of mutant phenotypes:

s	0.01	0.1	1
Q	10^{-4}	10^{-5}	10^{-6}

It is evident that even comparatively weak discrimination reduces the frequency of mutant phenotypes to low values. A relatively weak selection force suffices to keep the mutants at bay. However, for any given selection pressure it is impossible to reduce the mutant frequency below the equilibrium point.

Selection, even though weak at the borders of the frequency space, would eventually eliminate undesirable genes if mutation and/or migration did not bring them back.

A small disadvantage of heterozygotes will greatly reduce the equilibrium frequency of mutants. Since the great majority of mutant genes occur in heterozygotes, selection will be much more effective when these can be discriminated against. For example, assume $s = 1$, $\mu = 10^{-5}$, and $h = 0.02$. The equilibrium gene frequency will be 5×10^{-5}, in contrast to $1/316$ when the gene is completely recessive. Reduction of the adaptive value of heterozygotes to 98% of that of homozygotes will thus reduce the frequency of recessive lethal individuals to $1/4000$ of the value for completely recessive mutant genes.

The frequency of homozygotes is increased in small populations, where some inbreeding is unavoidable. This makes selection more effective and the selection–mutation equilibrium is reduced. In a population with inbreeding coefficient F (Chapter 3) the equilibrium frequency of a completely recessive undesirable gene will be

$$[q^2(1 - F) + Fq]s = \mu$$

Assuming $\mu = 10^{-6}$, $s = 1$, and $F = 0.01$, we find that the expression yields an equilibrium gene frequency of 10^{-4} instead of 10^{-3} under panmixis. An inbreeding coefficient of 0.10 will reduce the equilibrium frequency further to 10^{-5}.

The expression "genetic load" was coined by Muller, but Haldane, in writing of the costs of natural selection, had nearly the same concept in mind. Genetic load is the proportion by which the average adaptive value of a population is reduced by detrimental genes (Crow, 1958; Crow and Denniston, 1981):

$$L = \frac{w_{\max} - \bar{w}}{w_{\max}}$$

where $w_{\max}$ denotes the adaptive value of the optimal genotype and $\bar{w}$ the average adaptive value. This concept has been much discussed in population genetics. Today it tends to have a more neutral meaning and denotes the variability of adaptive values present in a population, either open (panmictic load) or hidden (inbred load) to natural selection. Load can be due to several factors. The two most important kinds are caused by recurrent mutations (mutation load) or by superiority of heterozygotes and thus inferiority of homozygotes (segregation load). Other causes of load can result from dam–fetus incompatibility, substitution of genes, and meiotic drive.

Mutation and segregation loads can be differentiated by comparing their extent in panmictic and in inbred populations. In a panmictic population, most rare lethal genes will be in heterozygotes. Therefore the reduction in average adaptive value of the population load will be $L_P = 2shpq$. The equilibrium frequency is $q = \mu/sh$, and therefore $L_P = 2\mu p$ and, since q will be very small, p approximates unity and $L_P = 2\mu$. Upon inbreeding without selection, the fraction q should become homozygous recessive, and $L_I = sq = \mu/h$. The ratio $L_P/L_I = 2/h$ will be large because h is small for lethals (for example, for *Drosophila* lethals the degree of dominance h is estimated as 0.02).

King (1967) has expressed the mutation load as $L_P = 3U/(N - m)$, where U denotes the total (gametic) mutation rate, N the mean number of mutants in individuals removed by selection, and m the average number of mutants per individual before selection.

Where undesirable homozygotes result from overdominance, $L_P = s_1 p^2 + s_2 q^2$. The equilibrium frequency is $q = s_1/(s_1 + s_2)$. Therefore,

$$L_P = s_1 \left(\frac{s_2}{s_1 + s_2} \right)^2 + s_2 \left(\frac{s_1}{s_1 + s_2} \right)^2 = \frac{s_1 s_2}{s_1 + s_2}$$

Upon inbreeding,

$$L_I = s_1 p + s_2 q = \frac{2 s_1 s_2}{s_1 + s_2}$$

and $L_I/L_P = 2$.

Therefore, while the ratio of inbred load to panmictic load is large when it is caused by mutations, it is small when heterozygote superiority is the cause (double, or, when k alleles occur at the locus, k times L_P).

The ratio L_I/L_P permits one to differentiate between mutation and segregation as causes for the load (Morton *et al.*, 1956; Crow, 1958). Of course, several assumptions must hold in order for this method to be valid. Most investigations of human populations strongly suggest that recurrent mutations are the principal cause of the load. In contrast, some research on cattle (Conneally *et al.*, 1963) revealed that this inbred load for return rate and fetal death is less than double the panmictic load. Therefore, part of the reproductive wastage in cattle may be caused by segregation load. The loci should be independent, i.e., the adaptive value of an individual should be the product of the adaptive values of individual loci ($w = w_1 w_2 \cdots w_{n-1} w_n$). Levene (1963) found in *Tribolium* indications of epistasis for adaptive value and differences between L_P and L_I of their various populations which could not be explained. Further, for the inference to be valid it must be assumed that environmental and genetic components of the load are independent and that the population on which the load ratio is estimated is in equilibrium between mutation and selection, or the opposing selection. Haldane and Jayakar (1965) have pointed out that the load is frequently estimated for parts of the selective value only. When the total selective value is considered, overdominance may be present even though it is lacking for any one of its components.

Two facts should be pointed out. First, in the case of segregation load, $L_P = 2\mu$, which is twice the load caused by complete recessives. In recessives two genes must be present to manifest damage, but one gene suffices in the case of dominants. Second, most of the segregation load is caused by the better homozygotes. Assume $s_1 = 1/6$ and $s_2 = 1/2$. The equilibrium gene frequencies will be 3/4 and 1/4 and the load caused by the better homozygote is 2/32, which is three-fourths of the total load.

Looked at from another angle, the genetic variability in a population will be much larger when it is maintained by overdominance than when it is due to the mutation–selection balance. Mukai (1964) has estimated both the total mutation rate and the load for lethals, viability polygenes, and polygenes affecting abdominal bristles in *Drosophila*. He found that in the case of the former two classes of genes, the load was about 70–80 times the mutation rate per generation but that the genetic variance for abdominal bristles extant in the lines was about 800 times the variance created by mutation in one generation. Probably, the load of lethals and of viability polygenes is caused by recurrent mutations and the high variance of bristle numbers by heterozygote superiority.

The expression "lethal equivalents" has been introduced in connection with the load concept. It equals $\Sigma\, sq$, i.e., the sum of the products between selection coefficients and gene frequencies. If $s = 1$, as for lethal and sterility genes, the lethal equivalent is the sum of such genes in a gamete. In diploids it must be doubled. It gives the number of genes an individual carries that, if doubled, would cause genetic death. It is approximated by the inbred load. For humans it is estimated that on average each individual carries one lethal equivalent. Pisani and Kerr (1961) estimated from perusal of the literature that White Leghorn chickens carry 4–5 lethal equivalents and Poland China pigs 1.6 for intra-uterine death and 0.4 for postnatal mortality.

2.4. Random Drift

In large populations mutation, migration, and selection influence gene frequency in principle in a predictable fashion in terms both of direction and of speed of change. Gene frequencies should reach, under the influence of these systematic pressures, equilibrium points. However, populations, particularly those of domestic animals, are frequently of limited size. In such populations gene frequencies may change randomly. Therefore a stochastic model is required to describe gene frequency changes in a finite population.

Mendel showed that distribution of genes into gametes is random. Therefore, the distribution of n progeny with n, $n - 1$, $n - 2$, etc., genes follows a binomial probability distribution. If in a small population most offspring will have received by chance a particular gene rather than its allele, the gene frequency of the progeny generation will be changed, and if the gene is neutral, sampling of gametes proceeds from the new gene pool.

Assume a population of four heterozygous individuals. Gene frequency equals 1/2 and this is the expected frequency of the offspring generation. The probability of four offspring being (a) homozygotes of one type, (b) heterozygotes, and (c) alternative homozygotes is given by

$$\frac{4!}{a!\,b!\,c!}\, P^a Q^b R^c$$

where P, Q, and R are the F_2 probabilities 1/4, 1/2, and 1/4, respectively. The probability that four offspring comprise two homozygotes of one type and two heterozygotes equals 6/64, nearly 10%. Their gene frequency is 0.75, a change of 0.25 from the parent generation. The probabilities of offspring of each type in the next generation depend on P', Q', and R', which are changed to 9/16, 6/10, and 1/16, and therefore the probability of fixation of gene A_1 is considerably increased.

Genetic drift involves random gene frequency changes from generation to generation and their accumulation. Ultimately this should lead to loss or fixation of alleles.

Populations of domestic animals have a hierarchic structure with comparatively small numbers in the top layer. This has the consequence that random gene frequency changes may assume importance. A large change in gene frequency reported by Zurkowski and Bouw (1966), where the blood-group allele I_2 increased from below 1% to 11% in a mere two generations, is probably a case of random drift. Its immediate cause was the extensive use of male progeny of one cow in Dutch Friesians. The occurrence of translocation, such as the 1/29 translocation in Swedish Red-and-White cattle (Gustavsson, 1969) and in Brown Swiss (Zwiauer *et al.*, 1980), is probably also an example of random drift.

A patent example of random drift as it causes lines to become distinct has been given by Cooper *et al.* (1967). Table 2.6 gives frequencies of transferrin genes in Australian Merinos. Frequencies in lines R, M, and CR deviated from line A, considered to be representative of most of the Australian Merino population. The deviation from A is particularly pronounced for lines M and CR, which have been separated from the main population for a long time.

The direction of drift cannot be predicted, as the expression itself implies, but its extent can be estimated. Assume that a population consists of N animals. Therefore, it has $2N$ genes at one locus. Assume two alleles A_1 and A_2 with frequencies p and q. If the gene is neutral, no change of gene frequencies is expected on average, but the random fluctuations have the variance

$$\sigma^2_{\Delta q} = \frac{pq}{2N}$$

The extent of drift is inversely proportional to the size of the population. The measurement of population size is frequently difficult, but gene frequency changes, or rather their variance, in populations can be used to gauge the genetically effective size (variance effective; Crow, 1954), which is defined as the size of a panmictic population giving rise to the variance of gene frequency changes actually observed.

Variances of gene frequency changes in populations of various size are given in Table 2.7. Frequencies in generation zero were 0.5. When $N = 5$, the standard deviation is 0.16. Therefore, the gene frequency of progeny of such lines should lie in the range between 0.34 and 0.66, but 5% of the progeny should lie beyond the limits set by the means ± 2 standard deviations, i.e., either below 0.18 or above 0.82. Thus in one out of 20 progeny groups of five individuals one expects the gene frequency to deviate by more than 0.32. In comparison to the small changes

TABLE 2.6
Random Drift of Transferrin Genes in Australian Merinos[a]

Line	A	B	C	D	E	N_e	Age of line, years
A	16	21	2	60	0.2	470	—
R	26	20	—	52	2	100	17
M	—	69	—	31	—	?	110
CR	34	—	—	49	18	?	80

[a] Source: Cooper *et al.* (1967).

TABLE 2.7
Variance Caused by Random Drift of Gene Frequencies
and Fixation Probability[a]

	Variance			
Generation	$N = 2$	$N = 5$	$N = 10$	$N = 50$
1	62.5	25	12.5	2.5
10	235	163	100	24
20	249	220	160	45
		u		
$s = 0.1$	0.559	0.668	0.881	1.000
$s = 0.01$	0.510	0.525	0.550	0.731

[a] Variance × 10^3 between progeny lines. Frequency in base generation $p = q = 0.5$. N, population size. u, Probability of gene fixation. s, Selection coefficient in favor of gene.

which usually happen under systematic pressures, random drift can assume large proportions in small populations.

Genetic drift results from small population sizes, from the random nature of the Mendelian process, and from the accumulation of the random deviations over the generations in persistently small populations. The characteristics of such populations will fluctuate over a wide range. The variance that accumulates by drift over t generations among populations of size N is

$$\sigma_q^2 = pq\left[1 - \left(1 - \frac{1}{2N}\right)^t\right]$$

which for populations that are not very small can be approximated by

$$\sigma_q^2 = pq(1 - e^{-t/2N})$$

A high value for t will lead to fixation or loss and the variance becomes pq. Table 2.7 gives variances of gene frequencies expected after 10 and 20 generations of isolation in lines with N of various sizes. Regardless of their size when populations are homozygous, the variance will be 0.25.

With increasing number of generations the probability of fixation of a gene increases. Kimura (Crow and Kimura, 1970) has approximated the probability of fixation by

$$1 - 6pqe^{-t/2N}$$

For the examples of Table 2.7 this gives as fixation probability of lines with $N = 2$ in generation ten (full sibs) 88%, of lines with $N = 5$, 45%, and for lines of ten individuals, 9%. Therefore, 9% of lines composed of ten individuals should have become homozygous for neutral loci up to and including generation ten. The approximation is really only permissible for larger values of t. It cannot be applied to $N = 50$ and $t = 10$. For $t = 50$ the fixation probabilities for the four population sizes given in Table 2.7 are, respectively, 100, 99, 88, and 9%. Random drift causes the original symmetric distribution of gene frequencies in the early generations of isolates from a parent population to degenerate to a rectangular distribution. Further drift degenerates to a U-type distribution, where the gene in the majority of isolates is either lost or fixed. When the rectangular distribution is established neutral genes in $1/(2N)$ of isolates become fixed or are lost in each generation. The discussion has referred only to one locus in N populations. The same results are expected when several neutral independent loci are assumed in one population of size N.

When certain genes are advantageous a larger fraction will be fixed. The probability of ultimate fixation for neutral genes equals the gene frequency. For genes with selection coefficient s, Kimura has shown that their fixation probability depends also on both their initial frequency p and population size N:

$$u(p) = \frac{1 - e^{-4Nsp}}{1 - e^{-4Ns}}.$$

In Table 2.7 fixation probabilities are given for genes with advantages of 1 and 10% in their adaptive values. Note that the probability of fixation of an allele with a given advantage increases with increasing size of the population. For example, an allele with an advantage of 1% in a population with $N = 50$ has the probability of its ultimate fixation increased to nearly 3/4, while in a population of ten individuals $u(p)$ is increased, relative to a neutral gene, by only 50–55%. Looked at from another point of view, the probability of loss of advantageous alleles is fairly large in small populations. For example, in full-sibling lines ($N = 2$), even strongly beneficial alleles ($s = 0.1$) with starting frequencies of 50% will be lost in 40% of the lines. Therefore, small and/or moderately sized populations which have been isolated for many generations are expected to be homozygous at a large fraction of their loci.

In the case of a newly arisen advantageous mutant, $p_0 = 1/(2N)$, and

$$u(p_0) = \frac{1 - e^{-4Nsp}}{1 - e^{-4Ns}} \approx \frac{2s}{1 - e^{-4Ns}}$$

When Ns is large the formula reduces to $2s$. The ratio of the probability of its fixation to that of a neutral mutant $1/(2N)$ is $4Ns$, i.e., it increases linearly with the size of the population, provided s is small and N is reasonably large. However, in absolute terms the fixation probability does not change very much in small populations where $Ns \geqslant 2$ because the fixation probability of even a neutral gene is relatively large.

One may ask whether there are such things as neutral genes. From the point of view of population genetics one may consider a gene to be neutral when drift is more important than selection in causing the gene frequencies to change. Depending on what criteria the judgement is based on, the size of s compatible with operational neutrality varies somewhat. King and Jukes (1969) consider a gene as neutral when $s < 1/N_e$. Another criterion could be that the probability of fixation of a neutral gene should be larger than for an advantageous allele, i.e., $u(p_0) = p$. Now $e^{-4Ns} < 1 - 4Ns$ when $Ns < 1/16$. If this holds,

$$u(p) \approx \frac{1 - (1 - 4Nsp)}{1 - (1 - 4Ns)} = \frac{4Nsp}{4Ns} = p$$

and $s \ll 1/(16N)$. Therefore a gene may behave as neutral in a small population while acting as a beneficial gene when N is large.

Random drift can easily be simulated by throwing coins, drawing random digits, etc. In Fig. 2.5 and Table 2.8 the results of a classroom exercise with cards are given. Forty-eight populations with $N = 4$ were

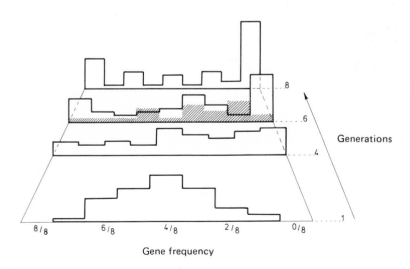

FIGURE 2.5. Simulated random drift.

TABLE 2.8
Means and Variances of Simulation Experiment on
Random Drift[a]

Generation	$\bar{p}$	$\sigma^2 \times 10^4$	$E(\sigma^2) \times 10^4$
0	0.500	—	—
1	0.508	311	312
2	0.461	540	586
3	0.425	690	825
4	0.414	1011	1034
5	0.388	1166	1216
6	0.406	1306	1378
7	0.382	1512	1518
8	0.378	1618	1640

[a] $N = 4$. Forty-eight populations.

simulated through eight generations. It will be noticed in Fig. 2.5 that one line with $p = 7/8$ and two with $p = 1/8$ had occurred already in the first generation. In generation four the distribution of gene frequencies had become rectangular and it was decidedly U-shaped in generation eight, even though somewhat asymmetric. Expected $[1 - (1 - 1/2N)^t]$ and observed variances between lines are given in Table 2.8. Their agreement is, not surprisingly, very close. The shaded area in Fig. 2.5 represents the frequency distribution of the amylase C gene in 24 European cattle breeds. It resembles the distribution in one of the middle generations of the simulated lines. Therefore, one may regard the amylase distribution as a result of drift in, of course, larger populations over longer time periods (however, there are some indications that the amylase genes are not neutral).

In small populations homozygosity will be increased because drift will have moved gene frequencies toward fixation and thus decreased heterozygosity, even though within each isolate genotypes are in Hardy–Weinberg equilibrium. This was noticed by Wahlund (1928) more than 50 years ago and is demonstrated with a simple example in Table 2.9. The following identity is used:

$$\sigma_p^2 = \overline{p^2} - (\bar{p})^2$$

i.e., the variance is given by the difference between the mean squares of quantities and their squared mean. The average frequency of homozygotes is $\overline{p^2} = (\Sigma\, p^2)/n$, i.e., the mean square of gene frequencies, and it is found from

$$\overline{p^2} = \sigma_p^2 + (\bar{p})^2$$

TABLE 2.9
Genotypes in Small Populations

Subpopulation	p_i	p_i^2	$2p_iq_i$	q_i^2
1	0.9	0.81	0.18	0.01
2	0.7	0.49	0.42	0.09
3	0.3	0.21	0.42	0.49
4	0.1	0.01	0.18	0.81
Total population	0.5	0.35	0.30	0.35
$E\ (p_ip_j)$ without subdivision		0.25	0.50	0.25
Δ		0.10	−0.20	0.10

$$\sigma_p^2 = \frac{(\Sigma\ p^2)}{n} = \overline{p^2} = 0.35 - 0.25 = 0.10$$

$$\overline{p}^2 + \sigma^2 p = 0.25 + 0.10 = 0.35$$

$$2\overline{pq} - 2\sigma^2 p = 0.50 - 0.20 = 0.30$$

$$\sigma_p^2/\overline{pq} = \frac{0.10}{0.25} = 0.40$$

The frequency of homozygotes is increased by an amount equal to the variance of gene frequencies. This has been called Wahlund's rule. Both types of homozygote are increased by the same absolute quantity. It follows that the heterozygotes are diminished:

$$2\overline{pq} = \frac{2\ \Sigma\ p(1-p)}{n} = 2\left(\frac{\Sigma\ p}{n} - \frac{\Sigma\ p^2}{n}\right) = 2\ (\overline{p} - \overline{p}^2 - \sigma_q^2)$$

$$= 2\ (\overline{pq} - \sigma_p^2) = 2\overline{pq}\left(1 - \frac{\sigma_p^2}{\overline{pq}}\right)$$

The ratio of the proportion of heterozygotes in later generations to their proportion in the base generation is given by

$$\left(1 - \frac{\sigma_q^2}{\overline{pq}}\right) = \left(1 - \frac{1}{2N}\right)^t \approx e^{-t/2N}$$

The quantity $\sigma_p^2/\overline{pq}$ is often called Wahlund's variance, or the standardized variance, of gene frequency since it expresses the observed variance in terms of the maximal variance of gene frequency when populations were homozygous. It is a measure of the deviation from panmixis and it is identical to the inbreeding coefficient (Chapter 3). Even though genotypes may be in equilibrium within any one isolate, homozygotes are increased and heterozygotes diminished in a population consisting of many isolates.

The smaller the isolates and/or the longer the separation has lasted, the greater the proportion of homozygotes and the larger the variance accumulated by drift. However, the average gene frequency of the whole population remains unchanged. The variance among gene frequencies of the isolates is a measure of their differentiation and is proportional to the time of isolation and to the inverse of the population size. A surplus of homozygotes in a large population is an indication of lack of homogeneity of the population, which may consist of several subpopulations (lines, strains).

Random genetic drift increases variability between isolates. This diversifying action of drift is counteracted by the systematic forces of mutation, migration, and selection, which seek to establish the same equilibrium gene frequencies in all subpopulations. These forces can be considered as exerting linear pressure. Their combined effect has been called the stabilizing coefficient (*coefficient de recall*) by Malécot. The gene frequency change caused by these three factors will be $b = m' + u^* + s$, where u^* denotes the combined mutation pressure $(u - v)$, m' the net migration rate (difference between immigration and emigration), and s the net selection pressure, in the case of selection for heterozygotes approximately $s_1 s_2 / s_1 + s_2$. Malécot's designation of b as *coefficient de recall* underlines the fact that the systematic forces tend to retrieve gene frequencies that drifted away from the equilibrium.

With small linear pressures the variance between populations of size N is given by

$$\sigma_q^2 = \frac{pq}{1 + 4Nb}$$

When $4Nb < 2$, drift is more important and the gene frequencies of isolates will be distributed in a U-shaped fashion, as appeared in round eight of the classroom experiment (Fig. 2.5). Alternatively, if one gene is preferred the distribution will be J-shaped. In large populations with significant systematic influences ($4Nb > 2$) the variance will be small and the distribution will tend toward a normal curve.

If sampling of gametes were random, each potential gamete would have an equal probability of coming from any potential parent. The distribution of parents with respect to progeny numbers would correspond to a Poisson series. If population size is static, the mean number of offspring per parent surviving to breeding age is one, or, under biparental reproduction, each parent averages two "successful" gametes

(gametes that are fertilized). Under such conditions the parents would be distributed with respect to progeny number as follows:

Number of offspring	Fraction of parents
0	0.135
1	0.271
2	0.271
3	0.180
4	0.090
5	0.036
6	0.012
≥7	0.004

This distribution clearly shows that the equal probability does not imply that all parents will produce equal numbers of offspring. If the distribution of progeny numbers corresponds to this Poisson series, the variance of gene frequency could be estimated from the formula given previously. However, among both natural and domestic animals the variance of family size is much larger than that of a Poisson series.

The large variability of family size under natural conditions is frequently due to the correlated fate of family members. If a predator discovers a nest of young birds, usually all of them will be destroyed; if they escape detection, the probability is fairly large that all or most of them will reach breeding age. In breeding domestic animals, selection leads to large differences in family size. Infertility is a common factor that contributes to the variation in family size within a population.

Gene frequencies vary more, therefore, than the actual number (census number) of breeding animals in the formula for the variance of gene frequencies would lead us to expect. The method of estimating the "effective size" of a population differs depending on whether that size is used for estimating the variance of gene frequency distribution or the decrease of heterozygosity in small populations (Crow, 1954). However, when the population size is nearly static, which implies an average of two successful gametes per individual, and when the distribution of family sizes follows a Poisson series the "effective numbers" for either purpose are about equal. The effective size of a population may be computed from the following formula:

$$N_e = \frac{4N - 2}{2 + \sigma_f^2}$$

where N_e is the effective size and σ_f^2 is the variance of family size. When family size follows a Poisson distribution its variance is 2 and the effective population size will roughly equal its actual size. If, however, the variance of family size is larger than that, as in natural and domestic populations, N_e will be smaller, in many cases much smaller, than N. Variance of gene frequency changes and therefore random drift will be correspondingly larger.

The (variance) effective population size can be estimated also from gene frequency differences between populations when their time of separation is known and also from frequency changes of neutral alleles over time.

This latter approach can be demonstrated with blood-group data from German Simmental and German Gelbvieh which were published in 1962 and again in 1974. Assuming that the genes are neutral, the observed differences Δ in their frequencies between the two sampling periods is caused by drift and by sampling error. Therefore $\sigma_{\Delta q}^2 = \Delta^2 - \Sigma s^2$, where Σs^2 is the sum of the two sampling error variances. Table 2.10 gives the frequencies. The time interval of 12 years corresponds to roughly $2\frac{1}{2}$ generations. The effective population size is found from

$$\frac{\sigma_{\Delta q}^2}{pq} = 1 - \left(1 - \frac{1}{2N_e}\right)^t$$

and in the case of Simmentals $[1 - 1/(2N_e)]^{2.5} = 0.990808$ and $N_e = 135$, which is roughly similar to estimates from breeds where the increment of the inbreeding coefficient can be used to estimate the (inbreeding) effective size.

At constant family size the variance becomes zero and the effective number is twice the actual number. Consequently, the variance of gene frequency changes and random drift decreases. This becomes important when a control population is to remain constant over long periods of time. Gowe and his colleagues (1959) suggested that relatively small populations comprising families of constant size are suitable as controls in poultry breeding programs. Table 2.11 shows effective sizes of control populations under different systems of mating and with different actual numbers.

In domestic animals the numbers of male and of female breeding animals usually are quite different and the effective population size can be computed from

$$N_e = \frac{4N_M N_F}{N_M + N_F}$$

TABLE 2.10
Estimation of Effective Population Size from Gene Frequency Changes[a]

	German Simmental				German Gelbvieh			
	$q \times 10^3$				$q \times 10^3$			
Locus[b]	1962 $N = 1000$	1974 $N = 1177$	Δ	$\sigma^2_{\Delta q} \times 10^6$	1962 $N = 480$	1974 $N = 147$	Δ	$\sigma^2_{\Delta q} \times 10^6$
J	101	112	−11	−3,092	66	170	−104	85,687
L	150	146	4	−2,999	132	154	−22	−9,300
M	17	7	10	−30,196	107	50	51	7,054
Z	380	480	−100	39,718	546	465	81	21,662
FV	770	775	−5	−784	839	806	33	2,843
a	522	636	−114	52,507	445	595	−150	21,522
Average				9,192				21,578
N_e [c]				135				57

[a] Here $\dfrac{\sigma^2_{\Delta q}}{pq} = \dfrac{\Delta^2 - \sum s^2}{pq} = \dfrac{121 - (225 + 189)}{94,764} = -\dfrac{293}{94,764}$, for $pq = \dfrac{101 + 112}{2} - 894 = 94,764$.

[b] Sampling variance $\dfrac{(1 - q^2)}{4N}$, except for FV, for which it is $pq/2N$.

[c] N_e, Effective population size. $t = 2\frac{1}{2}$ generations.

TABLE 2.11
Effective Population Size in Two Different Mating Systems

Actual number of parents		Effective population size	Increase in homozygosity per generation, %	Maximal random deviation after 20 generations, eggs per hen housed
♂	♀			
		Random mating		
25	25	50	1.00	±25
100	100	200	0.25	±13
50	250	167	0.30	±14
100	300	300	0.17	±10
		Pedigree control, family size constant		
25	25	100	0.50	±18
100	100	400	0.12	± 9
50	250	250	0.20	±11
100	300	480	0.10	± 8

[a] Source: Gowe et al. (1959).

where N_M is the number of breeding males and N_F is the number of breeding females. Assume that a small herd-book society has 1000 registered cows and 20 registered bulls, and that replacements come from the same population, i.e., the breed is maintained as a closed population. From the formula just given the effective size of such a population is 78. In other words, the variance of the gene frequency (the original frequency is assumed to be 0.5) and the extent of random drift have the same magnitude as in a population of 78 animals all of which have an equal chance to reproduce. The variance of a gene frequency of 0.5 is $0.25/(2 \times 78) = 0.0016$. Each of the homozygote classes will increase by this fraction and the heterozygotes will decrease by 0.0032 in each generation. When artificial insemination allows the reduction of bulls to four, say, the effective size of the population will decrease to 16 and the variance of the gene frequency will increase to 0.0008—almost fivefold. The increment of each of the homozygotes per generation will be 1.6%. Thus the rate of inbreeding and the accompanying disadvantages, such as reduction of genetic variability and increase of undesirable homozygotes, would increase to that of a population of 16 individuals in which each individual has an equal chance to reproduce.

When population size varies over time, that is, from generation to generation, the effective number of animals can be estimated from the following expression:

$$\frac{1}{N_e} = \frac{1}{t}\left(\frac{1}{N_1} + \frac{1}{N_2} + \cdots + \frac{1}{N_t}\right)$$

where t is the number of generations. A large decrease in population size in any one generation will have a large effect on the effective number, as any example with assumed numbers will readily demonstrate. This is quite important in nature, where, for example, populations of insects are severely reduced in winter. The genetic composition of future populations will depend largely on the genotypes of the few animals that survive to reproduce in spring. In domestic populations such drastic fluctuations are rare, although economic pressure may force severe reductions in a breeding population's size, which may then be kept small for some generations. Such a "bottleneck" may have a lasting effect on the future gene pool of a breed. When domestic breeds were founded, the first generations consisted of relatively few animals. Consider, for example, the history of the Shorthorn cattle breed.

It must be emphasized that applying the Hardy–Weinberg law to describe the distribution of genotypes presupposes that the gene frequencies of the sexes are equal. As mentioned already, populations of domestic animals usually have only a few breeding males. Thus, a difference in gene frequency between the breeding females (p_1) and the sires (p_2) could obtain. Robertson (1965) showed that a surplus of heterozygotes over the expected proportion exists if p_1 and p_2 are different. The actual heterozygote frequency in such a population will be

$$Q = p_1 q_2 + p_2 q_1$$

The heterozygote frequency expected on the basis of equilibrium is

$$Q' = 2 \frac{p_1 + p_2}{2} \left(1 - \frac{p_1 + p_2}{2} \right)$$

The heterozygotes will be in excess by

$$Q - Q' = \tfrac{1}{2}(p_1 - p_2)^2$$

that is, by half of the squared difference between the gene frequencies. When the total population is assumed to be distributed randomly into many small herds consisting of N_M males and N_F females each, the gene frequencies of the herds have a variance of $pq/(2N_M + 2N_F)$. Now the above difference $\tfrac{1}{2}(p_1 - p_2)^2$ equals twice the variance of the gene frequency, so the excess of heterozygotes can be expressed as $pq(\tfrac{1}{4}N_M + \tfrac{1}{4}N_F) = 2pq(\tfrac{1}{8}N_M + \tfrac{1}{8}N_F)$. Therefore, the actual proportion of heterozygotes is larger by $(\tfrac{1}{8}N_M + \tfrac{1}{8}N_F)$ of the value expected on the basis of the overall gene frequency. In the above example with four bulls and 1000 cows heterozygotes will show a surplus of 1/32 + 1/8000 above their fraction expected with average gene frequencies.

3

Inbreeding

The genotypic structure of a population is not determined exclusively by gene frequencies and factors influencing them, but is also affected by the system used in mating parents. A mating system may deviate from panmixis and the partners may be chosen according to (a) relationship or (b) phenotypic resemblance. When partners are related to each other more closely than randomly chosen individuals, the mating system is called inbreeding, and when they are less closely related, we speak of outbreeding. Choosing partners according to phenotypic likeness is called assortative mating. If the partners are more similar in morphological traits or performance than members of the population chosen at random, it is positive assortative mating. If the partners are unlike, it is negative assortative mating. A breeder practicing compensatory matings in an attempt to make the population more uniform, for example, mating large bulls to small cows and *vice versa,* is practicing negative assortative mating. Positive assortative mating would pair small bulls with small cows and large bulls with large cows.

Inbreeding means mating of relatives. Two related partners have common ancestors and may have received from them some genes that are alike. Offspring of a mating between relatives may receive, therefore, two genes at a locus that are identical, that is, they are both copies of the same gene that was carried by one of the common ancestors.

Breeders have long been aware of the effects of inbreeding, which on the one hand are detrimental with regard to many traits but very often also "fix" traits in lines which otherwise show much variation. Attempts to measure inbreeding in a quantitative sense were not very successful. However, for humans the Catholic Church, in granting marriage licenses, used a measure for the determination of the degree of relationship between prospective marriage partners which paralleled the inbreeding of the potential offspring of such couples.

The measure used almost universally today is the inbreeding coefficient, introduced by Wright (1921). Wright denoted it by F and defined it as the correlation between genic values of gametes, which will be discussed in Chapter 4. In Table 3.1 genic values are symbolized by a and b, respectively, for the male and female gametes.

Somewhat easier to visualize and leading to the same results is Malécot's (1948) definition. Basically, two genes, say x_1 and x_2 can be equivalent in two ways. The two genes can have the same function, that is, in terms of molecular genetics they have the same nucleotide sequence. For example, in a certain cow both genes at the blood-group locus F can be F alleles. The two genes are alike in state. If two mutations create the same gene, we would say that these are alike in state. Two genes, however, can also be identical because they are both copies of the same gene in a common ancestor. Naturally, they have the same function and the same nucleotide sequence (unless a mutation has occurred). Such genes are identical by descent. Some or most of the genes that we consider to be identical in state are also identical by descent, since they originated in a remote ancestor, perhaps several hundred generations back. Usually, however, only a few generations of the pedigree are considered. Inbreeding is relative to a base generation (the last generation of the pedigrees used). Genes present in the base generation that were identical in state are not considered identical by descent.

Cotterman (1940) made a similar distinction. He used the term autozygous genes for genes identical by descent and allozygous to denote identical in state.

TABLE 3.1
Inbreeding Coefficient as Correlation between Genic Values of Gametes

♂ a	♀ b	Frequency	ab	a^2	b^2
$A_2\ 0$	$A_2\ 0$	$q^2(1 - F) + qF$	0	0	1
$A_2\ 0$	$A_1\ 1$	$pq(1 - F)$	0	0	1
$A_1\ 1$	$A_2\ 0$	$pq(1 - F)$	0	1	0
$A_1\ 1$	$A_1\ 1$	$p^2(1 - F) + pF$	1	1	0

Covariance $= p^2(1 - F) + pF - p^2 = Fpq$
Variance $= pq$
$r = Fpq/(pq \times pq)^{1/2} = F$

Malécot defines the coefficient of inbreeding as the probability that two genes at the same locus are identical by descent. At a particular locus the genes may or may not be identical by descent. However, considering all loci, the inbreeding coefficient means a proportion of loci equal to F are expected to carry genes that are identical by descent. In other words a proportion F of loci that were heterozygous in the base generation have become homozygous. The same reasoning can be applied to a single locus in the whole population. Assuming the base generation to be an F_2 with 50% heterozygotes, an inbreeding coefficient of 40% means that 40% of the heterozygous loci in the population have become homozygous, or, in absolute terms, the inbred population is homozygous at 70% of the loci.

The computation of the inbreeding coefficient F can be explained by Fig. 3.1. Assume that ancestor A transmits gene a_1 to his offspring V. The probability that offspring Y inherits the same gene a_1 from A is 1/2. The probability that V, if it has gene a_1, passes it on to W is 1/2, and that W transmits the gene to X is also 1/2. The probabilities of the inheritance of a_1 in each generation are conditional. Therefore, the probability that both X and Y receive a copy of the same gene a_1 from A is

$$\tfrac{1}{2} \times \tfrac{1}{2} \times \tfrac{1}{2} = (\tfrac{1}{2})^3$$

The probability that Y passes the gene to X also equals 1/2. The total probability that X receives a copy of the same gene a_1 from both parents is $(1/2)^4 = 1/16$. Furthermore, the probability exists that X received gene a_1 from one parent and gene a_2 from the other, and that both genes already were identical by descent in ancestor A. The probability that they were identical by descent in A is F_A, the inbreeding coefficient of ancestor A. Therefore, the total probability that the genes are identical by descent in X is $(1/16)(1 + F_A)$. This can be generalized as follows: Assume that there are n_1 generations between a parent and a common ancestor, and n_2 generations between the other parent and the same

FIGURE 3.1. Path of gene from ancestor to offspring.

ancestor, and assume that the coefficient of inbreeding of the ancestor
is F_A. Then the inbreeding coefficient of the offspring is

$$F_X = \left(\frac{1}{2}\right)^{n_1 + n_2 + 1} (1 + F_A)$$

An individual may have several ancestors, for each of which there is a
separate probability that the individual has pairs of genes which are
copies of the same ancestral genes. These separate probabilities are added
to find the total probability of genes identical by descent in the individual.
Wright's formula for the inbreeding coefficient becomes, therefore,

$$F = \Sigma\left[\left(\frac{1}{2}\right)^{n_1 + n_2 + 1} (1 + F_A)\right]$$

The summation refers not only to each common ancestor, but also to
each possible connection between the two parents that goes through a
common ancestor. Each such path must touch the ancestor only once
and must not zigzag. Figure 3.2 demonstrates the computation of the
inbreeding coefficient. It is assumed that the inbreeding coefficient of
individual E is $F_E = 1/8$. In the analysis common ancestors are under-
lined.

The inbreeding coefficient gives the probability that two genes at a
locus are identical by descent, or autozygous, to use Cotterman's term.
Malécot considered the probability that a random gene in one individual
is identical by descent to a random gene in another, and called this
identity measure "*coefficient de parenté*," kinship coefficient, or coefficient
of coancestry. In population genetics this coefficient and the inbreeding
coefficient are frequently referred to simply as F coefficients. In our
further discussion we shall denote it by C, for coefficient of coancestry.
It gives the probability of autozygosity of random genes in two individuals. It is identical to the inbreeding coefficient of a potential offspring

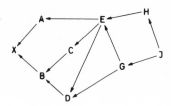

FIGURE 3.2. AECB: $(\frac{1}{2})^4(1 + \frac{1}{8}) = 9/128$. AEDB:
$(\frac{1}{2})^4(1 + \frac{1}{8}) = 9/128$. AEGDB: $(\frac{1}{2})^5 = 1/32 = 4/128$.
AEHJGDB: $(\frac{1}{2})^7 = 1/128$. Sum: 23/128.

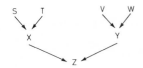

FIGURE 3.3. Generalized pedigree.

of these two individuals. Consider Fig. 3.3. Let the two genes at a given locus of individual X be denoted by a and b, and likewise the two genes at the homologous locus of Y be denoted by c and d. Let $P(a = b)$ denote the probability that genes a and b are identical by descent. The coefficient of coancestry between X and Y is

$$C_{XY} = \tfrac{1}{4}[P(a = c) + P(a = d) + P(b = c) + P(b = d)]$$

It can be shown easily that the coefficient of coancestry between two individuals equals the average of the coefficients of coancestry between their respective parents, or the average of the kinship coefficients of one individual and the parents of the other.

Inbreeding coefficients of individuals in small populations can be computed from the kinship coefficients between individuals. This requires the computation of the coefficient of coancestry of an individual with itself as a formal quantity. By definition it is the probability that any one random gene is identical by descent to a random gene of the individual. The probability of identical descent of a gene with itself is one, of course, and with the second gene at the locus the probability is F, the inbreeding coefficient. The average of the two probabilities equals the coefficient of coancestry of an individual with itself:

$$C_{XY} = \tfrac{1}{2}(1 + F_X)$$

This quantity equals the inbreeding coefficient of an offspring through selfing, and the inbreeding coefficient of the individual is then

$$F_X = 2C_{XX} - 1$$

Inbreeding and coancestry coefficients are closely related concepts. The latter gives the probability of identity of genes in two individuals, the former in one. The coefficient of coancestry between two individuals equals the inbreeding coefficient of their potential offspring:

$$F_{XY} = C_{XY}$$

Computation for the line given in Fig. 3.2 is illustrated in Table 3.2. Animals that are unrelated to their mates will be denoted by O and O'. The computation starts with the first ancestors:

$$C_{II} = C_{HH} = C_{GG} = 1/2$$

$$C_{IH} = C_{I.I \times O} = \tfrac{1}{2}(C_{II} + C_{IO}) = \tfrac{1}{2}(\tfrac{1}{2} + 0) = \tfrac{1}{4} = C_{IG}$$

$C_{I.I \times O}$ is the coancestry coefficient of an offspring with its parent I as a result of the mating I × O and it equals the average of the coefficient of the parent with itself and the coefficient between the two parents. The next animal E in the pedigree is an offspring of G and H:

$$C_{HG} = C_{I \times O.I \times O'} = \tfrac{1}{4}(C_{II} + C_{IO} + C_{O'I} + C_{OG'})$$

$$= \tfrac{1}{4}(\tfrac{1}{2} + 0 + 0 + 0) = \tfrac{1}{8} = F_E$$

The kinship coefficient of two animals equals the average of the coefficients among the four possible parents, or alternatively, the average of the coefficients between one of them and the two parents of the other individual:

$$C_{HG} = C_{H.I \times O'} = \tfrac{1}{2}(C_{HI} + C_{HO}) = \tfrac{1}{8}$$

Eventually one finds

$$F_X = C_{AB} = 23/128$$

in agreement of course with the direct computation. The method is appropriate for keeping a chart of current inbreeding and coancestry

TABLE 3.2
Coefficients of Coancestry between Animals of Fig. 3.2 (× 128)

A	B	C	D	E	G	H	I	
							64	I
						64	32	H
					64	16	32	G
				72	40	40	32	E
			84	56	52	28	32	D
		64	28	36	20	20	16	C
	78	46	56	46	36	24	24	B
64	23	18	28	36	20	20	16	A

coefficients in small lines, herds, etc., generation by generation. Other computing methods exist [see, for example, Emik and Terrill (1949); Cruden (1949); and Döring and Walter (1959)]. Mayerhofer (1979) has developed a computer program for handling large populations. The methods are basically alike but differ in detail.

The concepts just explained permit the estimation of inbreeding coefficients of regular mating systems in which in each succeeding generation relatives of the same degree of relationship are mated. Inbreeding coefficients resulting from various numbers of generations are given in Table 3.3.

Selfing is the most intense form of inbreeding, but not the quickest way to achieve homozygosity. This can be done in one generation by doubling the haploid chromosome set, as it is performed by anther doubling in experimental plant breeding. Selfing is impossible in bisexually reproducing higher animals, but it is common of course in lower animals and in plants. However, mating of drones to the mother queen is equivalent to selfing, as is the mating of a rare parthenogenetic turkey male to his dam (Olsen, 1962). The following relations express the general formula for the regular inbreeding system of selfing (generation zero is designated by X, generation one by Y, generation two by Z):

generation 1 $\quad F_Y = r_{XX} = \frac{1}{2}(1 + F_X) = \frac{1}{2}$ if $F_X = 0$

generation 2 $\quad F_Z = r_{YY} = \frac{1}{2}(1 + F_Y) = \frac{1}{2}(1 + \frac{1}{2}) = \frac{3}{4}$

TABLE 3.3
Inbreeding Coefficients of Regular Mating Systems

Inbreeding generation	Self-fertilization	Full-sib	Parent–offspring[a]	Parent–offspring[b]	Half-sib
1	0.500	0.250	0.250	0.250	0.125
2	0.750	0.375	0.375	0.375	0.219
3	0.875	0.500	0.500	0.438	0.305
4	0.938	0.594	0.594	0.469	0.381
5	0.969	0.672	0.672	0.484	0.449
.	.	.	.	.	.
.	.	.	.	.	.
.	.	.	.	.	.
∞	1	1	1	0.500	1

[a] Mating to younger parent.
[b] Mating to constant parent with $F = 0$.

Using t to represent the number of generations, we may write the general equation

$$F_t = \tfrac{1}{2}(1 + F_{t-1})$$

The panmictic index P is the probability that genes at a single locus are unlike by descent: that is,

$$P_A = 1 - F_A$$

$$P_Y = 1 - F_Y = 1 - \tfrac{1}{2}(1 + F_X) = \tfrac{1}{2} - \tfrac{1}{2}F_X = \tfrac{1}{2}P_X$$

$$P_Z = 1 - F_Z = 1 - \tfrac{1}{2}(1 + F_Y) = \tfrac{1}{2}P_Y = \tfrac{1}{4}P_X$$

Generally

$$P_t = (\tfrac{1}{2})^t P_0$$

Consequently, heterozygosity decreases in each generation of selfing by half of the amount present in the previous generation.

Full-sib mating is one of the most powerful inbreeding systems available in higher animals. If in Fig. 3.3 the letters S and V stand for the same animal A and the letters T and W for the animal B, it follows that

$$F_Z = r_{XY} = \tfrac{1}{4}(r_{AA} + r_{BB} + 2r_{AB})$$

Then

$$r_{AA} = r_{BB} = \tfrac{1}{2}(1 + F_{t-2}) \qquad \text{if} \quad F_A = F_B$$

since they refer to the grandparental generation. Furthermore,

$$r_{AB} = F_{t-1}$$

Because they are full-sibs, the inbreeding coefficient of the parents equals the coefficient of coancestry of their parents. The equation for Z can be generalized to

$$F_Z = \tfrac{1}{4}(1 + F_{t-2} + 2F_{t-1})$$

or, in terms of the panmictic index,

$$P_Z = \tfrac{1}{2}P_{t-1} + \tfrac{1}{4}P_{t-2}$$

The recurrence relation between F or P values of succeeding generations can be simplified further. After relatively few generations the decrease in heterozygosity per generation will be nearly constant:

$$\frac{P_2}{P_1} = \frac{P_1}{P_0} = l$$

Dividing both sides of the equation by P_{t-1} yields

$$\frac{P_z}{P_{t-1}} = \frac{1}{2} + \frac{1}{4}\frac{P_{t-2}}{P_{t-1}} = l$$

Multiplying by l and putting all quantities on one side of the equation yields a quadratic equation:

$$l^2 - \tfrac{1}{2}l - \tfrac{1}{4} = 0$$

the roots of which are 0.809 and -0.309, respectively. When the number of generations is large, since the second root is much smaller than the first, it becomes unimportant and can be neglected. Under continuous full-sibbing succeeding generations will have 80.9% of the heterozygosity of their previous (parent) generation, and

$$P_t = 0.809^t P_0$$

Heterozygosity is decreased by 19.1% in each generation of full-sibbing.

Parent–offspring mating can take two forms. The offspring can be mated back to the younger parent. Here the recurrence relation for the inbreeding coefficient is equal to that of full-sibs. Conversely, the offspring can be repeatedly backcrossed to one parent:

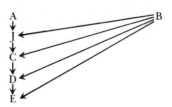

If the parent that is repeatedly mated is not inbred, the recurrence relation gives

$$F_t = \tfrac{1}{4}(1 + 2 F_{t-1})$$

or

$$P = \tfrac{1}{4} + \tfrac{1}{2}P_{t-1}$$

The inbreeding coefficient can never exceed 1/2 as long as this parent is not inbred.

Half-sib mating is possible in several forms, depending on how the dams are related to each other. If they are half-sibs to each other as well as to the sire, the recurrence relation gives

$$F_t = \tfrac{1}{8}(1 + F_{t-2} + 6F_{t-1})$$

or, for the panmictic index,

$$P_t = \tfrac{3}{4}P_{t-1} + \tfrac{1}{8}P_{t-2}$$

The decrease in heterozygosity from one generation to the next is 11% and any one generation has 89% of the heterozygosity of the immediately previous one when the mating system has been in operation long enough. Such a mating system occurs in practice when a breeder never buys a sire but continually uses one homebred male per generation.

A number of regular inbreeding systems are possible and recurrence relations for the majority of them were worked out by Wright. In some systems the initial rate of inbreeding is slow, but in the long run they may lead to a greater loss of heterozygosity than other systems with somewhat higher initial inbreeding increments.

The various regular mating systems may be compared by calculating the relative increment of homozygosity or loss of heterozygosity. One may compute the number of generations of mating necessary for a comparable loss in heterozygosity:

$$l_S = l_F^t = l_H^{t'}$$

where S, F, and H indicate selfing, full-sib mating, and half-sib mating, respectively, and t and t' are the numbers of generations. The loss of heterozygosity through one generation of self-fertilization, i.e., 1/2, is equivalent to 3.27 generations of full-sibbing or 5.95 generations of half-sib mating. One generation of full-sib mating is equivalent to 1.82 generations of half-sib mating.

Wright defined the coefficient of relationship R as the correlation between the genic values of two individuals—in contrast to F, which is the correlation between genic values of gametes forming a zygote. Here the covariance equals $\frac{1}{2}(1 + F_Z)$. The relationship coefficient refers to a whole zygote with two genes per locus in each individual, so that the covariance must be doubled. The variance of the average genic value of an individual X with genes 1 and 2 is

$$\tfrac{1}{2}(g_1^2 + g_2^2 + 2g_1g_2) = g^2(1 + F_X)$$

and

$$R_{XY} = \frac{\Sigma(\tfrac{1}{2})^{n_1 + n_2}(1 + F_A)}{[(1 + F_X)(1 + F_Y)]^{1/2}}$$

Under panmixis the coefficient of relationship equals twice the coefficient of coancestry.

When whole populations (e.g., lines) are to be examined for average relationship or inbreeding, approximations are useful. If one may assume that X and Y are equally inbred and that F_A can be neglected, R and F can be readily interconverted:

$$R = \frac{2F}{1 + F} \quad \text{and} \quad F = \frac{R}{2 - R}$$

The genic covariance is important for estimation of both the genic variance and the heritability and for estimation of breeding values. It is the product of the genic variance V_A and the probability of identity by descent of the genes in relatives, ie., $2C \times V_A$, which is equivalent to the product of the numerator of the relationship coefficent and the genic variance. The former has therefore been called the numerator relationship and Henderson (1975) has developed matrix procedures for rapid calculations.

The computation of inbreeding coefficients is fairly simple for individuals or for small populations. It becomes burdensome when a pedigree is followed back for several generations or when the average inbreeding coefficient for a larger herd or breeding population is desired. Wright and McPhee (1925) proposed for such cases an abbreviated method that uses only a sample of the pedigree yet permits good approximations. In their work they used two-line pedigrees, as shown here:

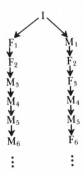

Starting with the two parents, it is decided by chance whether the male or the female ancestor of the preceding is to be considered. On the sire side in this example, the sequence of the ancestors selected by chance is F F M M F, that is, parental granddam, great-granddam, her sire, his sire, his dam. On the dam side, the sequence is F M M M M. All other ancestors are ignored. If there is an ancestor common to both lines, the probability of homozygosity in the descendant due to this "tie" is $(1/2)^{n_1 + n_2 + 1}$. The tie is between ancestors that are n_1 and n_2 generations, respectively, back from the parents. In generation n_1 the individual will have 2^{n_1} ancestors, in generation n_2 there will be 2^{n_2}. The total possible number of ancestral lines therefore equals $2^{n_1 + n_2}$. Consequently, the inbreeding coefficient of an individual with one tie in a two-line pedigree is $\frac{1}{2}$ or, more accurately, $\frac{1}{2}(1 + F_A)$, where F_A is the inbreeding coefficient of the common ancestor of the parents. The number of generations cancels out in this expression. This method would be inaccurate for an individual animal, but it suffices for estimating the average degree of inbreeding in a whole population. The average coefficient of inbreeding in a whole population equals half the proportion of ties. Using two-line pedigrees

$$ F = \frac{\text{number of ties}}{2 \times \text{number individuals investigated}} $$

The standard error of this estimate is given by $[p(1 - p)]^{1/2}/(n - 1)$, where p denotes the fraction of ties and n the total number of random lines compared (e.g., 100 when two lines and 50 animals each are drawn). The accuracy can be increased by drawing complete pedigrees to the grandparents or even further, and then sampling from there. Wright has shown that in pedigrees which are complete for k generations beyond the parents the inbreeding coefficient can be estimated from

$F = (1/2)^{2k+1}(1 + F_A)$ for each tie. By starting the sampling from grandparents, $k = 1$, and from great-grandparents, $k = 2$. Then the inbreeding coefficients of the populations are $\frac{1}{8}(1 + F_A)$ times the proportion of ties for four-line pedigrees, and $\frac{1}{16}(1 + F_A)$ for eight-line pedigrees. When ties occur between the complete and the random part of a pedigree, the inbreeding coefficient is

$$F = (\tfrac{1}{2})^{n_1 + k + 1}(1 + F_A)$$

where n_1 is the generation of the common ancestor and k the generation in which sampling starts, both counted from the parents. For example, assume that in a four-line pedigree the sire (A) occurs also in the random portion: This tie adds

$$(\tfrac{1}{2})^{0+1+1}(1 + F_A) = \tfrac{1}{4}(1 + F_A)$$

to the inbreeding. The method was used extensively by McPhee and Wright and by Lush and his students for investigating breed structure in the U.S. Robertson and Asker (1951) and Wiener (1961), among others, have investigated British breeds of livestock, Sciuchetti (1935), Langlet and Gravert (1961), and Fehlings et al. (1982) continental cattle and horse breeds, and Bohlin and Rønningen (1974) the breed structure in Scandinavia.

Inbreeding may occur in various different ways. For one, inbreeding in the usual sense results from the mating, intentionally or unintentionally, of more or less closely related animals. One may call it diffuse inbreeding. On the other hand, all animals are related, to some degree at least, in a finite population and therefore some inbreeding becomes unavoidable. In such a situation homozygosity will increase even though the finite population may have panmixis. An extreme but valid example is provided by lines with continuous full-sib mating, where $N = 2$. The degree of inbreeding attained, or, expressed somewhat differently, the level of homozygosis, once reached in such finite populations cannot be reduced as long as the populations remain closed and no genes are introduced. The connection between the finite size of a population and inbreeding is intuitively obvious to most breeders. The exact relationship between them can be most easily clarified for populations that permit selfing. Assume such a population of unrelated individuals, of size N. Upon mixing sperm and ova, the probability that uniting male and female gametes will derive from the same zygote equals $1/N$. For selfing

$F = \frac{1}{2}(1 + F_A)$. Therefore the increase in the inbreeding coefficient from the parent to the progeny generation will be

$$\Delta F = \frac{1 + F_A}{2N}$$

The average inbreeding of the progeny generation will be $F_0 + \Delta F = F_1$. Subscripts denote generations. The inbreeding present in the parent generation remains in the fraction $1 - 1/N$ of the progeny originating from the union of gametes that were derived from different individuals. The average inbreeding coefficient of the progeny generation then will be

$$F_1 = \frac{1 + F_0}{2N} + \left(1 - \frac{1}{N}\right)F_0 = \frac{1}{2N} + (1 - \frac{1}{2N})F_0$$

This relation makes it obvious that in isolated finite populations the degree of inbreeding will inevitably increase.

In domestic animals selfing is impossible so that a slight modification of the equation is required. Assume that in generation zero the population consists of $N/2$ males and $N/2$ females, from which $N/2$ full-sib families arise, each of which comprise one male and one female. Upon completely random mating the fraction $2/N$ of all matings will be between full-sibs, the progeny of which have a degree of inbreeding of $\frac{1}{4}(1 + F_0 + 2F_1)$. The fraction $1 - 2/N$ in generation two derives from matings between unrelated animals and this generation has a level of inbreeding equal to the coefficient of coancestry of the parent generation. Putting the two together gives

$$F_2 = \frac{1 + 2F_1 + F_0}{2N} + \left(1 - \frac{2}{N}\right)F_1 = \frac{1 + F_0}{2N} + \left(1 - \frac{1}{N}\right)F_1$$

If we can assume that $F_1 = F_0$, i.e., that there is little change in inbreeding between the generations, we get the same equation as given for the case where selfing is permitted.

The panmictic index for such a population is given by

$$P_2 = 1 - F_2 = 1 - \frac{1}{2N} - \left(1 - \frac{1}{2N}\right)F_1$$

$$= \left(1 - \frac{1}{2N}\right)P_1 = \left(1 - \frac{1}{2N}\right)^2 P_0$$

In generation t it will be

$$P_t = \left(1 - \frac{1}{2N}\right)^t P_0 = P_0 e^{-t/2N}$$

and

$$F_t = 1 - e^{-t/2N}$$

From the discussion heretofore it is evident that the inbreeding of a population is composed of two parts: the increment ΔF from one generation to the next, and the inbreeding prevalent in the previous generation:

$$F_t = \Delta F + (1 - F)F_{t-1}$$

with

$$\Delta F = \frac{F_t - F_{t-1}}{1 - F_{t-1}}$$

which makes it obvious that the increase in homozygosity due to inbreeding is measured relative to the panmictic index. Further, it is evident that in a closed population homozygosity cannot be diminished even if inbreeding is stopped. A breed or strain which has experienced intensive inbreeding during its formation will retain the same level of homozygosity even when it expands in later generations to a larger panmictic population. Examples include Shorthorn cattle and many lines of inbred mice which may have F values in excess of 99%.

When a breed is subdivided into strains—subpopulations (S)—animals within such groups will be related to each other more closely than to other animals in the breed. Matings of otherwise unrelated animals within such subgroups result in progeny with a degree of inbreeding F_{ST} in which F_{ST} equals the probability of identity of genes in an animal (or in two different animals) relative to the probability in the breed as a whole. Take, for example, the many strains of Friesians—Dutch, Holstein, New Zealand, and Danish. The probability that two Holstein animals possess identical genes is much larger than the probability of identity between a Holstein bull and a Dutch Friesian relative to the Black-and-White population of the mid 1800s. Animals may be inbred within a subpopulation and F_{IS} denotes the degree of inbreeding relative to the subpopulation. F_{IS} would describe the extent of what has been called diffuse inbreeding. For example, assume that for a progeny from a half-sib mating in the U.S. Brown Swiss, $F_{IS} = 1/8$. Yoder and Lush (1937) have estimated F_{ST} for the Brown Swiss breed of the 1930s to the whole breed around 1890 to be 4.3%. The total inbreeding F_{IT} of a calf out of a half-sib mating in the Brown Swiss breed would be

$$1 - F_{IT} = P_{IT} = P_{IS}P_{ST} = (1 - 0.043)(1 - 0.125) = 0.837$$

and

$$F_{IT} = 0.163$$

F_{ST} is identical with the Wahlund variance:

$$F_{ST} = \frac{\sigma_p^2}{pq}$$

While F_{ST} will always be positive, within subpopulations F_{IS} may be negative when animals are paired that are less related than average, as, for example, when substrains are crossed. In such situations the definition of F as a correlation is easily comprehended.

Cockerham (1973) has expressed the F statistics as intraclass correlations. The gametes were assigned genic values 1 and 0, so that the total score of an individual assumes values of 0, 1, 2.

The total variance is the sum of three variance components, σ_S^2 among subpopulation means, σ_I^2 among individuals, and σ_W^2 among gametes within individuals. The F statistics are defined as

$$F_{ST} = \frac{\sigma_S^2}{\sigma_S^2 + \sigma_I^2 + \sigma_W^2}, \quad F_{IT} = \frac{\sigma_S^2 + \sigma_I^2}{\sigma_S^2 + \sigma_I^2 + \sigma_W^2}, \quad F_{IS} = \frac{\sigma_I^2}{\sigma_I^2 + \sigma_W^2}$$

Between inbreeding increment and population size the relationship is fairly simple:

$$\Delta F = \frac{1}{2N_e}$$

Therefore the increment in the degree of inbreeding can be used to estimate the effective size of a population (inbreeding effective size; Crow, 1954). This is the size of a panmictic population that has the same ΔF as observed in a real population. In polygamous domestic animals, the numbers of breeding animals of each sex differ widely, so the effective size can be approximated from

$$\Delta F = \frac{1}{8N_M} + \frac{1}{8N_F} = \frac{N_M + N_F}{8N_M N_F}$$

Since in general there are fewer males than females, an even simpler approximation is

$$\Delta F = \frac{1}{8N_M}$$

In our example where 20 bulls and 1000 cows were assumed, ΔF will be 0.0064, or, when applying the last formula, 0.00625. When males are reduced to four, the increment in the inbreeding coefficient will be 0.03125 by both approximations.

The increment in the degree of inbreeding can be used instead of N_e to estimate the variance of the gene frequency change

$$\sigma_{\Delta q}^2 = \frac{pq}{2N} = pq \, \Delta F$$

and the variance of gene frequencies in later generations

$$\sigma_q^2 = pqF = pq\left[1 - \left(1 - \frac{1}{2N}\right)^t\right] = pq(1 - e^{-t/2N})$$

The variance increases with increasing t and will ultimately reach pq. The distribution of genotypes in small populations is given in Table 3.1. The proportion $1 - F$ of the population has allozygous, independent genes, where the probability of homozygosity is given by the product of two independent probabilities, i.e., p^2. In the fraction F of the population genes are identical by descent. It should be emphasized that this is valid for populations that have diffuse inbreeding or are structured, i.e., where subpopulations occur. It is not valid for any subpopulation considered by itself which may be inbred i.e., $F_{ST} \neq 0$, but where panmictic conditions will prevail.

In Table 3.4 increments of inbreeding rates, population sizes, and effective numbers of sires are given for a number of breeds. In most cases inbreeding was estimated by the sample pedigree method outlined earlier. It is striking that the increment of inbreeding coefficient ΔF varies little between breeds even though the actual sizes of the breeds vary by orders of magnitude—while the Holstein-Friesian breed accepts some 300,000 registrations per year, the yearly Brown Swiss registrations

TABLE 3.4

Inbreeding Increment and Population Size[a]

Breed	F, %	$E(F)$, %	R, %	t	ΔF, %	N_e	N_M	Ref.
Hereford U.S.	8.1	4.6	8.8	12.9	0.63	79	20	Willham (1964)
Holstein U.S.	4.1	1.7	3.4	10	0.41	122	30	Lush *et al.* (1936)
Brown Swiss	3.8	2.2	4.3	8.3	0.46	109	27	Yoder and Lush (1937)
Haflinger	5.3	5.7	10.8	5.9	0.52	96	24	Fehlings *et al.* (1983)
Bavarian Draft Horse	2.3	3.6	6.8	4	0.63	79	20	Fehlings *et al.* (1983)
Trotter FRG	2.9	3.1	6.0	7	0.41	122	30	Fehlings *et al.* (1983)

[a] $E(F) = R/(2 - R)$. t, Number of generations used in estimating F. $N_e = 1/(2\ \Delta F)$. $N_M = 1/(8\ \Delta F)$.

number around 10,000, and in the case of the Bavarian Haflinger horse, a few hundred. The large inbreeding increment must certainly be ascribed to the dominant position of some elite breeding herds, i.e., to the hierarchic structure within breeds of domestic animals. It is of interest also that the actual inbreeding of some breeds, notably the U.S. breeds, is larger than expected from the degree of relationship between random animals of the breed. This indicates the existence of subpopulations within which the probability of mating is increased. Quite the opposite can be observed for some of the horse breeds, where the degree of inbreeding is less than expected, indicating that breeders desire to avoid inbreeding.

Loss of genetic variability, rate of inbreeding, etc., are important to the quest of maintaining small rare breeds. In such breeds N_e should be kept as large as possible, a requirement facilitated by keeping family size constant. Robertson (1964) and S. Wright demonstrated that genetic drift precedes loss of heterozygosity by one generation.

If the proportion of heterozygotes in generation t is designated by H_t and drift variance by V_t, the increment of the drift variance can be expressed as

$$V_{t+1} - V_t = \frac{H_t}{4N} = \frac{2pq(1 - C'_{t-1})}{4N}$$

where C'_{t-1} denotes the coancestry coefficient of the mating partners in generation $t - 1$. The variance ultimately accumulated will be pq when $C = 1$. Therefore

$$V_\infty - V_t = pq(1 - C_t)$$

The ratio of drift variance to the total possible increment of variance can be described by

$$\frac{V}{V_\infty - V_t} = \frac{1}{2N} \frac{1 - C_{t-1}}{1 - C_t}$$

The ratio will be smaller if $C'_{t-1} > C_t$, which is expected when mating partners are closely related. In other words, in such a situation the loss of heterozygosity is smaller. Extrapolating, it turns out that drift can be avoided when the population is split into completely inbred lines. For animal breeding this implies that conservation of genetic variability can be accomplished efficiently by splitting the population into several independently propagated lines. In such situations drift will be reduced at the cost of increased homozygosity. However, heterozygosity can be reconstituted at any time by crossing the lines.

Several mating systems have been proposed which should minimize the increase in homozygosity. However, this can be only a temporary, though in practice a quite long-lasting, achievement since in finite populations partners inevitably will become related. In "circular" mating plans partners are rotated among four families, for example. Therefore, only after four generations will partners originating from the same family be mated. In such a setup C'_{t-1} refers to the second preceding generation and therefore it is smaller. This implies greater drift, but at the same time smaller loss of heterozygosity. In later stages, loss of heterozygosity will become greater than in panmictic or even in hierarchically structured populations of similar size. One example is provided by continuous mating of double cousins, where the initial inbreeding increment is smaller than expected for a population of size four. In contrast, with half-sib mating the initial (for 12 generations) loss of heterozygosity is larger, although it will be surpassed later by the loss occurring with the system of double cousin mating (Crow and Kimura, 1970).

4

Quantitative Genetics

Nearly all economic traits of domestic animals display continuous variability. Lactation milk yield, for example, varies from a few hundred kilograms to more than 10,000, and in large populations one may expect yields in each 10-kg interval. Such a population cannot be divided automatically into two or a few groups, as is possible in the case of traits, such as coat color and lethals, with discontinuous distributions. Therefore, traits with a continuous distribution from the lower to the upper limit are referred to as traits with continuous variability. Since individual observations differ quantitatively, such traits are also called quantitative and one refers correspondingly to quantitative variability and the genetics thereof.

How can quantitative variability arise when the underlying distribution of genes is basically discontinuous? At any one locus of an animal there are either none, one, or two alleles of a particular kind. Classical Mendelian characters are distributed in a discontinuous fashion, which incidentally facilitated the discovery of the laws of inheritance, since only such traits lead to clear segregation ratios. In contrast, the biometric approach dealing with continuously varying traits was not nearly as successful in uncovering the mechanism of inheritance. In the early periods of genetics and sporadically into the 1950s opinions existed which proposed that there was a basic difference between inheritance of Mendelian traits and that of quantitative traits. However, already in 1908 Wilhelm Weinberg showed that Mendelian or, in his expression, "alternative" inheritance is compatible with the inheritance of quantitative traits and that there is no need to invoke separate "blending inheritance."

Two facts might explain continuous distribution. First, a trait might be influenced by a large number of genes, with a single gene influencing the total expression relatively little; and second, a number of nongenetic influences, usually called environment for lack of a better term, modify the final expression of a trait. The term "environment" is common and in many cases also correct, particularly when it includes the "*milieu interne*" in which genes must express themselves. For example, the age of cows affects the expression of genes influencing milk yield. However, the term "environment" includes such disparate things as disturbances and irregularities of development, measurement errors, etc., many of which are barely attributable to environment. A more precise term would be nongenetic influences, but environment has become generally used and will be adhered to also in this text. Nevertheless, one should keep in mind that the expression denotes more than environmental influences proper.

In quantitative genetics the meaning of phenotype and genotype differs from their meaning in Mendelian genetics. Phenotype is understood as a quantity, a performance, etc., for example, 7000 kg milk, or 250 eggs. The phenotype is considered to be a function of genotype and environment: $P = f(G,E)$. The real function is surely very complicated and therefore approximations must suffice. The simplest function is additive,

$$P = G + E$$

In most cases one also assumes that genotype and environment are independent of each other, i.e., that good genotypes occur in all environmental niches and *vice versa*. It simplifies further treatment when the quantities are considered as deviations from a general mean.

In classical genetics genotype is understood as the sum of genes an individual possesses, and one may describe it by a list of gene symbols, or by the library of DNA. In contrast, genotype in quantitative genetics is the quantity of milk, to use an example, produced by a cow with particular genes, i.e., with the genotype in the Mendelian sense, under the prevailing conditions. One may assume that all individuals of an inbred line have an identical genotype in the Mendelian sense. They are distributed to environmental niches in proportion to the size of the niches and to their own numbers relative to the total population size.

The average performance of the inbred line, or its deviation from the overall mean, corresponds to its genotype in the sense used in quantitative genetics. The environmental influence should be understood as complementary, i.e., it is the performance of all animals in a particular environmental niche. "Environment" also can be thought of as the mod-

ification of the "genotype" in a particular environmental niche. Both
views are really identical: The expected value of the phenotype of animals with a particular set of genes distributed proportionately over the
whole environment is the genotype, and the expected value of the group
of animals in one environment which contains proportionately all genotypes is the "environmental influence." In this statement it is assumed
that environmental influences between animals are independent, which
implies, of course, that genotype and environment are also independent.

Let P_{ij} denote the phenotype of animals possessing genotype i (in
the Mendelian sense) which perform in environment j. The quantitative
genotype is

$$G_i = \sum_j P_{ij} = \sum_j (G_i + E_j)$$

since $\sum_j E_j = 0$. Similarly,

$$E_j = \sum_i P_{ij} = \sum_i (G_i + E_j), \qquad \sum_i G_i = 0$$

Quantitative traits must be measured, weighed, counted. Consequently,
measurement errors are unavoidable. So the model must be extended:

$$P_{ijk} = G_i + E_j + R_{ijk}$$

where R_{ijk} denotes the measurement error.

The model can be illustrated by results of a feeding trial where
heifers from four breeds were fed at three feed energy levels (Table
4.1). The genotype of Brown Alpine heifers equals their average performance over all three feeding levels and the environmental influence
of the feeding level I is given by the average performance of all four
breeds, or the deviation from the overall average, under this feeding
regime.

In many situations the additive model will be only a rough approximation and possibly not permissible. It may be that in environment I
differences between genotypes are small or even nonexistent, for example, when environment I represents feeding at a starvation level, while
genotypic differences in environment II under liberal feeding are clearly
distinguishable. When the additive model is obviously insufficient, the
model can be extended and a term for the interaction between genotypes
and environment can be introduced. Interaction is to be understood in
purely statistical terms. It implies that the sum of the two main effects,

TABLE 4.1
Influence of Genotype and Environment (Feeding Level) on Milk Production

	Milk yield[a]				
	Brown Alpine	Simmental	Pinzgau	Murbodner	∅ FL
Environment[b]					
FL 1	1000	1000	810	680	870
FL 2	1100	1100	920	840	990
FL 3	1220	1110	1010	810	1040
Mean genotypes	1110	1070	910	780	970

[a] Least square means of 16-week milk yields, rounded to 10 kg.
[b] FL, Feeding level: (1) 1 kg, (2) 3 kg, (3) 5 kg concentrate/day.

genotype and environment, is insufficient as a description of the phenotype, because of mutual modification, enhancing or impeding each other's effects. These interactions will be discussed later. Here it need only be pointed out that differences in scale, when, for example, genotypic effects are of different magnitudes in different environments, can frequently be handled by transformation of the data. In contrast, when genotypes cause no differences in some environments or when their effects are reversed, transformation cannot retrieve the simple additivity. In general, one observes that the magnitude of interaction increases with increasing size of genetic and/or environmental differences.

Not infrequently, objections, sometimes even heated ones, are raised against the assumption of additivity. It leads probably to a poor approximation when factors vary over a wide range, as for example (Fig. 4.1) feeding levels from starvation to overfeeding. However, in agriculture only a fairly small section is relevant and over this range additivity is a practical and acceptable approximation.

Figure 4.2 demonstrates how increasing the number of loci influencing a trait changes the distribution of the phenotypes. With increasing number of loci the distribution approaches the normal curve or normal distribution. When constructing these curves the favorable genes were assumed to be dominant at all loci. Further, the effects of the genes were assumed to be equal at all loci and the effects of all loci to be additive. One locus with at least one dominant gene increases the phenotype by one unit, *Aa, bb* will be equal in phenotype to *AA bb, aa Bb, aa BB*, etc. The height of the columns indicates the proportion of individuals with the particular phenotype. When two loci are involved, more than 56%

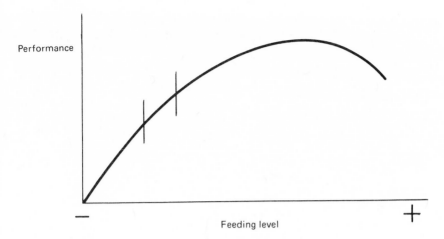

FIGURE 4.1. Relationship between performance and feeding level.

of all individuals are in the phenotypic class of two units and the distribution is strongly discontinuous (top left curve). With four loci acting, the distribution is still discontinuous even though individuals are distributed over five phenotypic classes. For eight loci the distribution clearly approaches normality. The curve at the bottom right was drawn under the assumption that in addition to eight loci, three environmental factors influence phenotype. One factor affects the phenotype as a dominant gene does, the second is neutral, and the third depresses the phenotype by one unit. Each factor affects one-third of the individuals of the same genotype. As illustrated by the figure, the environmental modification of the genotypes causes a further smoothing of the distribution and a closer approximation to the normal curve.

Naturally, the assumptions made here simplify the real situation. A very much larger number of genes probably influences traits as complex as those of economic importance to the cattle breeder. Individual genes do not have equal effects, but are of varying importance in determining a particular trait. Environmental influences vary from one extreme to the other. Furthermore, it is often difficult to measure a phenotype exactly. Measurement errors might be large enough to make it difficult to differentiate between phenotypes belonging to neighboring genotypic classes. When all of these forces act together the distribution of the phenotypes becomes truly continuous, that is, it is impossible to divide the population into separate classes.

Some model traits exist where additivity of gene action and the approximation of a continuous distribution can be directly demon-

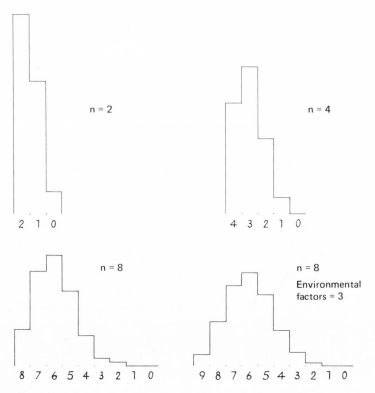

FIGURE 4.2.   Distribution of phenotypes when two, four, or eight loci are segregating; three environmental factors and genes from eight loci influence phenotype.

strated. Gahne and his colleagues (1970) found four alleles at the locus for cholinesterase of horses. They found the enzyme activity in Swedish horses to be distributed as shown by the histogram of Fig. 4.3. The distribution is not normal, but it probably could be normalized by a transformation.

The example demonstrates that even multiple alleles of a single locus may be the cause of a continuous distribution of a trait. An example where quantitative differences between isoalleles exist is also provided by Gahne (1961). Aryl esterase alleles of pigs, which cannot be distinguished by electrophoresis, can be differentiated when the enzyme activity is measured quantitatively. A number of major genes act on different traits as minor genes. For example, the gene for human

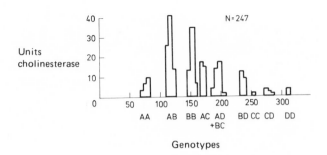

FIGURE 4.3. Cholinesterase in horses. (Source: Gahne *et al.*, 1970.)

phenylketonuria (PKU) affects the enzyme level, with tragic consequences when homozygous, but also other characters, such as hair color and head length as a minor gene with effects which cause overlapping distributions with normal genotypes. In animal genetics quantitative effects of a number of marker genes are known, even though only few of them are significant statistically. Pani (1974), for example, found that chicken embryos with the genotype $i^s i^s c^s c^s$ (colored) are much less susceptible to leucosis infection than $IIc^s c^s$ (white) embryos. Hartmann (1972) showed that the pea-comb gene affects the frequency of breast blisters and the growth rate of broilers.

4.1. Gene Effects

To explain the concept of gene effect we may assume as a simplified model that the character is influenced by two alleles at a single locus and is expressed independently of the environment. The mean of the two homozygotes O is the basis of measurement, or the point of origin. The deviation of the homozygous dominant phenotype A_1A_1 is the genotypic value and is denoted by a. The genotypic value of the recessive homozygote is $-a$. The genotypic value of the heterozygote is denoted by d and has a value between zero and a or, with overdominance, larger than a. With complete dominance $d = a$; when inheritance is intermediate the genotypic value of the heterozygote is zero, that is, $d = 0$. Table 4.2 and Fig. 4.4 illustrate these concepts with the egg-laying performance of three genotypes. The population mean does not equal the mean of the two homozygotes. It would only do so if gene frequencies were 0.5 and dominance was absent. The population mean is a weighted average

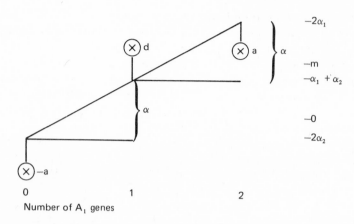

Number of A_1 genes

FIGURE 4.4. Gene effects. α_1, α_2, average gene effects; α, effect of gene substitution; a, d, genotypic values; O, point of origin.

of the genotypic values. Therefore, it is computed by adding the products of the genotypic values and their frequencies:

$$m = p^2a + 2pqd + q^2(-a) = a(p - q) + 2pqd$$

The population mean is determined by the genotypic values and gene frequencies. When dominance is absent the population mean is influenced by gene frequencies and genotypic values of homozygotes. In the example of Table 4.2 the population mean value, expressed as a deviation from the point of origin, is computed as follows:

$$m = (0.6 - 0.4)14 + 0.48 \times 16 = 10.48$$

TABLE 4.2
Egg-Laying Performance of Blood-Group Genotypes[a]

Genotype	n	Frequency	Percent production	Genotypic value[b]	
A_1A_1	54	0.38	38	$a =$	14
A_1A_2	63	0.44	40	$d =$	16
A_2A_2	25	0.18	10	$-a =$	-14

[a] Source: Modified from Briles and Allen (1961).
[b] Deviation from point of origin 24 [$= (38 + 10)/2$].

In absolute terms the population average is $m + O = 34.48$. This is a bit larger than the weighted average (33.84) computed from Table 4.2, because the genotypic frequencies in the table differ a little from the exact equilibrium frequencies. The formula assumes Hardy–Weinberg equilibrium, so the computed m is that expected in a large population with $p = 0.6$ and $q = 0.4$. When the frequency of the favorable gene A_1 is increased to 0.9 the population average increases to 38.68%.

How will introducing additional A_1 alleles affect the population? In a part p of the population an A_1 allele pairs with another A_1 allele; in another part q, however, an A_1 allele pairs with an A_2 allele. In the first part an A_1A_1 individual results with the genotypic value a, in the second an A_1A_2 heterozygote with the genotypic value d. The population average changes by

$$pa + qd - a(p - q) + 2pqd = q[a + d(q - p)] = \alpha_1$$

Fisher (1918) termed this change α_1, meaning the average effect of the gene A_1. Similarly, the average effect of gene A_2 can be computed:

$$\alpha_2 = -p[a + d(q - p)]$$

These two quantities in the example of Table 4.3 are $\alpha_1 = 4.32\%$ and $\alpha_2 = -6.48\%$. Introducing additional A_1 genes causes the population average to increase by 4.32% while its allele reduces the average by 6.48%. Thus, gene effects are expressed as the changes in the population mean caused by the introduction of the respective genes. In individuals, introducing a particular gene to replace another may not affect the phenotype. In our example, replacing the A_2 gene in a heterozygote with an A_1 gene decreases performance, but the same replacement in an A_2A_2 homozygote increases performance by 30%. Regardless of whether dominance is complete or incomplete, the different alleles have "average" effects in the sense that the population mean rises or drops when one allele is replaced by another. Consequently, even genes with complete dominance or overdominance have "average" effects.

Average gene effects are connected closely to the breeding value of an animal. An individual of genotype A_1A_1 will introduce only A_1 genes into the population. Therefore, the progeny of such an individual will deviate from the population mean by the average effect of gene A_1. The offspring receive only half their genes from a particular parent. Consequently, if the offspring differ from the population mean in performance by a certain amount, the breeding value of the parent must be twice that amount. It is assumed that the other parent or parents are a

TABLE 4.3

Breeding Value, Dominance Deviations, and Genetic Variance of Blood-Group Genotypes

	A_1A_1	A_1A_2	A_2A_2	
Genotypic value	$a_{11} = 2q(a - dp)$	$a_{12} = a(q - p) + d(p^2 + q^2)$	$a_{22} = -2p(a + dq)$	
Breeding value	$2\alpha_1$	$\alpha_1 + \alpha_2$	$2\alpha_2$	
Dominance deviation	$d_{11} = -2q^2d$	$d_{12} = 2pqd$	$d_{22} = -2p^2d$	
$p = 0.6$, $m = 34.48\%$:				
a_{ij}	3.52	5.52	−24.48	
α_i	4.32	—	−6.48	
$\alpha_i + \alpha_j$	8.64	−2.16	−12.96	
d_{ij}	−5.12	7.68	−11.52	
$p_ip_j(\alpha_i + \alpha_j)^2$	26.87	2.24	26.87	55.9872[a]
$p_ip_jd_{ij}^2$	9.44	28.31	21.33	58.9824[b]
$p_ip_ja_{ij}^2$	4.46	14.63	95.88	114.9696[c]
$p = 0.9$, $m = 38.08\%$:				
a_{ij}	−0.08	+1.92	−28.08	
α_i	0.12	—	−1.08	
$\alpha_i + \alpha_j$	0.24	−0.96	−2.16	
d_{ij}	−0.32	2.88	−25.92	
$p_ip_j(\alpha_i + \alpha_j)^2$	0.047	0.166	0.047	0.2592[a]
$p_ip_jd_{ij}^2$	0.083	1.493	6718	8.2944[b]
$p_ip_ja_{ij}^2$	0.005	0.664	7885	8.5536[c]

[a] Genic variance.
[b] Dominance deviation.
[c] Total variance.

random sample of the population and that their contribution is the same as the population average. The breeding value of an animal corresponds, therefore, to the sum of the average effects of all genes affecting the trait. Table 4.3 gives breeding values and genotypes used earlier.

The concept of average gene effect describes the changes in the population mean when, for example, an A_1 gene replaces another gene at the A locus. If the A_1 allele does replace an A_2 allele, the population mean will change by the difference between the average effects of the two genes, that is, by $\alpha_1 - \alpha_2$. This quantity was first defined by Fisher as the effect of gene substitution. We shall denote it by α without a subscript, in contrast to α_1 and α_2. Consequently,

$$\alpha = \alpha_1 - \alpha_2 = a + d(q - p)$$

or, conversely,

$$\alpha_1 = q\alpha, \qquad \alpha_2 = -p\alpha$$

The effect of gene substitution can be estimated directly from the genotypic values. The A_2 alleles are paired with A_1 alleles in a proportion p of the population and with other A_2 alleles in a proportion q. If A_1 is substituted for A_2 in all pairs, the heterozygotes A_1A_2 will be changed to homozygotes A_1A_1 and the homozygotes A_2A_2 to heterozygotes A_1A_2. Using the frequencies as weights, we find that the effect of substituting gene A_1 for A_2 is

$$\alpha = p(a - d) + q(d + a) = a + d(q - p)$$

the same expression as given before. In the example this is 11.2%, that is, substituting gene A_2 for A_1 increases the average laying performance by 11.2%.

When dominance is complete, then $d = a$, and population mean, gene substitution effect, and average gene effect are, respectively, $(1 - 2q^2)a$, $2qa$, and $2q^2a$. When heterozygotes are intermediate, then $d = 0$, and the three quantities are given by $(p - q)a$, a, and qa.

Gene effects may be analyzed by the method of least squares. The regression coefficient of the phenotype on the genotype, or, more precisely, on the number of favorable genes, equals the gene substitution effect. The model of genotypic effects of the three genotypes at a locus is

$$A_1A_1:a_{11} = \mu + \Sigma\alpha_1 + \delta_{11}$$

$$A_1A_2:a_{12} = \mu + \alpha_1 + \alpha_2 + \delta_{12}$$

$$A_2A_2:a_{22} = \mu + 2\alpha_2 + \delta_{22}$$

The square of deviations of observed (a_{ij}) from expected values ($\mu + \alpha_i + \alpha_j$) is minimized. Thus dominance deviations are, analogous to the "errors" of regression estimates, the difference between observed and estimated additive values. The estimation involves differentiation of the squared differences between a_{ij} and the additive value and subsequent minimizing by setting the differential equal to zero ($a_{ij}^* = a_{ij} - \mu$):

$$\frac{\delta Q}{\delta \alpha_1} = \frac{\delta[p^2(a_{11}^* - 2\alpha_1)^2 + 2pq(a_{12}^* - \alpha_1 - \alpha_2)^2]}{\delta \alpha_1}$$

and correspondingly $\delta Q/\delta \alpha_2$.

These equations lead to

$$4p^2(a_{11}^* - 2\alpha_1) + 4pq(a_{12}^* - \alpha_1 - \alpha_2) = 0$$

$$4pq(a_{12}^* - \alpha_1 - \alpha_2) + 4q^2(a_{22}^* - 2\alpha_2) = 0$$

These can be rearranged:

$$\alpha_1 + p\alpha_1 + q\alpha_2 = pa_{11}^* + qa^{*12}$$

$$\alpha_2 + p\alpha_1 + q\alpha_2 = pa_{12}^* + qa_{22}^*$$

The weighted sum of deviations from the mean equals zero:

$$p^2 2\alpha_1 + 2pq(\alpha_1 + \alpha_2) + q^2 2\alpha_2 = p\alpha_1 + q\alpha_2 = 0$$

The average gene effects are

$$\alpha_1 = pa_{11}^* + qa_{12}^*$$

$$\alpha_2 = pa_{11}^* + qa_{22}^*$$

This result is plausible: The average effect of a gene is equivalent to the sum of the effects of genotypes in which the gene occurs weighted by the gene frequencies. When numerator and denominator are multiplied by the frequency of the gene, the resulting expressions sum the performance of all genotypes in which the gene occurs, weighted by the number of alleles in them:

$$\alpha_1 = \frac{p^2 a_{12}^* + pq a_{12}^*}{p}$$

$$\alpha_2 = \frac{pq a_{12}^* + q^2 a_{22}^*}{q}$$

Deviations of genotypic values from the breeding value are denoted as dominance deviations. Their weighted sum equals zero. In our example the mean, computed as deviation from midpoint, equals 10.48%, $\alpha_{11} = (14 - 10.48)\% = 3.52\%$, $a_{12} = 5.52\%$, etc. We have

$$\alpha_1 = 0.6 \times 3.52 + 0.4 \times 5.52$$

$$= (0.36 \times 3.52 + 0.24 \times 5.52)/0.6 = 4.32$$

$$\delta_{11} = a_{11}^* - 2\alpha_1 = 3.52 - (2 \times 4.32) = -5.12$$

So far only one locus has been considered. However, when individual loci can be considered to act independently, their effects can be added and the genotypic value corresponds to the sum of the additive effects and the dominance deviations of the individual loci affecting a trait:

$$A + D = \sum_l (\alpha_i + \alpha_j) + \sum_l \delta_{ij}$$

The independence of average gene effects is implicit in their definition: It is the change of the population mean when the gene is introduced, i.e., it is the average of the effects in all genotypes, and these may be positive or negative, depending on the rest of the genotype.

However, the individual loci will influence the effects of other loci, in other words, epistasis will be present. The concept of epistasis in quantitative genetics differs from the classical understanding of it, where, for example, the albino locus is epistatic due to its action in suppressing the effects of other loci. In quantitative genetics epistasis denotes the lack of additivity between the effects of the loci. It is a statistical interaction between loci—interallelic (more precisely interlocal)—while dominance is an intra-allelic (or intralocal) interaction.

The approach of quantitative genetics consists first of an attempt to explain as much of the variation as possible by additive gene effects. When this proves insufficient, the residual variation is ascribed as far as possible to dominance deviations. Only insofar as this, too, still leads to unsatisfactory results is epistasis invoked to account for the residual variation. It must be emphasized that epistasis in the classical sense may be present but the quantitative variation may be explicable to a large degree by additive gene action. Therefore quantitative genetic analysis applies methods based on statistical concepts and can be considered as an example of reductionism.

One simple example is the interaction of color genes in rabbits. In the presence of C genes, B animals are black, bb animals brown. When C genes are absent, all animals are albinos regardless of which genes they possess at the B locus. Therefore, the action of the B genes depends on what genes are present at the C locus (considering for the moment only the B and the C loci) and, conversely, the C gene acts differently on B or bb genotypes. Assume now that black animals have the score 2, brown animals the score 1, and albinos the score zero (for example, these could be color intensities). Assume further that the frequency of the brown gene is 0.7 ($= p_B$) and that of the C gene is 0.6 ($= p_C$). Both the B and C genes are completely dominant. First the genotypic values are computed separately by the formulas shown in Table 4.5. The action of the gene at one locus depends on the genotype at the second locus. Therefore the genotype frequencies have to be considered. Changing bb to B changes the coat color in C animals from brown to black, that is, by one unit on our scale. C animals (CC and Cc) constitute 84% of the population. In cc animals B has no effect. The average difference between BB and bb is therefore $(0.84 \times 1) + (0.16 \times 0) = 0.84 = 2a_B$. Similarly, the genotypic values at the C locus can be computed: $(0.91 \times$

TABLE 4.4
Phenotypes P and Frequencies f of Rodent Coat Color

		CC	Cc	cc	$\emptyset\, B$
BB	P	2	2	0	1.68
	f	0.1764	0.2352	0.0784	0.49
Bb	P	2	2	0	1.68
	f	0.1512	0.2016	0.0672	0.42
bb	P	1	1	0	0.84
	f	0.0324	0.0432	0.0144	0.09
Mean C		1.91	1.91	0	1.6044
		0.36	0.48	0.16	1

TABLE 4.5
Genetic Effects and Variances for Two Loci C and B[a]

	Effects				Variances	
	C	B			C	B
Genotypic						
$a\ (=d)$	0.420	0.955	σ_T^2		0.4903	0.05779
Additive						
$\alpha = 2qa$	0.764	0.252	$\sigma_A^2 = 2pq\alpha^2$		0.2802	0.02667
Dominance						
$d_{11} = -2q^2d$	−0.0356	−0.0756	$\sigma_D^2 = 4p^2q^2d^2$		0.2101	0.03112
$d_{12} = 2pqd$	0.4584	0.1765				
$d_{22} = -2p^2d$	−0.6876	−0.4114				

Genotype	Absolute	Phenotype P	f	G_A	G_D	G_I
$CCBB$	2	0.3956	0.1764	0.7264	−0.3812	0.0144
$CCBb$	2	0.3956	0.1512	0.5104	−0.1292	0.0144
$CCbb$	1	−0.6044	0.0324	0.2584	−0.7172	−0.1456
$CcBB$	2	0.3956	0.2352	−0.0016	0.3828	0.0144
$CcBb$	2	0.3956	0.2016	−0.2536	0.6348	0.0144
$Ccbb$	1	−0.6044	0.0432	−0.5056	0.0468	−0.1456
$ccBB$	0	−1.6044	0.0784	−0.7856	−0.7632	−0.0756
$ccBb$	0	−1.6044	0.0672	−1.0176	−0.5112	−0.0756
$ccbb$	0	−1.6044	0.0144	−1.2696	−1.0992	0.7644
Population mean	1.6044	0	1	0	0	0

Variances (directly computed)

$$\sigma_A^2 = \Sigma fG_A^2 = 0.30684, \ \sigma_D^2 = \Sigma fG_D^2 = 0.241229, \ \sigma^2 = \Sigma fG_I^2 = 0.01099$$

$$G_A(CcBb) = 0.3056 + (-0.4584) + 0.0756 + (-0.1764) = -0.2536$$

$$= (\alpha_C + \alpha_C) + (\alpha_B + \alpha_b)$$

$$G_D(CcBb) = 0.4584 + 0.1764 = 0.6348 = d_{C12} + d_{B12}$$

$$G_I(CcBb) = 0.3956 - (-0.2536) - 0.6348 = 0.0144$$

[a] α, Gene substitution effect. d_{C12}, Dominance deviation of Cc. α_C, Average effect of gene C. G_A, G_D, G_I, Additive dominance, epistatic values of genotype. σ_A^2, σ_D^2, σ_I^2, Corresponding variances. σ_T^2, Total variance.

2) + $(0.99 \times 1) = 1.91 = 2a_C$. The genotypic values of Table 4.5 simplify with complete dominance to $2aq^2$ for heterozygotes and homozygous dominants and to $-2ap(1 + q)$ for homozygous recessives. These genotypic values are the sum of the additive effects and the dominance deviations. Where the genotypic effects at both loci are summed, a difference remains between the sum and the phenotypic value. This difference is caused by the interaction of both loci. The phenotype is more

(or less) than the sum of the genotypic effects of the single locus. This difference is called the epistatic effect or gene interaction.

For example, the four average gene effects of the double hetero-zygotes sum to -0.2436 and the two dominance deviations to 0.6348, so that the sum of the single-locus estimates is 0.3812. This is less than the genotypic value (0.3956). The difference between the two, the de-viation of the observed from the expected (on the basis of single-locus effect) value, corresponds to the epistatic or interaction effect. For the double heterozygote it is 0.0144. Cockerham (1954) has shown how to break the epistatic effect down into effects due to additive $\times$ additive, additive $\times$ dominance, and dominance $\times$ dominance effects, in strict analogy to the analysis of interactions between main effects in factorial experiments. The present example has been chosen in part to demon-strate how even for an example with epistasis in the classical sense, most of the variation can be explained by invoking a model where genetic effects are considered independent between loci.

In quantitative genetics, epistasis or interallelic gene interactions signify that additive effects and dominance deviations are insufficient to account completely for the expression of a trait (not considering envi-ronment for the moment). The genotype G in the sense explained can be defined as

$$G = M + A + D + I$$

where M, A, D, and I denote, respectively, population mean, additive effects, dominance deviations, and epistatic deviations. In the example of Table 4.5, A and D are combined in G_B and G_C for the two loci, and I arises from the interaction of the B and the C loci.

We may never be able to analyze a complex trait such as milk yield or egg number using such references to gene frequencies and gene effects. We must assume, nevertheless, that a complex trait is the result of the sum of additive effects, dominance, and epistatic deviations of all genes that affect the trait. That the breeding value equals the sum of all additive effects and a very small part of the epistatic effects will be shown later.

4.2. Variances

The various genetic and environmental effects cause variances. In common language, variability is associated with inaccuracy, or with in-ability to make a precise measurement. In quantitative genetics variance is associated with information. It permits the estimation of the importance of genetic or nongenetic influences reflected in the magnitude of the variance caused by the particular influence. This variance may be taken as that which disappears when the factors are held constant. Where the simple additive model of the phenotype is applicable, the observed phenotypic variance is composed correspondingly:

$$V_P = V_G + V_E$$

The three quantities denote, respectively, phenotypic, genotypic, and environmental variance. Note that the latter two are the portions of V_P that are caused by differences in genotypes and environment, respectively, or, stated differently, which disappear when they are prevented from varying.

The variances given in Table 4.6 were computed from results of Kronacher's classic twin investigations. The total variance for head length is 16.05 cm². Differences between partners of individual monozygotic twin pairs contain no genetic component. Therefore, the intrapair variance, 1.2 cm², is caused by environmental differences and measurement errors. We cannot accomplish the reverse, i.e., keep the environment constant to the same degree as we can the genotype and exclude all measurement errors. However, when the model is satisfactory and when genetic and nongenetic influences are independent, the phenotypic variance between unrelated animals will contain both V_G and V_E and subtracting the intrapair variance from the total yields the genetic variance:

$$V_P - V_E = V_G$$

$$16.05 - 1.2 = 14.85$$

The genetic variance is composed of entities which are caused by the three types of genetic effects. Additive-genetic or genic effects give rise

TABLE 4.6
Genetic and Environmental Variance of Head Length of Cattle Twins[a]

Source of variance	df	MS	E(MS)
Pairs of twins	21	30.9	$\sigma_E^2 + 2\sigma_G^2$
Twins/pairs	22	1.2	σ_E^2
Total	43	15.2	$\sigma_E^2 + (42/43)\,\sigma_G^2$

[a] Source: Kronacher (1931). df, Degrees of freedom. Variance of a population of unrelated animals = $\sigma_E^2 + \sigma_G^2 = 14.05$.

to the corresponding variance V_A, dominance deviations lead to the dominance variance V_D, and epistasis leads to the epistatic or interaction variance V_I. Thus the genetic variance has the following composition:

$$V_G = V_A + V_D + V_I$$

Knowledge of genetic effects and frequency of underlying genotypes permits direct computation of the variances. In the example of Table 4.5 the genic variance equals the sum of the squares of genic deviations G_A times the corresponding frequencies, and similarly for the dominance and epistatic variances.

When all the quantities—genetic effects and frequencies—are known, the variances can be computed without further assumption. However, this will rarely be so.

When additive gene effects are expressed as gene substitution effects, the additive-genetic variance or variance between breeding values is given by

$$V_A = p^2(2q\alpha)^2 + 2pq[(p - q)\alpha]^2 + q^2(-2p\alpha)^2 = 2pq\alpha^2$$

In the example of Table 4.3 this yields $0.48 \times 10.8^2 = 55.9873$, equal to the value computed directly in the table from $\Sigma f(\alpha_i + \alpha_j)^2$. The same can be done using average gene effects:

$$
\begin{aligned}
V_A &= p^2 4\alpha_1^2 + 2pq(\alpha_1 + \alpha_2)^2 + q^2 4\alpha_2^2 \\
&= 2(p^2 + pq)\,\alpha_1^2 + 2(p\alpha_1 + g\alpha_2)^2 + 2(q^2 + pq)\alpha_2^2 \\
&= 2p\alpha_1^2 + 2q\alpha_2^2 = 2\,\Sigma p_i\alpha_i^2 \\
&= 2p(q\alpha)^2 + 2q(-p\alpha)^2 = 2pq\alpha^2
\end{aligned}
$$

Again for the example of Table 4.3

$$V_A = 2 \times 0.6 \times 4.32^2 + 2 \times 0.4 \times 6.48^2 = 55.9873$$

The last formula is particularly handy for multiple alleles.

The dominance variance is computed in an analogous fashion:

$$V_D = p^2(-2q^2d)^2 + 2pq(pqd)^2 + q^2(-2p^2d)^2 = 4p^2q^2d^2$$

In our example $4 \times 0.36 \times 0.16 \times 16^2 = 58.9824$, the same as from $\Sigma f\delta_{ij}^2$. A slightly different formula displays more clearly the character of the intra-allelic interaction:

$$V_D = p^2 q^2 (a_{11} - 2a_{12} + a_{22})^2 = p^2 q^2 (a - 2d - a)^2$$

with, of course, the same result.

The covariance between genic and dominance effects equals zero, as is expected of a covariance between regression estimates and deviations from the regression line. This can be demonstrated with the example of Table 4.3:

$$\text{cov } AD = \Sigma f(\alpha_i + \alpha_j)(\delta_{ij})$$

$$= 0.36(2 \times 4.32)(-5.12) + 0.48(4.32 - 6.48)$$
$$7.68 + 0.16(2 \times 6.48)(-11.52)$$

$$= -15.925 - 7.963 + 23.888$$

$$= 0$$

The formulas indicate and the example illustrates that the sizes of the variances depend on the gene frequencies. Figure 4.5 shows the additive variance and the dominance variance of the example over the whole range of gene frequencies. The variances are zero at the extremes ($p = 0, 1$) and increase toward the middle of the range. At the extremes the genotypes are homozygous and no differences exist. As the gene frequencies move toward the middle, the genetic diversity increases. The maximum is reached by the additive variance at a gene frequency of less than 1/2 and by the dominance variance at 1/2. The maximum of the genic variance depends on gene action. With semidominance it also is at the maximum at $p = 0.5$. Of particular interest is the relation between genic and dominance variances. At low gene frequencies nearly all of the genotypic variance is additive and the dominance variance has very little significance even if dominance is complete. Conversely, at high gene frequencies the larger part of the genotypic variance is caused by dominance deviations and the additive variance is relatively small.

The variance reflects the contributions of additive effects and dominance deviations at various gene frequencies (Table 4.3). Selection in the usual sense acts on the additive gene effects, that is, only the frequencies of those genes causing additive effects are changed by selection. Therefore, the proportion of the total variance caused by genic effects is of paramount importance to predict selection progress. Figure 4.5

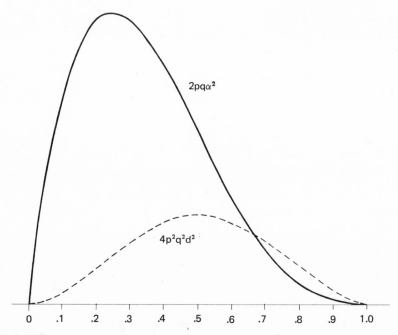

FIGURE 4.5. Additive and dominance variances at varying gene frequencies; $a = 14\%$, $d = 14\%$.

shows the dominance variance at higher gene frequencies to be the major part of the genotypic variance. Thus, selection will be less effective than at low gene frequencies, where most of the variation is genic in origin. If differences in phenotypes are caused mainly by additive effects, these will recur in the progeny. Conversely, dominance deviations are not correlated between parent and offspring. Gene combinations in the parents disintegrate at meiosis and recombine at random in the offspring. In our single-locus example, selection will be successful at low gene frequencies. Conversely, at higher gene frequencies, progress will be very slow because mass selection cannot differentiate between heterozygotes, the largest part of the population, and dominant homozygotes. Consequently, moving the gene frequencies to the extreme becomes very difficult.

Genic and dominance variances of quantitative traits are the sum of the respective variances caused by single loci—in the example of Table 4.5, the sum of the variances at loci B and C. To this has to be added the epistatic or interaction variance, which cannot be attributed to individual loci. Formally, it can be considered as a residual:

$$V_I = V_G - V_A - V_D$$

In polygenic traits the additive variance and the dominance variance are the sum of the respective variances caused by the effects at single loci. In addition, the epistatic or interaction variance is caused by differences in epistatic effects between individuals. The composition of the epistatic variance in terms of gene action was clarified by Cockerham (1954) and Kempthorne (1957). Epistatic effects caused by the simultaneous action of two additive effects at different loci cause the so-called additive-by-additive variance. An interaction caused by the simultaneous presence of an additive effect at one locus and a dominance deviation at another is an additive-by-dominance effect causing the additive-by-dominance variance ($A \times D$). Interactions of additive effects at three loci are, correspondingly, the source of $A \times A \times A$ variance, and so on. Similar to dominance combinations, epistatic combinations mostly disintegrate at meiosis, making selection for epistatic effects largely unsuccessful.

The environmental variance can be subdivided according to several criteria. One important criterion is whether the causes of the environmental variation are tangible, as are such factors as feeding level, type of management (open versus closed barns for cattle, floor or battery for laying hens), month of calving, age, and so on. These types of environmental influences sometimes are spoken of collectively as the macroenvironment.

Environmental influences common to groups of relatives such as families or progeny groups are of particular importance. Littermates have a common intrauterine environment, and in swine the milk yield and mothering ability of the sow provide a common postnatal environment for the piglets. They share an environmental niche from before birth until weaning. All these influences will tend to make such full-sibs more alike than if the only source of their likeness was their genetic relationship. Paternal half-sibs, if born at about the same time, may be subjected to similar environmental influences that differ from those experienced by other half-sib groups. This can create a similarity within half-sib groups over and above that caused by the common share of genes. Such nongenetic influences, which vary from one group of relatives to another, are often called C effects in animal breeding literature. Their importance and the trouble they cause in the analysis and interpretation of genetic data arise because they cause likenesses among family members that are often difficult to separate from the likeness caused by their common genes.

After all these facts are considered, a residual variation among animals still remains unaccounted for. Even monozygotic twins, for example, fed and managed under seemingly identical conditions differ somewhat in most traits. The causes for the residual variation are often called the microenvironment.

If the sources of environmental variance are tangible, it is often possible to correct data statistically to eliminate this variance. For example, it is customary to correct individual records of American dairy breeds for age effects and times of milking. Environmental variance can also be excluded by comparing only individuals that perform in the same macroenvironment, that is, animals of the same age in the same testing station. The residual variance caused by differences in the microenvironment, measurement errors, and simple chance deviations cannot be removed, and thus this variance conceals the genotype of an animal.

However, when a genotype can be repeatedly tested and when independent measurements are possible, the residual error can be reduced at will. This results from the relation between the variance of a mean of independent observations and the number of such observations ($\sigma_{\bar{x}}^2 = \sigma^2/n$). When n is large the variance approaches zero. This principle is widely used in estimating breeding values or the producing abilities of animals.

5

Repeatability

To understand the concept of repeatability and to quantify it, a different division of the variation is used. Repeatable traits include lactation milk yields, body measurements, litter size, and numbers of eggs laid in different periods. Some environmental effects are permanent and therefore influence performance in all periods. Other environmental effects are temporary and vary from one period to the next. Since the temporary effects vary in a random way from period to period, they are as likely to be positive as negative and should tend to average to zero over several periods. In contrast, permanent environmental effects together with the genotype determine an animal's performance potential during its whole life and are termed "real producing ability." The variance of an average of independent quantities equals the variance among the observations divided by the number of observations in the average. Only the temporary environmental factors vary in a random way from period to period, so only the variance caused by these is correspondingly reduced. This is illustrated in Fig. 5.1. The repeatability or the intraclass correlation r is an estimate of the proportion of the variation among observations caused by permanent differences among animals:

$$r = \frac{V_P}{V_P + V_T}$$

where V_P is the variance caused by permanent differences among animals and V_T is the variance caused by differences in the performance of one individual in various periods, that is, by temporary differences.

The permanent differences between individuals are caused by dif-

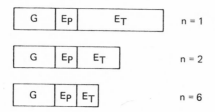

FIGURE 5.1. Distribution of variance with single records ($n = 1$) and averages of two records ($n = 2$) and of six records ($n = 6$). G, Genetic variance; E_P, variance due to permanent environmental effects; E_T, variance due to temporary environmental effects.

ferences in genotypes and in permanent environmental factors. An example of the latter could be udder damage incurred by a heifer that might decrease her milk yield throughout her productive life. A temporary environmental factor would be feed quality, which certainly varies from year to year. The variance caused by temporary environmental influences decreases in averages of n observations to V_T/n. The regression of the performance potential on averages of n observations becomes

$$b = \frac{V_P}{V_P + V_T/n} = \frac{nr}{1 + (n - 1)r}$$

When observations are repeated, the variance in the denominator decreases and the regression coefficient of the real producing ability on the actual average performance increases and the former may be estimated more accurately.

The quantity r, the intraclass correlation between performance records of an individual, is called the coefficient of repeatability, or simply repeatability, and indicates the extent to which observed differences in performance between animals will be repeated in future periods. Expressed differently, repeatability indicates the proportion of observed differences in performance between animals caused by differences in real producing ability.

Repeatability corresponds to a regression coefficient under conditions where variances are equal in different periods. When variances differ, as for milk yield in different age classes for which no age correction has been performed, application of repeatability coefficients for prediction requires multiplication by the ratio σ_2/σ_1, where σ_2 is the standard deviation of the performance to be estimated and σ_1 is the standard deviation of the "independent" yield.

Kögel (1976) found the correlation coefficients given in Table 5.1 from performance data of German Brown cattle. Repeatability as estimated from the correlation between first and second lactation yields is 0.58. The standard deviations of the selected cows (cows with two lac-

TABLE 5.1
Repeatability of Yields of Different Lactations[a]

		Lactation		
		1	2	3
Second lactation	a	0.73		
	r	0.58		
Third lactation	a	0.63	0.75	
	r	0.48	0.61	
Number of lactations		Standard deviation		
1		555		
2		542	628	
3		530	605	641

[a] Source: Kögel (1976). a, Absolute yield. r, Deviation from mean.

tations) were 542 and 628 for the first and second lactation yields. Therefore, the regression coefficient is $0.67 = 0.58 \times 628/542$ and a difference of 1000 kg in the first lactation should, on average, be followed by a difference of 670 kg in the second lactation. The regression coefficient can exceed 1 when the variance of future performance is much larger than the variance of the earlier period. In contrast, repeatability as defined above cannot exceed 1.

Repeatability of discontinuous traits can be estimated directly or by a method to be discussed in Chapter 13. When estimated directly, the repeatability gives the portion of the class differences repeated in future records, analogous to repeatability of continuous traits:

$$r = \frac{\Sigma \, kd}{\Sigma \, k}$$

where differences d are weighted by $k = n_0 n_1 / (n_0 + n_1)$.

Thus information from several classes of different size can be combined. For a single class repeatability can be computed as follows: Johansson et al. (1974) found that the twinning rate of Swedish Red-and-White cattle was 2.3%, but cows that had borne twins once had in later parities a twinning rate of 6.1%. Repeatability is then

$$r = 6.1\% - 2.3\% = 3.8\%$$

which is to be understood as an increase in the probability of twin birth of 3.8%.

Real producing ability, or performance potential, of an animal is affected by its genotype and by permanently acting environmental influences. However, repeatability is estimated generally from records of a limited number of periods. By definition, permanent influences act in all periods. However, contiguous periods have more in common than do more distant ones. Therefore, one may expect the repeatability estimated from few performance periods to be greater than that found when the total time span considered is longer. The correlations computed between consecutive yields in German Brown cattle are higher ($r_{12} = 0.58$, $r_{23} = 0.48$). A similar argument is valid for other traits and other species. However, compensation may occur between consecutive periods and introduce a negative correlation. For example, Johansson and Hansson (1940) found a negative correlation between dry period length and current lactation milk yield but a positive correlation between dry period and consecutive yield. These relationships should diminish the correlation between consecutive lactation yields.

The repeatability of a trait is not a constant. If measurement errors or, more generally, varying environmental conditions tend to increase V_T, repeatability tends to decrease. For example, Castle and Searle (1957) report a repeatability of 0.44 for lactation milk yield estimated from bimonthly milk controls and a repeatability of 0.49 from monthly controls. An example of the effect of a change in environmental conditions on repeatability was reported by Shrode and co-workers (1960). Their material showed repeatabilities for milk yield, butterfat yield, and fat percentage of 0.37, 0.32, and 0.70. After management was improved, the coefficients became 0.49, 0.44, and 0.71. Apparently the change in management removed some of the environmental variations that had masked the real producing abilities of the animals.

Repeatability generally refers to similarities in an animal's performance records in such a macroenvironment as a herd. When repeatability for dairy traits is estimated for a whole area containing many herds, the coefficient will be larger than that commonly found for repeatability in a herd. This is because herd differences will be included with differences in individual producing ability. In statistical terms the variance component for herds V_H will occur both in the numerator and denominator:

$$r = \frac{V_H + V_P}{V_H + V_P + V_T}$$

For example, correlations estimated from raw yield data of German Brown cattle were higher than estimates from data corrected for herd production class. Differences between animals in different herds or different herd production classes will be repeated to a much greater extent since the herd, etc., difference is repeated also.

Repeatability is important for estimation of breeding values from yield averages. It is directly applied, so to speak, when the performance potential or real producing ability of animals is to be estimated. For example, in a herd of cows, animals should be evaluated taking into account the different numbers of records. In reality, of course, the life expectancy of cows is taken into consideration also, and therefore real producing ability will not be the only criterion.

For an illustration, we can assume that cow A had a single lactation butterfat yield of 240 kg and cow B had a four-lactation average of 210 kg; that the herd average is 180 kg per lactation, $r = 0.4$, and that all observations are age-corrected. Which cow has the highest estimated real producing ability? The peformance of cow A is much higher than the average of cow B. But cow A only has a single observation and in that particular year conditions might have been unusually favorable. At any rate, an average of four lactations is a much safer estimate. By using repeatability, the real producing abilities of the two cows can be estimated and thus compared on an equal basis. For cow A, the regression coefficient b of real producing ability on the observation is equal to the repeatability (0.4). Her real producing ability is estimated to be 204 kg butterfat [0.4(240 − 180) + 180]. The regression coefficient for cow B is 0.73 and her probable performance potential is 202 kg butterfat. Only a small difference exists between the estimated real producing abilities of the cows. The real producing ability or performance potential of an animal P is estimated by the following equation:

$$P = b(O - M) + M$$

where M denotes the population (herd) mean and O the actual observed performance. The equation can be rearranged:

$$P = bO + (1 - b)M$$

The rearranged equation clearly shows that two sources of information exist for estimating the real producing ability of an animal, its own performance O and the population mean M. The larger the quantity b, the greater the weight given to the animal's own performance. If b is low, individual performance is of little value in predicting real producing ability. If no observation is available on the animal itself ($b = 0$), the best estimate of its performance potential is the population mean (neglecting the possibility of using information either on relatives or on other traits for estimating real producing ability).

Increased repeatability indicates the performance potential with higher accuracy. Consequently, a single record may be sufficient to estimate the real producing ability and repeated measurements are superfluous. Repeatabilities for a number of performance traits are given in Table 5.2. Several groups can be discerned. Differences in milk composition are highly repeatable and one lactation record may suffice to estimate the potential of a cow. Body measurements of adult animals show high repeatability. In contrast, repeatability of reproductive traits such as litter size of sows is low and a second or even a third litter may contribute valuable information.

Consideration of additional records implies a delay in decision-making and therefore a loss of time. One may ask whether the additional

TABLE 5.2
Coefficients of Repeatability $(r)^a$

Cattle	
Milk yield	35–55
Fat yield	35–50
Fat content	50–70
Persistency	15–25
Lactation length	15–25
Milk flow	60–80
Strip yield	10–20
Calving interval	1–10
Inseminations per conception	1–5
Gestation length	15–20
Weaning weight of calves	30–50
Weaning score	20–60
Swine	
Litter size	10–20
Sheep	
Fleece weight	40–80
Staple length	50–80
Weaning weight	20–30
Horse	
Trotting speed	60–80
Points for movement	30–40
Type score	30–80

^{a} The range of values comprises the majority of published values.

information is worth the additional time required to reach the culling decision. Probably the only practical situation in which such a question presents a problem occurs with large animals where selection may be postponed from the first to the second records. The progress by selection is proportional to the correlation between real producing ability and information.

$$(r_n)^{1/2} = \left[\frac{nr}{1 + (n - 1)r} \right]^{1/2}$$

The correlations to be compared are those involving one ($\sqrt{r_1}$) or two ($\sqrt{r_2}$) records. Assume that the second record will only be considered if the prediction accuracy of one record is less than 80% of the accuracy of two records. The following ratio permits the computation of repeatabilities that help in judging whether one should wait for another record:

$$\frac{r_1}{r_2} = \left[\frac{1 + (n - 1)r}{n} \right]^{1/2} \leq 0.8$$

It turns out that waiting is justified only when weakly repeatable traits with $r < 0.3$ are concerned.

Repeatability should be estimated from unselected data. In reality this is rarely possible, since poor cows, sows with small litters, etc., are not permitted to repeat their performance. Butcher and Freeman (1968) showed how original variances and covariances can be estimated from truncated data originally having a normal distribution. It is known that regression coefficients from normally distributed data remain unbiased when estimated from truncated data. Therefore, the regression coefficients permit the estimation of the original covariance; for example, from Table 5.1:

$$\text{cov } 12 = b_{21.1} \sigma_1^2 = 0.672 \times 555^2 = 206,992$$

$b_{21.1}$ denotes the regression coefficient of the second lactation yield on the first from cows that survived the first lactation. The mean squares due to deviation from regression are also unbiased. This permits reconstruction of the original variance of second records:

$$\sigma_2^2 = \frac{\Sigma d^2}{n - 2} + \frac{(\text{cov } 12)^2}{\sigma_1^2} = 261,725 + 139,098 = 400,823$$

The correlation as computed from corrected variances and covariances is 0.59, identical for all practical purposes to the correlation estimated from selected data. Butcher and Freeman also found in their results little difference between estimates from raw data and those computed with the method just described. The method can be extended to several consecutive records.

Repeatability coefficients also can be computed on observations and measurements that can be repeated consecutively. If the repeatability of a single observation is high, little can be gained by a second or even a third observation. In contrast, low repeatabilities indicate that measurements should be repeated. After studying repeatabilities, Touchberry and Lush (1950) decided that single measurements of cattle (girth, wither height, etc) were accurate enough for most practical purposes, and other investigators have reached the same conclusion (Weber, 1957). Similar measurements on pigs have a lower repeatability, and thus repeated measurements are justifiable (Fewson and Le Roy, 1959). Repeatabilities for determinations of milk constituents, as determined by Von Krosigk (1959) and Pirchner and colleagues (1960), can be used to judge how frequently fat or protein determinations should be made to achieve a desired accuracy in estimating real producing ability.

6

Heritability

The concept of heritability is of central importance to modern animal breeding theory. First, it is of considerable interest to know the extent to which traits are influenced by the genotype. Second, and possibly even more importantly to breeders, heritability provides, as the term implies, an estimate of the degree to which differences between animals are repeated in their progeny.

To discuss the broader concept first, breeders have been concerned since time immemorial with the question of how much observed differences are due to inheritance and how much to environment. Not infrequently the question is posed in "either–or" terms, as in the "nature or nurture" controversy in psychology. Of course the question cannot be posed in this form, and instead one must ask what proportion of the differences is caused by genes and what proportion is due to environment. The importance of each of the two classes of influence can be estimated by the portion of variance each causes. Of course the estimate arrived at is strictly valid only for existing conditions, and it may change when the environment changes or when a different population is tested. Fundamental problems connected with the justification and general significance of heritability have been discussed in human genetics in connection with the debate around the inheritance of intelligence. Feldman and Lewontin (1975) object to the concept of heritability on the grounds of its apparent lack of consistency. Purely genetic traits such as phenylketonuria (PKU) or galactosemia will cause no genetic variance when the diet is correctly modified. This could be extrapolated to animal production, where genetic differences in amino acid, trace mineral, and vitamin requirements can be masked by surplus provision of these micronutrients in the ration. In this case the genetic differences will have no consequences for the phenotype and therefore cause no genetic variation. However, a diet change where the level of critical nutrients is lowered may expose the heretofore hidden genetic variation. Against such arguments it is to be pointed out that estimates of heritability pertain to the environment to which the animals were exposed when the estimate was made.

The portion of the total variance caused by the total of genotypic differences was called by Lush "heritability in the broad sense":

$$h_w^2 = \frac{V_G}{V_G + V_E} = \frac{V_A + V_D + V_I}{V_A + V_D + V_I + V_E}$$

where the symbols in the denominator refer to variance caused by genic, dominance, epistatic, and environmental differences, respectively, and V_G is the variance caused by the totality of genotype differences. Heritability in the broad sense has been called coefficient of genetic determination in human genetics. It corresponds to the square of the correlation coefficient in regression theory. The size of this ratio is of interest insofar as it indicates the maximal portion of phenotypic variance that can be controlled when the whole genotype can be manipulated. This is possible by selection among certain types of populations and by clonal selection. The ratio is also of interest with regard to the possibility of environmental improvement. However, also in this respect one must take care concerning extrapolation. Inference is valid only insofar as the range of environments to which individuals are to be exposed is not altered. For example, under a normal diet PKU may have a high heritability and one may conclude that environmental measures would have little prospect of success. Yet changing the diet in an appropriate manner will prevent the disease.

Repeatability is larger than h_w^2 to the extent that it involves, in addition to genotype, permanent environmental effects. Repeatability contains in the numerator both V_G and $V_{E(p)}$, the variance caused by permanent environmental effects. If environmental effects can be neglected, both ratios will be the same; therefore $r \geq h_w^2$ and repeatability can be taken as the upper limit of heritability.

The estimation of heritability requires knowledge of the relationship between animals, or performance records from two generations, information which may be lacking in some situations. For example, in less developed countries the experimental herds assembled for investigation frequently lack in the beginning any information about the relationships of their animals. In such situations computation of repeatability is often possible and it yields initial information about the upper bounds of heritability.

Heritability in the narrow sense, defined by Lush as the ratio of genic variance to the total variance, is of yet greater, even central importance to breeding theory:

$$h_e^2 = \frac{V_A}{V_A + V_{E*}}$$

where V_{E*} denotes all nongenic variance, $V_{E*} = V_D + V_I + V_E = V_P - V_A$, since dominance and epistatic effects are largely nontransmissible.

Therefore, permanent selection gain is proportional to heritability in the narrow sense. It can also be conceived of as regression of breeding value on phenotype, since it indicates the change in breeding value when the performance changes by one unit:

$$h_e^2 = b_{AP} = \frac{\text{cov } AP}{V_P} = \frac{\text{cov } A(A + E^*)}{V_P} = \frac{V_A}{V_P}$$

where cov AP is the covariance between breeding value and phenotype, cov $AE^* = 0$ (breeding value and nongenic influences are independent), and cov $AA = V_A$.

Heritability can be estimated in two rather different ways: either from the ratio of selection gain to selection differential or from the phenotypic likeness or differences between relatives.

6.1. Likeness of Relatives

We know neither the number nor the kind of genes for any economic trait, as already mentioned. One possibility of measuring the importance of the genotype consists in comparing the fraction of variance between animals that remains when all genetic differences are removed. As pointed out the variance between monozygous twins is caused by environmental differences only, while the variance between unrelated animals (population variance) contains both environmental and genetic contributions. The difference between the population variance and the variance between monozygotic twins yields an estimate of the genotypic variance. Animals less related than monozygous twins have only part of the genotype in common. The variance between such relatives will then consist of the environmental contribution plus a fraction of the genetic variance which decreases with increasing relationship of the animals compared. The diminution of this intrapair variance with increasing closeness of relatives permits one to estimate the genetic variance. This approach is illustrated in Fig. 6.1 with data from a Scottish twin experiment (Watson, 1960).

In general, heritability is not estimated from variances between relatives, but from covariances (respectively, correlations) between them. The covariances (respectively, correlations) are identical to variance components relating to group means (respectively, intraclass correlations).

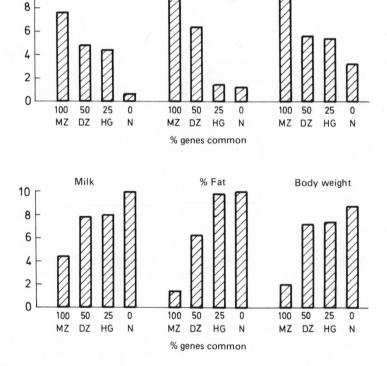

FIGURE 6.1. Correlations (above) and residual variances (below) between members of monozygous twin pairs (MZ), dizygous twin pairs (DZ), half-sib (HG), and unrelated pairs (N) of cattle. (Source: Watson, 1960.)

The variance of group means is given by

$$\frac{\sigma_w^2}{n} + \frac{n-1}{n} \, \text{cov} \, x_1 x_2 = \frac{\sigma_w^2}{n} + \frac{n-1}{n} \, \sigma_B^2$$

where σ_w^2, $\text{cov} \, x_1 x_2$, and σ_B^2 denote, respectively, the variance caused by differences between members of the group, the covariance between them, and the variance between true group means. The variance between means of very large groups is identical to the covariance between members of the group.

The covariance is caused by common elements in the group members. Insofar as genes only are the common elements, the covariance will be of genetic origin and thus provide an estimate of the genetic variance. Assume that the performance of individuals X and Y is described by the following model:

$$P(X) = x_1 + x_2 + dx_{12} + r$$

$$P(Y) = y_1 + y_2 + dy_{12} + r'$$

x_i and y_i are genic effects in individuals X and Y, dx_{12} denotes the dominance deviation, and r is the residual. Further,

$$E(x_1 \cdot dx_{12}) = E(x_1 \cdot dy_{12}) = E(rr') = 0$$

The covariance between X and Y is given by

$$E(XY) = E(x_1 y_1 + x_2 y_1 + x_1 y_2 + x_2 y_2 + dx_{12} dy_{12})$$

Now,

$$E(x_1 y_1) = P(x_1 = y_1) \, \Sigma \, p\alpha^2 = \tfrac{1}{2} C_{XY} \, \sigma_A^2$$

Note that $\Sigma \, p\alpha^2$ is the variance caused by effects of single genes. Therefore the covariance equals the product of the probability that the genes of X and Y are identical times half the genic variance. The probability of identity is given by the coancestry coefficient between X and Y. Common dominance deviations in both individuals can only arise when genes at both loci in both individuals are identical. Let x_1, y_1 denote paternal genes and x_2, y_2 maternal genes in the two individuals and let ϕ and ϕ' be the probabilities of identity of the paternal and the maternal genes, respectively, in the two relatives. The probability of identity of gene combinations in the two individuals is then $\phi\phi' = U_{XY}$, which can be described as a dominance relationship. It is zero when animals are related via one parent only. The covariance between relatives caused by common dominance deviations is given by

$$U_{XY} \Sigma\, p_i p_j \sigma_y^2 = U_{XY} V_D$$

When the one-locus model for the genotype is used, the covariance between the effects of individuals X and Y is given by

$$2C_{XY}V_A + U_{XY}V_D = \tfrac{1}{2}(\phi + \phi')V_A + \phi\phi'\, V_D$$

Of course there are unaccounted for epistatic effects, which Cockerham (1954) and Kempthorne (1957) have systematized. Let us consider the simplest case, where additive effects at two separate loci interact. When such interaction effects are common to two relatives a covariance exists. The probability that two individuals have two identical genes common at two different loci is given by

$$(2C_{XY})^2 = \left(\frac{\phi + \phi'}{2}\right)^2 = I_{XY}, \text{ say}$$

The covariance caused by common epistatic deviations between two loci is given by

$$I_{XY}V_{AA}$$

where V_{AA} is the variance due to interactions between additive gene effects at two loci. The general formula for covariance caused by interactions between additive gene effects at n loci and dominance deviations at m loci can be formulated as

$$(2C_{XY})^n\,(U_{XY})^m\, V_{AnDm}$$

The quantities $2C_{XY}$ and U_{XY} are less than one between all relatives except for monozygotic twins. Higher powers will make them very small. Also, numerous investigations have indicated that epistasis as it is understood in the regression model is a relatively unimportant cause of differences between animals within populations. Therefore, covariances between relatives caused by epistatic interactions are generally ignored, even though admittedly this is not a very satisfactory state of affairs. In general, covariances between relatives are interpreted by supposing that similarity is caused only by common possession of genic effects and dominance deviations:

$$\text{cov XY} = 2C_{XY}V_A + U_{XY}V_D$$

It must be pointed out that when estimating the genetic variances from differences between relatives of varying closeness, the contribution of variances due to higher order interactions increases with decreasing relationship. The variance between relatives is given by

$$V_E + (1 - 2C_{XY})V_A + (1 - U_{XY})V_D + [1 - (2C_{XY})^2]V_{AA}$$

The coefficients of the genic- and dominance-caused covariances between relatives of varying closeness are given in Table 6.1. The covariance between parents and offspring can be derived directly:

$$\text{parents} \quad x_1x_2 \qquad y_1y_2$$

$$\text{offspring} \quad z_1z_2$$

When z_2 derives from the second parent and random mating exists,

$$\phi = P(x_1 = z_1) + P(x_2 = z_1) = \tfrac{1}{2} + \tfrac{1}{2} = 1$$

$$\phi' = P(x_1 = z_2) + P(x_2 = z_2) = 0$$

$$\tfrac{1}{2}(\phi + \phi')V_A = 2C_{XY}V_A = \tfrac{1}{2}V_A$$

TABLE 6.1
Degrees of Relationship

Relationship	$2C_{XY}$	U_{XY}
Monozygous twins	1	1
Full-sibs	1/2	1/4
Half-sibs	1/4	0
First cousins	1/8	0
Double first cousins	1/4	1/16
Parent–offspring	1/2	0
Grandparent–grandchildren	1/4	0
Uncle–nephew	1/4	0

Progeny receive half of their genes from one parent. No correlation exists between the breeding values and dominance deviations of the parents nor between them and environmental effects [i.e., $E(AE) = E(AA') = 0$, where A' denotes the breeding value of the partner]. When epistatically caused similarity between parent and offspring is ignored, the covariance between the two is given by

$$\text{cov } PO = \left(\frac{A}{2} + \frac{A'}{2} + E^*\right)(A + E) = \frac{E(AA)}{2} = \frac{V_A}{2}$$

The estimation of the genic variance can be illustrated by the example given in Table 6.2. The variance component for the sire σ_S^2 equals the covariance between half-sibs which has the expectation cov HS $= \frac{1}{4}V_A$. Therefore $V_A = 4\sigma_S^2 = 0.0688$ cm^2. The difference cov FS $-$ cov HS yields the variance component between full-sibs within a half-sib group and it has the expectation

$$(\tfrac{1}{2} - \tfrac{1}{4})V_A + \tfrac{1}{4}V_D + V_M$$

where V_M is the variance caused by maternal effects. If these can be considered as unimportant, the dominance variance is given by

$$4(\sigma_D^2 - \sigma_S^2) = 4(\tfrac{1}{4}V_A + \tfrac{1}{4}V_D - \tfrac{1}{4}V_A) = V_D$$

$$= 4(0.0211 - 0.0172) \text{ cm}^2 = 0.0246 \text{ cm}^2$$

When half-sib covariances are estimated, it is generally assumed that the parents are unrelated and that no relationship exists among the sires or among the dams. These assumptions will be unjustified in many situa-

TABLE 6.2
Analyses of Variance for Full-Sib and Half-Sib Families[a]

Source of variance	Degrees of freedom	Mean square	Expected mean square
Sires	$s - 1 = 1125$	$V_S = 0.1464$	$\sigma_R^2 + k_2\sigma_D^2 + k_3\sigma_S^2$
Dams within sires	$d - s = 326$	$V_D = 0.1028$	$\sigma_R^2 + k_1\sigma_D^2$
Full-sibs within dams	$t - d = 1499$	$V_R = 0.0587$	σ_R^2

$$\sigma_R^2 + \sigma_D^2 + \sigma_S^2 = \sigma_T^2 = 0.0980, \quad \sigma_R^2 = \sigma_T^2 - \text{cov } FS = 0.0587$$
$$\sigma_D^2 = \text{cov } FS - \text{cov } HS = 0.0221, \quad \sigma_S^2 = \text{cov } HS = 0.0172$$
$$k_1 = k_2 = 2.00, \quad k_3 = 2.54$$

[a]Source: Jonsson (1959). s, Sires. d, Dams. t, Individuals.

tions. For example, a boar will be used on a herd of sows which are related to each other as full and half-sisters, etc. Hinkelmann (1971) investigated the problems of estimating genetic variance in situations where some relationship between parental animals may exist. However, the problems arising from related mating partners are not considered here, but in such situations the definition of genetic effects presents further problems. The simplest case, relationship ($2C$) between partners of a sire, will be discussed here.

When genes x_2 and y_2 come from dams the probabilities of identity of genes are $\phi = \frac{1}{2}$ and $\phi' = P(x_2 = y_2) = \frac{1}{4}C_{BC}$. Therefore the coefficient of the genic variance (Fig. 6.2) is

$$2C_{XY} = \tfrac{1}{2}(\tfrac{1}{2} + C_{BC}) = \tfrac{1}{4}(1 + 2C_{BC})$$

The offspring are related through both parents—same sire, related dams—so that some dominance-caused similarity exists: $\phi\phi' = \frac{1}{2}C_{BC}$. Therefore, the sire covariance becomes

$$\sigma_S^2 = \tfrac{1}{4}(1 + 2C_{BC})V_A + \tfrac{1}{2}C_{BC}V_D$$

The relationship between mated sows in one herd may amount to $2C_{BC}$ = 0.2 (Flock, 1970). Using this value results in a sire component of variance of approximately

$$\sigma_S^2 = \tfrac{1}{4} \times 1.2V_A + 0.2V_D$$

If dominance can be neglected, the genic variance of the example in Table 6.2 will be smaller than the value reached ignoring the similarity caused by related dams:

$$V_A = 4 \times 0.0172/1.2 = 0.0573$$

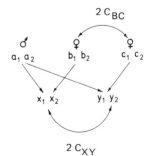

FIGURE 6.2. Kinship coefficients when mating partners are related.

Related dams cause the dam variance component to be biased downward by the same amount as the sire components are inflated:

$$\sigma_D^2 = \tfrac{1}{4}(1 - 2C_{BC})V_A + \tfrac{1}{4}(1 - 2C_{BC})V_D$$

If dominance deviations cannot be excluded *a priori*, the genic and dominance variances can be estimated by equating the respective variance components to their expectations and solving the two equations for their unknown. The results $V_A = 0.0309$ and $V_D = 0.0159$ differ considerably from those estimated without considering the relationship of the dams ($V_A = 0.0688$ and $V_D = 0.0194$), which illustrates the bias possible when applying a model insufficient to account for the data structure.

Similarity can also be caused by sex-linked genes. Only in poultry do sex chromosomes contain a larger part of the total genome (about one-seventh of the larger chromosomes). Nevertheless, the following discussion presupposes the situation in mammals, where females are homogametic. In the heterogametic sex, genes are hemizygous. Therefore no dominance deviations can arise and the sex-linked genic variance is $V_{A_s} = \Sigma \, p\alpha_i^2$.

Of course epistatic interactions can arise, but between sex-linked genes only interactions of the $A \times A$ type are possible. In the homogametic sex the genic variance caused by sex-linked genes is no different from autosomal variance: $V_A = 2 \Sigma \, p\alpha_i^2$. Covariances between relatives differ, depending on the sex of the relatives. Covariance between females is analogous to the autosomal genic covariance. The covariance between male relatives is a part of the sex-linked genic covariance. Coefficients of covariances for various combinations of relatives are given in Table 6.3. Two examples will be discussed.

TABLE 6.3
Covariance Caused by Sex Linkage (Mammals)

	♂	♀
Sire–offspring	0	V_{A_s}
Dam–offspring	V_{A_s}	$\tfrac{1}{2}V_A$
Paternal half-sibs	0	$\tfrac{1}{2}V_A$
Maternal half-sibs	$\tfrac{1}{2}V_{A_s}$	$\tfrac{1}{4}V_A$
Full-sibs	$\tfrac{1}{2}V_{A_s}$	$\tfrac{3}{4}V_A + \tfrac{1}{4}V_D$
Paternal uncle–nephew (niece)	$\tfrac{1}{4}V_{A_s}$	$\tfrac{3}{8}V_A$

The covariance between a sire with gene a_1 and daughters with genes x_1, x_2 at the sex chromosomes is

$$P(x_1 = a_1) \Sigma\, p\alpha_i^2 = V_{A_s}$$

All daughters receive the same gene from their sire. Obviously then the covariance between paternal half-sisters is also V_{A_s}. The covariance between maternal half-brothers X and Y with sex-linked genes x_1 and y_1 is $P(x_1 = y_1)V_A = V_A/2$ since their only X chromosome will be one or the other of the maternal X chromosomes. The influence of sex-linked genes is mainly noticeable in poultry, which, given the relative size of the sex chromosomes, is not surprising. For example, Hogsett *et al.* (1964) report sire and dam intraclass correlations for egg laying performance of 2.6 and 1%, respectively. In poultry, males are homogametic, and therefore the expectations of Table 6.3 must be exchanged. In poultry no sex-linked covariance should exist between maternal half-sisters, while for paternal half-sisters its expectation is $\frac{1}{2}V_{A_s}$. Neglecting other possible complications, such as maternal effects, the sex-linked genic variance can be estimated as $2 \times (2.6 - 1.0) = 3.2\%$, while the autosomal genic variance is $4 \times 1.0 = 4.0\%$.

In sex-limited traits sex-linked variance can be inferred from a comparison of maternally caused covariances with covariances caused by paternal relationship. For example, in mammals there is no sire–son covariance caused by sex-linked genes, while these cause a covariance of $\frac{1}{2}V_{A_s}$ between a male and his maternal uncle.

Overdominance of the adaptive value may influence the off-spring–parent and the half-sib covariances to different degrees (James, 1966). If natural selection acts on viability and therefore covariances are to be estimated on survivors' performance, the ratio of the half-sib to the offspring–parent covariance is approximately

$$\frac{\text{cov } HS}{\text{cov } PO} = \frac{1}{2}(1 - L_P)$$

where L_P is the weighted average of the segregation load at the loci involved (coefficient of homeostatic strength; Robertson, 1956), the weights being the genic variance caused by the respective loci. Epistasis affects the ratio in a similar way since the contribution of covariance caused by $A \times A$ interactions to parent–offspring covariance is greater ($\frac{1}{4}V_{A \times A}$) than to the half-sib covariance ($\frac{1}{16}V_{A \times A}$). It must be assumed that any connection between the adaptive value and performance will affect covariances between relatives to various degrees. The extent of the changes brought about thereby also depends on the stage of the life cycle at which natural selection acts and at which the performance is measured.

6.2. Estimation of Heritability

Heritability in the narrow genetic sense is of great importance to breed improvement. It primarily indicates to what extent genetic transmission of performance differences between animals can be expected and therefore how large the progress by selection will be. The parameters chosen help to decide whether a performance or progeny test is to be applied and, in an aggregate of traits, what weights the components should receive to arrive at efficient criteria for selection decisions. The concept has several limitations which will be discussed later.

At the start of any breeding program information is required about the approximate range of heritability of the traits to be improved and also of genetic correlations among them and with other important characteristics. It is desirable to estimate these parameters from the population that is to be improved. If this is not possible, the information must be taken from the literature.

The method of estimation is frequently dictated by the nature and structure of the available data. For example, full-sibs are very rare in cattle. Therefore, full-sib correlations cannot be used for estimating genetic parameters of cattle performance. Parent–offspring correlations can rarely be used for estimating genetic parameters of slaughter performance, etc. The various methods available have both specific advantages and disadvantages which make them well suited for certain purposes but unsuited for others. With the exception of the realized heritability, which will be discussed later, the parameters are estimated from covariances (respectively, correlations) among relatives and from regression of offspring on parents. The use of covariances between relatives requires the validity of the Mendelian laws (which are of course universally accepted) and the justification of some assumptions, such as absence of nongenetic correlations between relatives, random mating, etc. In the case of regressions of offspring on parents no assumption about genetic theory is really necessary. Heritability is used initially to estimate the selection gain. The offspring–parent regression yields directly the gain in offspring performance due to selection of parents with unit superiority, so that the detour via the somewhat abstract concept of heritability is really unnecessary.

In populations of domestic animals heritability is frequently estimated from the regression of offspring on dam, somewhat less frequently from the offspring–sire regression, and least of all by the regression of offspring on midparent. The expectation of the offspring–parent covariance is given in Table 6.2. It should be noted that the expectation contains, in addition to half of the genic variance, epistatic components as well. If these are not negligible, the heritability derived from parent–offspring regression will be slightly overestimated. The regression of offspring on midparent gives immediately the heritability. While the parent–offspring covariance is $V_A/2$, the variance of the average performance of two unrelated parents is $V_P/2$, so that

$$b_{O\bar{P}} = \frac{V_A/2}{V_P/2} = h^2$$

Not only may parent–offspring regressions be biased by epistasis, but maternal effects, to be discussed later, are another source of bias. Phenotypic assortative mating, which is much discussed in human genetics, is yet another source. Positive assortative mating, where good sires are mated to good dams, medium sires to medium dams, etc., leads to an increase in the genic covariance between dam and offspring (or sire and offspring), since genes of the mating partner accentuate in the offspring the effects of genes of the first parent. The reverse is expected with disassortative mating, where mating partners are negatively correlated for some phenotypic characteristic and where the genetic effects coming from one parent are partly neutralized in offspring by those from the other parent.

As will be discussed in Section 11.3, the offspring–parent regression remains unbiased by phenotypic selection among parents, provided that the regression is linear. However, when selection is based on information from relatives, the heritability estimates from offspring–parent regressions will be biased. However, in many instances selection will not be very accurate and the resulting bias will be negligible.

The estimation of heritability based on regression of offspring on parents selected in opposite directions is most efficient. This divergent selection increases the variance of the independent variable, i.e., the parental performance, which makes the variance of the regression coefficient very small (Hill and Nicholas, 1975).

When progeny number per parent varies, a weighting problem arises. Three solutions are possible: to repeat each parent with each progeny, to use the progeny average for each parent, or to weight each parent–progeny combination by the information it contributes. The last method was introduced by Kempthorne and Tandon (1953) and clarified in some respects by Reeve (1955). Ollivier (1974a) extended it to take care of structured progeny groups where several full-sib groups may occur per sire. Kempthorne and Tandon derived the weights for the contributions of each parent–offspring group to the regression coefficient:

$$w_i = \frac{n_i}{1 + n_i \tau}$$

$$\tau = \frac{t - \beta^2}{1 - t}$$

and

$$b = \Sigma \, w_i(x_i - \bar{x}) \, (y_i - \bar{y}) / \Sigma \, w_i(x_i - \bar{x})^2$$

where t is the intraclass correlation between sibs, β is the prior estimate of regression of offspring on parents, and n_i is the number of offspring of parent i.

The method shows its greatest advantage when progeny group size varies very much. Bohren *et al.* (1961) used the three methods for an analysis of poultry data and found little difference among the parameters estimated. Alsing *et al.* (1980) applied the weighted regression method in genetic analysis of litter size of sows where both daughter–dam and granddaughter–granddam regressions were computed.

The use of grandchild–grandparent correlations is recommended in human genetics as a means of circumventing problems caused by environmental contributions to child–parent correlations. Lörtscher (1937) has used this approach in an analysis of dairy records. Grandchild–grandparent correlations contain fewer epistatic contributions than offspring–parent correlations. For example, four times the granddaughter–granddam correlation leads to a heritability which contains 3/16 less of the correlation caused by $A \times A$ interactions in both relatives than does two times the daughter–dam correlation. Van Vleck and Bradford (1965) have computed heritabilities of dairy traits by both methods. However, the large errors of the estimates precluded definite conclusions. In such investigations one must assume that maternal effects are negligible.

Male breeding animals in nearly all domestic species have, or at least may have, large numbers of progeny. Therefore sire intraclass correlations are one of the, if not the most, popular of methods of estimating heritability. The sire variance components are frequently estimated from a one-stage hierarchic analysis of variance, an example of which is given in Table 6.4. The fourfold sire intraclass correlation estimates heritability. This simple form of analysis requires that each progeny has a separate dam, a condition which is nearly automatically met in monoparous species. The epistatic contribution to such a heritability is com-

TABLE 6.4

Analysis of Variance for Half-Sib Families[a]

Source of variance	df	MS Milk yield Absolute	MS Milk yield Deviation[b]	MS Fat-% deviation[b]	MP Milk yield, fat-%		Expected value[c]
Sires	$s - 1 = 92$	7,040,357	1,942,803	1.3210		−205.9727	$\sigma_R^2 + 126.6\sigma_s^2$
Residual	$t - s = 11{,}858$	516,786	326,338	0.0930		−22.8078	σ_R^2
σ_P^2		567,009	338,724	0.1024	cov_P	−24.2112	
σ_S^2		51,529	12,769	0.0097	cov_S	−1.4468	
h^2		0.36	0.15	0.38	r_G	−0.13	
s_{h^2} 1)[d]		0.058	0.027	0.060	r_P	−0.15	
2)		0.058	0.027	0.061	s_{r_G} 4)	0.208	
3)		0.031	0.020	0.032	5)	0.118	

[a] Source: Hindemith (1978). df, Degrees of freedom; MS, MP, mean square, mean product, respectively.
[b] Deviation from herd group–season average.
[c] For MP covariance components.
[d] See text for explanation of 1–5.

parable to that in estimates derived from grandchild–grandparent correlations. Van Vleck and Bradford (1965) have used the different expectations of half-sib-derived heritabilities and heritabilities from daughter–dam analyses to estimate the importance of epistasis. With similar objectives Allaire and Henderson (1958) compared values derived from dam–daughter covariances with half-sib estimates. Neither workers achieved much success, because of too large errors, but the possible effect of natural selection on the covariances between the various relatives was not considered. Apart from the rather small contribution of epistasis, heritabilities computed from half-sib correlations are largely unbiased. Their principal disadvantage is the necessity of multiplication by four, which automatically quadruples all errors and biases. These could occur, for example, when environmental correlations between half-sibs exist which are neglected in the analysis. Such correlations may be present even in an AI population, for example, when progeny groups perform in different time periods. The heritability of absolute milk yield in the example of Table 6.4 is 36%, but when records were corrected for year, season, and herd level, it dropped to 15%.

Rittler *et al.* (1968) have compared heritabilities estimated from the total data set available to them with heritabilities calculated from the same data where only one record occurred in any one herd. They concluded that the latter approach removed environmental contributions and led to estimates much closer to the true value than if the restriction had not been made. Similar inferences were drawn by Bereskin and Lush (1965) from investigations of U.S. dairy records. Carryover of environmental effects from herds of origin may even inflate sib correlations from test station data and seriously inflate heritabilities derived therefrom. This bias may be aggravated by correlations due to contemporarity, since sibs often enter test stations simultaneously. Using extensive material from German swine-testing stations, Flock (1970), for example, found that heritabilities estimated from various sib correlations were all considerably larger than heritabilities derived from relatives one generation removed (for example, uncle–nephew). Likewise, Rutzmoser and Pirchner (1979) found relatively large estimates for beef traits from correlations between half-sibs in test stations, but much smaller heritability values when these were based on uncle–nephew correlations.

Mating to groups of dams, distinct due to either different selection or relationship, may inflate heritability. The latter cause of bias has been discussed in the previous chapter. In contrast, heritability may be seriously underestimated when the progeny groups are sired by highly and accurately selected males, as may be the case in AI breeding (see Section 11.3).

Data from species such as pigs and poultry where full-sibs are common permit handling by a two-stage hierarchic analysis of variance as shown in Table 6.2. Sire and dam variance components can be derived. The dam variance component is due to the additional covariance between full-sibs within half-sib groups. In addition to one-fourth of the genic covariance, it contains one-fourth of the dominance covariance and the whole of the maternal environmental covariance. Therefore, the dam variance component usually is larger than the sire variance component.

The difference between the two is considered as being caused by maternal influence and/or dominance deviations, which cannot be distinguished in a hierarchic design (several females mated to one male only). When other information permits the neglect of maternal effects, the difference between the sire component and the dam-within-sire component of variance allows an estimation of the importance of the dominance variance:

$$\frac{V_D}{V_P} = 4 \left(\frac{\sigma_M^2}{\sigma_P^2} - \frac{\sigma_S^2}{\sigma_P^2} \right)$$

Conversely, when dominance can be assumed to be unimportant, the relative contribution of maternal influences to the phenotypic variance can be estimated:

$$\frac{V_M}{V_P} = \frac{\sigma_M^2}{\sigma_P^2} - \frac{\sigma_S^2}{\sigma_P^2}$$

Full-sib correlations must be used in estimates of heritability from data on monogamous species such as pigeons. Also, in many cases test station data may consist largely of full-sib information when only full-sib groups are provided from single herds. The expectation of the full-sib covariance is given in Table 6.2. Heritability is estimated by doubling the full-sib correlation. It is only equivalent to heritability in the narrow sense when the nongenic influences on the correlation can be excluded. Contemporaneity contributes to likeness between litter mates; the effect of this on full-sib correlations is apparent in the investigation on the growth rate of swine by Johansson and Korkman (1950), where a correlation of 0.21 was found between litter mates, whereas between full-sibs from different litters it was only 0.13.

Factorial mating designs in which dams are mated with several sires and *vice versa* have been introduced by the Danish geneticist Johannes

Schmid to distinguish paternal from maternal contributions to the offspring genotype. Schmid called such designs diallels, and they will be
discussed in greater detail in Chapter 16. When matings within a breed
are arranged as diallels, maternal and dominance influences can be separated. Jerome *et al.* (1956) applied such mating designs to poultry and
Lauprecht *et al.* (1966) found similar mating patterns in swine. The
expectations of variance components estimated from such data are

$$\sigma_S^2 = \tfrac{1}{4}V_A, \qquad \sigma_D^2 = \tfrac{1}{4}V_A + V_M, \qquad \sigma_{S \times D}^2 = \tfrac{1}{4}V_D$$

Jerome *et al.* found that dominance variance was three times the importance of genic variance in their material on laying performance.
Therefore there seems to be a considerable difference between heritability in the narrow and broad senses. Sib performance data are contemporary and therefore they often reflect the most recent state of technology. Initially genetic parameters of newly introduced traits can only
be estimated from sib covariances and usually slaughter or test station
data can only be analyzed by sib covariances. Residual correlations between sibs pose problems in many cases. The number of investigations
based on sib data is enormous. Genetic parameters estimated for the
same traits often differ widely. On rare occasions the causes for such
differences are evident or the differences can confidently be assumed
to reflect true population differences. Familiarity with the material and
the population providing it will in many cases permit judgement.

When data derived are from two generations and also possess a
known family structure, several methods of heritability estimation can
be applied simultaneously and the results combined. For example, heritabilities can be computed from half-sib covariances and from daughter–dam covariances. Hill and Nicholas (1975) showed that correlations
exist between such estimates. Therefore, their combination by simple
weighting with the reciprocal of the variance of heritability is not optimal.
The correlation between heritability from midparent–offspring regres-
sion (h_{bM}^2) and full-sib heritability (h_{FS}^2) is positive, while heritabilities from
daughter–dam regression and from half-sib correlations (h_{HS}^2) are negatively correlated. The regression of h_{bM}^2 on h_{HS}^2 is given by the relatively
simple formula

$$-\frac{h^2\,[4\,+\,(m\,-\,1)h^2]}{2m}$$

where m is the number of dams, and S is the number of sires. The
expression for the correlation is more complicated, but when m and s

are large, it is $-h^2/2m$. The negative correlation between h_{bM}^2 and h_{HS}^2 comes about because a large variance between dams causes a high regression, but at the same time a smaller variability between half-sib groups. Large differences between dams will tend to level out differences between half-sib groups. Heritabilities from offspring–dam and offspring–sire regressions are independent. Heritabilities from two different estimations are combined by means of the formula

$$h_C^2 = h_{HS}^2 + a(h_{bD}^2 - h_{HS}^2)$$

In some cases three estimators may be available, for example, offspring–dam, offspring–sire regression, and half-sib correlations. Hill and Nicholas compared the various estimators and found, in agreement with Robertson (1959a), that for poorly heritable traits half-sib correlations are more efficient estimators than are daughter–dam regressions and that the combined heritability is little inferior to maximum-likelihood estimates. The most efficient estimation is from offspring–parent regression when parents were selected in divergent directions, which, naturally, rarely will be possible in the breeding of domestic animals.

The combination of heritabilities from different estimation procedures is permissible only when the expectations of the estimated parameters are identical or at any rate very close. Revelle and Robison (1973) and Alsing et al. (1980) suggested that for sow fertility half-sib covariance and daughter–dam covariance have very different expectations (Section 6.4). Therefore, a combination of the heritabilities in the sense discussed above is not permissible. Equal reservations would be justified about combining the various heritabilities from German pig test station data reported by Flock (1970). Estimates derived from relatives separated by one generation were only 60% of the half-sib estimates for seven traits.

6.3. Realized Heritability

Heritability mostly serves to predict selection gains. Therefore a direct comparison of selection gain and selection differential suggests itself. It circumvents the indirect approach of estimating heritability via resemblance of relatives, with its various assumptions, i.e., Mendelian laws, panmixis, missing dominance, c^2 effects, etc. The direct empirical method looks attractive and is a priori optimal. However, it can be applied only after selection results have become available. The concept of realized heritability, introduced by Falconer (1955), is applied mostly to laboratory animals, where the generation interval is very short.

The realized heritability should correspond to heritabilities estimated from offspring–parent regression. However, when epistasis, dominance, and c^2 effects are unimportant, heritabilities from sib correlations should be similar.

It appears at first that this method of estimating h^2 is optimal. However, as will be shown, most of those few epistatic combinations that are inherited are transmitted in the first or in the first few generations. Therefore, if such gene combinations are important, selection gains will be larger in early than in later generations of selection. Furthermore, selection gains based on such epistatic gene combinations are not permanent: When selection ceases, the combinations disintegrate and the corresponding gains disappear. In other words, some selection pressure is necessary to hold the selected gene combinations together. Heritability estimated from selection gains in the first generations may be too large and in direct proportion to the transmitted additive interaction effects. Thus, it overestimates permanent selection gains by this amount.

In addition, maternal effects can influence selection gain, in positive and in negative directions.

Realized heritability can be estimated in various ways. Falconer suggested the regression of cumulative selection gain R on cumulative selection differential D:

$$h_R^2 = b_K = \frac{\Sigma_i D_i R_i}{\Sigma_i D_i^2}$$

D_i and R_i are, respectively, selection differential and selection response up to and including generation i. Selection differential d_i and response r_i of individual generations can also be utilized:

$$h_R^2 = b_I = \frac{\Sigma_i d_i r_i}{\Sigma_i d_i^2}$$

The simplest method computes the ratio of final response R to the total selection differential D:

$$h_R^2 = b_R = R/D$$

The three estimators of h_R^2 are about equally efficient (Section 6.4). When heritability is high and selection has lasted for many generations, b_I and b_R seem preferable, but differences are small.

Realized heritability can be estimated from a two-way selection experiment which neutralizes any environmental trend, provided no genotype–environment (generation) interactions arise. While divergent selection can be applied to laboratory populations, it appears in general hardly practicable in domestic animal improvement. Here, one-way selection is the rule and realized heritability must be estimated from response and selection differentials in one direction, with or without an unselected control population. These permit the recognition and neutralization, if present, of environmental trends. Where such a trend exists to the extent of U units per generation, it will overestimate the realized heritability by U/D if it is not recognized and taken account of. Control populations are common in selection experiments with laboratory animals. However, here environment can usually be well controlled, in contrast to selection endeavors with domestic animals, where control populations are difficult to maintain. However, environmental trends can also be recognized by repeat matings. When control populations are employed in two-way selection experiments, asymmetry of the selection response can be recognized and thus asymmetric heritabilities estimated. Of course the application of repeat matings also permits this, but it is considerably more complicated.

Asymmetry of selection response first became evident in selection experiments with mice and *Drosophila*. It also may be a problem of general importance in domestic animals. For example, Hetzer and Harvey (1967) have published a series of papers on a divergent selection for back fat in Duroc and in Yorkshire pigs (Table 6.5). Agreement was reasonable between offspring–parent heritability estimated in the base generation and realized heritability computed from the divergence of the lines. However, a strong asymmetry became apparent. In both breeds upward progress during the first five generations was much larger than downward response, while in generations 6–10 realized heritabilities were more equal. It should be noted that realized heritabilities for each direction and each time interval agree fairly closely with the off-

TABLE 6.5
Realized Heritability ($\times 100$) of Back-Fat Thickness[a]

Direction of selection	Generations	Duroc		Yorkshire	
		R	PO	R	PO
+	1–5	73	79	64	78
	6–10	30	34	46	51
−	1–5	48	40	13	24
	6–10	29	45	60	76

[a] Source: Hetzer and Harvey (1967). R, Realized; PO, parent-offspring.

spring–parent heritabilities estimated simultaneously within each selection line. Similar observations are reported by Meyer and Enfield (1974) from selection experiments with *Tribolium*. Realized heritabilities from the total response of the divergent lines agreed with the estimate from offspring–parent regressions (both 28 ± 20%). The asymmetric heritabililities observed in separate lines (16% in the downward, 28% in the upward selected lines) were reasonably predicted from offspring–parent regressions in the same lines (−5%, 15% in the minus, 47% in the plus lines). In contrast, Falconer (1973) reached the conclusion that asymmetry in a large-scale selection experiment with mice could have been due to random drift.

A method of estimation based on the same principle but using field records was suggested by Lush and Straus (1942). Hartmann (1959) used this method to estimate the heritability of milk yield in a herd of German Friesians. As shown in Fig. 6.3, the herd can be divided into a better half and a poorer half. The two daughter groups differ in performance because their breeding values differ by half of the difference of the breeding values of their dam groups (half, because the daughters receive only half their genes from their dams). If the difference between the two dam groups is denoted by $M_H - M_L$, where H and L denote high- and low-performance, respectively, and that between the two daughter groups by $D_H - D_L$, the heritability can be computed as

$$h^2 = 2\frac{D_H - D_L}{M_H - M_L}$$

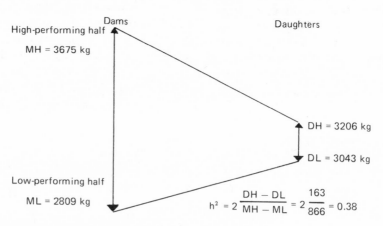

FIGURE 6.3. Estimation of heritability from selection gain. $h^2 = 2(D_H - D_L)/(M_H - M_L) = 2$ (163/866) = 0.38. (Modified from Hartmann, 1959.)

When the daughters are sired by several different animals, the influence of the sires must be excluded. Dividing each group of mates of a single sire into two halves (high- and low-producing) and computing the difference between their daughters accomplishes this. The differences between the half groups of mates and of daughters are summed over sires using the appropriate weighting factor for each group:

$$h^2 = 2 \frac{\Sigma w_i(D_{iH} - D_{iL})}{\Sigma w_i(M_{iH} - M_{iL})}$$

where M_{iH} denotes the average of high-performing dams mated to sire i, D_{iL} is the average of daughters of sire i out of low-performing dams, and w_i is the weighting factor of the difference between the animals of sire group i. Comstock (Kincaid and Carter, 1958) has suggested $(M_{iH} - M_{iL})/V$ as an optimal weighting factor, where V is the residual variance remaining when daughter differences are regressed on dam differences. The method can be used for other traits than dairy records, of course. Kincaid and Carter used this method to estimate the heritability of cattle growth rate.

An estimation procedure termed "linear heritability estimates," which is also based on comparing genetic and phenotypic effects, has been suggested by Abplanalp (1961). With effective heritability, response in the progeny generation is used to indicate the extent of genetic effects in the parents, and this is compared to the phenotypic selection differential. Linear heritability estimates use true sire and dam effects of the same generation as measures of genetic variability. True sire effects are estimated from the average deviation of a sire's progeny from the population mean. Such a deviation ΔP_S contains in addition to the sire effect ΔS, which is alike for all half-sibs, dam effects ΔD and random deviations $\Delta P_F = \Delta F$. Since these are assumed to be independent of each other and also independent from one half-sib to another, the true sire effect can be estimated by subtracting the averaged dam effect, which is $(\Delta P_D)/d$ for a sire mated to d dams, from the sire deviation: $\Delta S = \Delta P_S - (\Delta P_D)/d$. Likewise the true dam effect ΔD can be estimated by subtracting from the dam deviation the averaged random deviation: $\Delta D = \Delta P_D - (\Delta P_F)/n$. Heritability is estimated by the ratio of genetic effects to the sum of all effects:

$$h^2 = \frac{2(\Delta S + \Delta D)}{\Delta S + \Delta D + \Delta F} \quad \text{or} \quad \frac{4 \Delta S}{\Delta S + \Delta D + \Delta F}$$

This kind of analysis can be performed on the end segments of the population. Any asymmetry in the heritability coefficients, or, rather, in the expected response, should be apparent.

The concept of estimating genetic parameters from the ratio of selection response to selection differential has been further developed and generalized. It will be discussed in more detail in Section 12.2.3.

6.4. Maternal Influences

Common environment increases likeness and thus the covariance among relatives. The maternal environment, which causes maternal effects, is of particular significance. Formally similar to the familial environment of humans, maternal effects cause nongenetic "inheritance," the maternal phenotype influencing directly offspring phenotypes. Breeders know of its existence and are familiar with the phenomenon. Its significance to animal breeding was brought to light by Hammond's classic experiment of reciprocal crosses between Shire horses and Shetland ponies. Judicious choice of experimental material permitted the unequivocal demonstration of the existence, duration, and extent of maternal effects. Similar experiment were later performed with South Devon and Dexter cattle and, using embryo transfer, with both large- and small-bodied rabbits and dwarf and normal pigs (Haring *et al.*, 1966).

Maternal effects affect the likeness of maternally related animals such as full-sibs, maternal half-sibs, and dam–daughter groups. The biological effects embrace a whole spectrum of different causes, such as differences in intrauterine and postnatal nutrition, as exemplified in Hammond's experiment, infections of offspring through the dam, maternal antibodies, and last but not least cytoplasmic inheritance. Maternal effects touch predominantly but not exclusively young animals.

Dickerson (1962) and Willham (1964) proposed a model which includes such traits as those resulting from the action of the individual's own genotype and of the directly experienced environment (direct effects, indicated by the subscript D) on the one hand and the action of the maternal phenotype (maternal effects, indicated by the subscript M) on the other:

$$P_X = A_{DX} + R_{DX} + A_{MW} + R_{MW}$$

which describes the performance of offspring X from dam W, where A and R refer, respectively, to genic and residual (dominance, epistatic, environmental) effects. Performance of relative Y out of dam Z can be described in similar terms. The covariance between X and Y is given by

$$E(P_X P_Y) = E(A_{DX} A_{DY} + A_{DX} A_{MZ} + A_{MW} A_{DY} + A_{MW} A_{MZ})$$

$$= 2C_{XY} V_{AD} + 2\sigma_{AD} \sigma_{AM} r_{DM}(C_{XZ} + C_{YW}) + 2C_{WZ} V_{AM}$$

r_{DM} is the genic correlation between direct and maternal effects. It is assumed that no correlations exist between the various residual effects. The model is illustrated in Fig. 6.4 and the expectations of covariances between various relatives are given in Table 6.6.

The maternal phenotype, or, more precisely, the phenotype for maternal influences, manifests itself in the offspring and not in the dam herself. Therefore, males cannot have a phenotype for maternal influences affecting the offspring, but, of course, they possess genes for it which will influence the maternal effect in their grand-progeny. The genotype for maternal effects in males must be inferred from their female relatives. This model has been used to analyze traits such as gestation length, for example, for Arabian horses (Rollins and Howell, 1951), or growth and weaning performance of beef cattle (Koch and Clark, 1955).

Another interesting model, which considers a trait as uniform, was introduced by Falconer (1964). In a selected mouse population offspring–parent regression was zero, the full-sib correlation 0.10, and the long-term realized heritability 0.24. These results can be explained by a maternal effect on litter size of the daughters. Mice out of large litters produce fewer young, while mice from small litters tend to have above

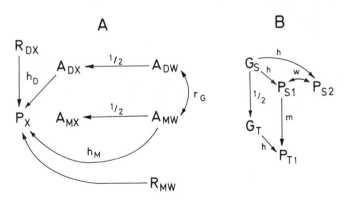

FIGURE 6.4. Maternal effects. A, Dickerson–Willham model; B, Falconer model.

TABLE 6.6
Covariance of Composite Traits[a]

	V_{AD}	$A_D \cdot A_M$	V_{AM}
Sire–offspring	1/2	—	—
Dam–offspring	1/2	5/4	1/2
Full-sibs	1/2	1	1
Maternal half-sibs	1/4	1	1
Paternal half-sibs	1/4	—	—
Granddam–grandchildren	1/4	1	1/4

[a] A_D, A_M, Genic values of direct and maternal effects, respectively. $A_D \cdot A_M$, Covariance between genic values of direct and maternal effects.

average litter size. Falconer's suggestion was to consider the maternal effect M as a function of dam's phenotype P': $M = mP'$. Phenotypes of dam and offspring can be described as

$$P' = A' + M' + R' \quad \text{(dam)}$$

$$P = A + M + R \quad \text{(daughter)}$$

where A, M, and R represent, respectively, breeding value (genic value), maternal effects, and residual effects. The dam–daughter covariance is given by (letter combinations symbolize covariances):

$$PP' = AA' + AM' + MP'$$

Covariances involving R and R' are considered to be zero. Now $M = mP'$; therefore $MP' = mV_{P'}$, and further,

$$AM' = Am(A'' + M'' + R'') = \frac{m}{2} V_A \left(\frac{1}{2} + \frac{m}{4} + \frac{m^2}{8} + \cdots \right)$$

The quantity within the parentheses is a geometric series with the factor $m/2$. Therefore V_A is to be multiplied by $m/[2(2 - m)]$ and

$$PP' = \frac{V_A}{2 - m} + mV_{P'}$$

The models differ in various ways. In the Dickerson–Wilham model the phenotype is conceived of as being composed of a direct and a maternal component and maternal effects are assumed to be caused by the genotype and nongenetic factors of the dam. Eisen (1981) amended the model by admitting an environmental covariance between relatives raised by the same dam. In the Falconer model maternal effects are caused by the phenotype of the dam, which in turn is caused, naturally, not only by the maternal genotype but also by the grandmaternal phenotype and so on. Insofar as in this model nongenetic influences of more distant generations can have an effect, it can be said to contain a "Lamarckian" component.

Maternal effects have most impact upon juvenile traits and it may be assumed *a priori* that at later ages maternal effects will have faded. For example, Baker *et al.* (1943) reported that for weight the variance caused by litter environment comprised 50% and genic variance 5% of the total for 3-week-old piglets, while at 1/2 year of age each factor caused 25% of the total variance. Maternal effects may be balanced by compensatory growth, as demonstrated with mice by negative correlations between growth in the prepuberal and in the postpuberal stages (Monteiro and Falconer, 1966).

Nevertheless there are numerous examples of carryover of maternal effects into adult or at least preadult stages. In cattle maternal effects on wither hight can be discerned in 3- to 4-year-old animals. In Falconer's results maternal effects on litter size of daughters was evident. Revelle and Robinson (1973) and Alsing *et al.* (1980) applied Falconer's model to litter size of pigs. The regression of gilt litter size on litter size of her dam in which she was born was much smaller than the regression on another (independent) litter of the same dam. The regression of gilt litter size on granddam's was about equal to the daughter–dam (litter of origin) regression, apparently signifying a heritability twice as large. The results can be explained by postulating an environmental correlation between size of birth litter and gilt litter size of about -0.1 (Fig. 6.4A). The heritability of litter size, free from any maternal (nongenetic) contribution, amounts to about 30%. The repeatability of litter size in the data set was around 10%. The deficit relative to the probable real heritability can be explained by a negative environmental correlation between adjacent litters.

An interesting explanation of likeness between nonrelated animals was forwarded by Skjervold (1977), in which an influence of the fetus on the milk yield of the cows was proposed. Such an effect would cause correlations between mating partners of a bull as well as an increase in repeatability if the same bull is used in successive years. Skjervold found correlations between performance of mating partners of bulls of the order of 0.10, but Van Vleck (1978) failed to find similar values in U.S. dairy data.

Environmental influences on young animals other than maternal effects may have consequences for later performance. For example, Lauprecht and Walter (1960) and Jonsson and King (1962) found in data from Danish swine-testing stations that for growth to 90 kg live weight equal fractions of variance were to be attributed to herd of origin and to sire breeding value (about 7% each), and similar results were found by Puff (1976) in German test station data. Since boars are largely confounded with herds, importance of genotype will be grossly overestimated if the carryover effect of the herd environment cannot be taken into account.

Environmental influences need not act only to increase similarity, just as maternal influences are not necessarily positive. Limited space or limited feeding of groups of relatives, such as a litter, may lead to competition and therefore increase the variance between relatives. For example, Jonsson (1959) reported from Danish swine-testing station data that under group feeding of litter test groups the residual variance between full-sibs was five times the variance computed from singly fed pigs. Competition among full-sibs neutralized similarity caused by common genes and common environmental influence, and reduced litter mate correlations to near zero. Under single feeding the correlation between full-sibs within half-sib groups was found to be about equal to the half-sib correlation, as is expected when dominance and maternal influences are missing.

6.5. Variance of Heritability

Variance and standard deviations of heritability are required for comparisons of the accuracy of estimates. These comparisons have less application to significance tests, since the existence of genetic variability and therefore of some nonzero value for heritability can be accepted *a priori*.

The expression of the variance appropriate to the regression estimator of heritability is fairly straightforward. It can be derived from the variance of a regression coefficient:

$$V_{h^2} = 4V_b = \frac{4}{(N-2)\,\Sigma\,x^2} \left(\Sigma\,y^2 - \frac{(\Sigma\,xy)^2}{\Sigma\,x^2} \right)$$

y and x are about equally variable, so the formula can be approximated:

$$V_{h^2} \approx \frac{4}{N-2}\left(1 - \frac{h^4}{4}\right) \approx \frac{4}{N} = \frac{8}{T}$$

where N denotes the number of offspring–parent pairs and T the total number of animals.

Intraclass correlations from one-stage hierarchic analyses of variance are frequently used to compute heritability. With half-sibs $h^2 = 4t$, and with full-sibs $h^2 = 2t$. The variance of intraclass correlations given by Fisher (1941) is

$$V_t = \frac{2[1 + (n-1)t^2(1-t)^2]}{n(n-1)(N-1)}$$

where N refers to the number of subclasses and n to subclass size. Since the denominator contains n^2, small subclass size causes a large variance.

For multistage analyses of variance several approximations have been suggested. Osborne and Patterson's (1952) development is based on equal subclass size. Dickerson (1960) proposed the formulas given in Table 6.7.

On the assumption of normal distribution, the variance of a mean square (MQ) is $2(MQ)^2/\text{df}$ (degrees of freedom). The sire variance components are estimated from $(MQ_S - MQ_R)/k$. The variance of this difference (respectively, ratio) is the quantity A^2 of Table 6.7. The intraclass correlation is a ratio of variance components, or, respectively, of func-

TABLE 6.7
Variance of Heritability[a]

h^2	σ	
$\dfrac{4\sigma_S^2}{\sigma_P^2}$	$\dfrac{4A}{\sigma_P^2}$	$A = \left[\dfrac{2}{k_3^2}\left(\dfrac{MQ_S^2}{n_S} + \dfrac{MQ_D^2}{n_D}\right)\right]^{1/2}$
$\dfrac{4\sigma_D^2}{\sigma_P^2}$	$\dfrac{4A'}{\sigma_P^2}$	$A' = \left[\dfrac{2}{k_1^2}\left(\dfrac{MQ_D^2}{n_D} + \dfrac{MQ_R^2}{n_R}\right)\right]^{1/2}$
$\dfrac{2(\sigma_S^2 + \sigma_D^2)}{\sigma_P^2}$	$\dfrac{2(A + A' + 2C)^{1/2}}{\sigma_P^2}$	$C = -\dfrac{k_2}{k_3}\left[(A')^2 - \dfrac{2MQ_R}{n_R k_1}\right]$

[a]MQ_S, Mean square between sires. n_S, Degrees of freedom for sires.

tions of mean squares. When subclass sizes are balanced and assuming a normal distribution, a first approximation to the variance of a ratio is

$$V_{x/y} = \frac{V_x}{y^2}$$

Heritability based on sire variance components is $4\sigma_S^2/\sigma_P^2$, with a variance $16A^2/\sigma_P^4$.

In the example of Table 6.4 the standard error is

$$s_{h^2} = \frac{4}{338,724} \left[\frac{2}{126.6^2} \left(\frac{1,942,803^2}{92} + \frac{326,338^2}{11,858} \right) \right]^{1/2} = 0.027$$

This formula is approximated by one given by Robertson (1959a):

$$s_{h^2} = \left(h^2 + \frac{4}{k} \right) \left(\frac{2}{n_V} \right)^{1/2}$$

which yields the same value (0.027) as the more complete formula. Heritability in the example given in Table 6.2, where a two-stage analysis of variance is shown, equals $4 \times 0.07 = 0.28$. Its standard error is

$$\frac{4}{0.098} \left[\frac{2}{2.542} \left(\frac{0.1464^2}{1125} + \frac{0.1028^2}{326} \right) \right] = 0.163$$

The simpler approximation gives a rather different result, $s_{h^2} = 0.096$, probably due to the unfavorable structure of the data (small subclass sizes). It should easily be possible to form replicates when data are numerous and subclasses large, as, for example, in the data used for Table 6.7. In such cases the variance could be computed from the estimates of each replicate and no approximation is necessary.

As mentioned above, combining heritability estimates in an optimal way from offspring–parent regressions and sib analyses requires consideration of covariances between the two estimates. Hill and Nicholas (1975) have developed formulas for covariances relating to several important combinations. The covariance between heritabilities estimated from daughter–dam regression and half-sib correlations (s denotes sires, m is the number of dams/sire, n is the number of progeny/dam, and c^2 is the nongenic correlation among half-sibs) is

$$\text{cov } h_b^2 h_S^2 = -\frac{h^2}{2sm(m-1)n} [4 + (m-1)h^2]$$
$$\times [4 - 2h^2 + nh^2(1-h^2) + 4(n-1)c^2]$$

The variance of the combined estimate is minimized when the weight a (p. 132) is

$$a = \frac{V(h_S^2) - \text{cov } h_b^2 h_S^2}{V(h_S^2) + V(h_b^2) - 2 \text{ cov } h_b^2 h_S^2}$$

The computation can be demonstrated by an investigation of udder infections by *Staphylococcus aureus* (Leithe, 1972). The data come from 96 cows which belong to 19 bull progeny groups. Both half-sib correlation and daughter–dam regression was used for heritability estimation (the method requires that information is available for each daughter–dam pair; in the data used as example this was only approximately so): $h_S^2 = 0.01 \pm 0.30$, $h_{bD}^2 = 0.65 \pm 0.20$, and average subclass size is 5. The computed covariance is -0.0115 and the combined heritability is

$$h_c^2 = 0.01 + \frac{0.09 - (-0.015)}{0.09 + 0.04 - 2(-0.0115)} (0.65 - 0.01) = 0.44$$

Its variance is

$$V(h_c^2) = \frac{1}{0.04} + \frac{1}{0.09} = 0.0277, \qquad s_{h^2} = 0.17$$

When sib correlations are estimated from data with optimal structure ($n = 1/t$), the variance roughly equals $V_t = 8(1-t)^2 t/T = 8t^2/N$. From this the variance of heritability computed from the half-sib correlation can be approximated by

$$V_{h^2} = 32h^2/T$$

and if computed from full-sib correlations it is

$$V_{h^2} = 16h^2/T$$

When full-sib groups are nested within half-sib groups, the optimal family size should be about $n = 2/h^2$ and $d = 3$–4. For the example of Table

6.4 the approximate standard error $s_{h^2} = (32 \times 0.15/11950)^{1/2} = 0.02$, smaller than the one computed by the more complicated formula. The justification for the short formula rests on the assumption of optimal subclass sizes, which was not true in the example ($k = 126.6$ instead of $4/h^2 \approx 4/0.16 = 25$). Hill and Nicholas arrive at similar approximations. When heritability is to be estimated from both daughter–dam and half-sib covariances, the optimal group size can be derived from $dn = 4/h^2$. When only males or only females can be measured, this is changed to $dn = 3/h^2$ and $dn = 5/h^2$, respectively.

When information about the probable size of the heritability is lacking, Hill and Nicholas recommend two offspring per dam and six dams per sire. This should be nearly optimal when both sexes or when males only can be measured. When measuring is restricted to females there should be 12 dams per sire.

Experimental plans for estimating heritability are robust and serve well over a wide range of actual heritability values. Even rather large deviations from optimum in subclass sizes result in comparatively small losses of efficiency.

Estimation of heritability from half-sib correlations is inefficient for high heritabilities. When the total number measured equals T, variances from daughter–dam regressions ($8/T$) and half-sib correlations ($32h^2/T$) should be equal when $h^2 = 1/4$. When $h^2 < 1/4$ the half-sib estimation should be more efficient than the daughter–dam regression and *vice versa* when $h^2 > 1/4$.

The variance of the realized heritability is inflated by the drift variance. The variance of means of unselected lines is

$$\sigma^2 \left[\frac{h^2/4}{N_M} + \frac{(h^2/4) + c^2}{N_F} + \frac{1 - (h^2/2) - c^2}{N_F n} \right]$$

The line mean of a given generation will vary around the mean breeding value of the previous one, which has a variance

$$V_{BV} = V_A \left(\frac{1}{4N_M} + \frac{1}{4N_F} \right) = \frac{V_A}{N_e}$$

This can be defined as genetic drift variance. The error variance is, correspondingly,

$$V_P \left[\frac{c^2}{N_F} + \frac{1 - c^2 - (h^2/2)}{N_F n} \right]$$

The genetic drift variance accumulates with generations. At generation t its contribution to the variance between line means is $(t - 1)V_A/N_e$. The line mean of generation t has the additional variance given above as V_{VB}. Therefore the total variance of line means in generation t will be $V_{x_t} = t\sigma_{DR}^2/N_e + \sigma_E^2$ and it is evident that with increasing number of generations the genetic drift variance becomes the dominating component.

In selected populations the range of phenotypes and genotypes is restricted. Therefore the breeding value of the selected group of animals has a variance $h^2(1 - h^2)/N$. As the number M measured is limited, the variance contributed is $h^4\sigma^2/M$. Therefore the variance of the mean breeding value of the selected group is

$$\sigma_P^2\left[\frac{h^2(1 - h^2)}{N} + \frac{h^4}{M}\right]$$

DR is drift, M is the number of animals measured, N is the number selected, N_F is the number of breeding females, and n is the number of progeny per female. The expression would be appropriate for the first generation in selection experiments. Notter *et al.* (1976) extended the use of the formula to selection for family mean performance.

These formulas assume that the variance within lines remains reasonably stable, when in fact it will decrease due to drift and linkage disequilibrium and also either increase or decrease as gene frequencies change. As these influences are difficult to quantify and may largely cancel each other, it has been suggested that use be made of the formula for unselected lines. The variance of the realized heritability is

$$V_{b_c} = \frac{6}{t(t + 1)(2t + 1)}\left(\frac{2t^2 + 2t + 1}{5}\sigma_D^2 + \sigma_R^2\right)$$

For large t the coefficient for the genetic drift variance approaches $1/t$, while the coefficient for the error variance is $1/t^3$. Therefore, after a sufficient number of generations, V_{b_c} will be influenced almost only by drift.

Realized heritability estimated as the ratio of final response to total differential has the variance

$$V_{b_R} = \frac{\sigma_{DR}^2}{t} + \frac{\sigma_E^2}{t^2}$$

which again shows the diminishing contribution of the error variance. When response and differential of individual generations are used as estimator, the variance is similar to V_{b_R}.

The h^2 values from the swine selection experiment reported by Hetzer and Harvey (1967) are given in Table 6.5. These authors give the following quantities for Durocs: $\bar{d} = 0.27$ cm, $\sigma_P = 0.56$ cm, $h^2 = 0.48$, $M = 60$, $N = 16$. Use of these figures yields the following results when all ten generations of selection are utilized: $\sigma_{DR}^2 = 0.0122$, $\sigma_R^2 = 0.0054$, and $\sigma_{h^2} = 0.02$. Comparison of the two estimates makes clear the importance of drift in selection experiments.

In one-way selection σ_{DR}^2 is the same size as in two-way selection; but σ_R^2 is only half as large if no c^2 effects exist. The variance of the heritability estimate is larger than the regular variance of the regression coefficient:

$$V_{b_c} = V_b + \frac{2(3t + 4)}{5d(t + 1)(t + 2)} \sigma_{DR}^2$$

Realized heritability estimation from divergent selection experiments can be very efficient even over few generations. Meyer and Enfield (1974) found good agreement between empirically estimated variances with those computed from the formulas of Hill and Nicholas.

It may be concluded that in many cases one must, at the beginning of an experiment, be satisfied with published heritability values. However, optimal plans are relatively robust. Therefore, an experiment can be planned on the basis of such values without fear of undue loss of efficiency. The greatest efficiency accrues from divergent selection and assortative mating, which nevertheless is hardly a practical proposition for domestic animals. When parents are unselected, estimation by combining offspring and parental information is relatively efficient.

6.6. General Remarks about Heritability

The predictive value of heritability is limited, since it relates to the population for which it has been estimated. It can be extrapolated to other populations which are similar in genetic structure, history, etc., and which are exposed to a similar environment. Experiments and field investigations have shown that heritability is reasonably stable over the span of a few generations, where "few" means half a dozen to a dozen. For large domestic animals this implies a rather long time period during which breeding goals and/or environmental conditions may change, requiring a new estimation. Heritabilities of several performance traits are given in Table 6.8.

The coefficient of heritability is a ratio, and therefore it changes when either numerator or denominator changes. The numerator is the genic variance, which changes with changing gene frequencies. However, when a trait is influenced by n additively acting genes of equal effect and equal frequencies, then $\sigma = (2npqa^2)^{1/2}$. A change in frequency at one locus will be proportional to $(1/2n)^{1/2}$, which is low in value when n is large. In general, then, changes of the genic variance caused by gene frequency changes brought about by selection are negligible and heritability should be little affected. However, selection increases linkage disequilibrium, decreasing the genic variance, which may influence heritability, as will be discussed in more detail in Section 9.2. Likewise, genetic variance decreases in closed lines and this, too, can affect heritability.

TABLE 6.8
Heritability Coefficients (in Percent)

	Field data	Test station data	Twins
Cattle			
Milk yield	20–40	60–70	75–90
Fat-%	30–80	70–80	90–95
Protein %	40–70		
Solids, nonfat %	40–70		
Persistency	15–30		
Milk flow rate	50–80	45–80	
Percent milk from fore udder	10–50	25–75	
Feed efficiency milk	—	20–40	
Calving interval	0–5		
Nonreturn rate	20–50		
Gestation length (calf)	25–45		
Twin birth	1–3		
Dystocia: calf	1–5		
dam	1–3		
Mastitis resistance	10–40		
Eyelid carcinoma	20–40		
Wither height	50–70		
Heart girth	30–60		
Conformation score	20–30		
Body weight gain	10–30	25–50	
Feed efficiency gain	—	20–40	
Percent valuable cuts	20–50		
Percent carcass fat	20–50		
Area of longissimus dorsi muscle	20–50		

Continued

TABLE 6.8 *(Continued)*

	Field data	Test station data	Twins
Sheep			
Fleece weight	30–60		
Clean fleece weight	30–60		
Staple length	30–60		
Fineness	20–50		
Face cover	40–60		
Crimp	20–40		
Body weight	20–40		
Litter size	10–30		
Swine			
Weight gain[a]: individual feeding	10–50		
group feeding	10–25		
Feed efficiency: individual feeding	15–50		
group feeding	20–30		
Carcass length	30–70		
Back-fat thickness	30–70		
Side-fat thickness	20–40		
Area of longissimus dorsi muscle	20–60		
Meat–fat ratio	30–70		
Ham score	30–60		
Meat color	30–40		
Litter size[b]	10–15		
Poultry			
Eggs/hen housed	5–15		
Eggs/hen day	15–30		
Age at first egg	20–50		
Body weight	30–70		
Egg weight	40–70		
Albumen height	30–50		
Specific weight	30–50		
Resistance to Marek's disease	5–20		
Fertility	5–15		
Hatchability	5–20		
Horses			
Racing speed	30–60		
Handicap rating	35–40		
Trotting speed	20–40		

[a] Low values from stations with restricted feeding.
[b] Maternal influence not considered.

In some selection experiments heritability changes in a rather abrupt fashion, an event difficult to explain under a polygenic model. Asymmetry of realized heritabilities is a fairly common experience in two-way selection experiments. It may be explained by various phenomena, such as asymmetric gene frequencies, directional dominance, maternal effects, and inbreeding depression. Explanation by asymmetry of gene frequencies assumes implicitly a gene number much smaller than commonly believed. The explanation is not relevant to the *Tribolium* experiment of Mayer and Enfield discussed in Section 6.3, since the lines originated from an F_2 of inbred lines and therefore had gene frequencies of 1/2. In this experiment, as in *Drosophila* experiments, realized heritability seemed to decrease with decreasing selection intensity, which cannot be explained by our present theory. Falconer pointed out that the asymmetry of realized heritabilities is rarely statistically significant in many reported experiments when the standard errors suggested by Hill (1980) are used for testing and that rather large-scale experiments would be necessary to detect a modest asymmetry. Nevertheless, the many cases of asymmetry, or the dependence of heritability on selection intensity, are observations which cannot easily be accommodated within our present quantitative genetic theory.

Population differences in heritabilities can be considered from two viewpoints. One is spatiotemporal, i.e., populations have dissimilar breeding histories with consequences to their genotypes, and populations experience a variety of environments, which may cause both different genetic and different environmental variances. The second viewpoint on different heritabilities is more causal and relates to differing plasticity of traits, about which little is known in spite of innumerable publications on the subject. One explanation involves the proximity of a trait to Darwinian fitness which, according to Fisher's Theorem of Natural Selection, would have no genic variance. Now, reproductive traits are not synonymous with adaptive value, but they should be closer to such value than, say, milk fat content or back-fat thickness of pigs. However, as is evident from Table 6.8, the heritability of some reproductive traits is not all that small—consider gestation period, nonreturn rate of bulls, or even litter size when maternal effects are taken into consideration. For calving interval the cause of low heritability may well be the "all-or-none" character of a single service and therefore more a problem of accuracy of timing than of fitness. When the random error can be minimized, as for nonreturn rate of bulls, heritability becomes moderate. Reeve and Robertson (1952) concluded from their extensive investigations with *Drosophila* that size of heritability is influenced by the complexity of a trait; the more complex it is, the lower its heritability. It may be hoped that a satisfactory explanation of the differences in heritabilities can be provided by a causal analysis, i.e., the identification of the relevant loci which influence a trait and the norm of reaction of the various genotypes to different stimuli.

Heritability seems to be difficult to influence. Of course, heritability increases when records can be made more accurate, for example, by higher sampling frequency of milk yield or fat content, or by use of deviations from analogous (same herd, year, season) instead of from contemporary (same herd, year) dairy records (Barker and Robertson, 1966). A few other indications exist for influencing factors—higher heritability of first lactations compared to later lactations, higher heritability of dairy performance in herds above average, at least in Europe, higher heritability for rate of gain in cattle in medium age groups—although many exceptions exist. Heritability is frequently lower for very high records, probably due to the special treatment extended to high-producing animals. Taylor and Craig (1967) showed that heritabilities for cattle measurements are maximum in the adult stage, and that heritabilities of measurements on juvenile and preadult cattle increase as a function of age. This was confirmed by Hintz and Van Vleck (1978) for horses. Although a few other differences between heritabilities can be explained, or at least plausible explanations can be constructed, the good agreement between values computed at different times and on different populations seems to be more surprising than the few deviations. However, the good agreement may also simply reflect the lack of sensitivity of the parameter estimated to considerable changes sometimes found in the population and/or the environment.

A causal analysis of phenotypic variability is no doubt very desirable, but it may be very protracted. Therefore the concept of heritability will continue to occupy a central position in the theory of animal improvement.

7

Genetic Correlations

The smallest unit breeding can manipulate is the animal, excluding the future possibilities of handling genes by recombinant techniques. The fact that all breeding activities concern at least an animal implies that selection for one trait cannot be considered in isolation but will have consequences for other traits. Direction and extent of correlated selection response depend on the sign and size of genetic correlations. Genetic correlation between two traits is defined as the correlation between genetic effects that influence the two traits. As discussed in Chapter 4, the phenotype P is considered to be a function of both the genotype G and environment E. Therefore the observable correlation between two traits r_P will arise from correlations between genetic and environmental effects that influence the two traits, as shown in Fig. 7.1. There is also the possibility of correlations between genetic effects on the one hand and environmental effects on the other, but such correlations are usually neglected even though in reality they may be of some importance. A genetic correlation is manifested by simultaneous inheritance of two or more traits, or expressed differently, by the joint appearance of two or more traits in relatives.

The causes of genetic correlations are threefold: Permanent, so to speak, genetic correlations are caused by pleiotropy. For example, the gene that causes malignant hyperthermia when pigs are exposed to halothane improves meatiness and impairs the meat color. Therefore a negative genetic correlation between the two traits will be observed in a population where the gene occurs.

Genetic correlation may be caused by linkage disequilibrium. Such a correlation will be transient, in theory at least, and will disappear when

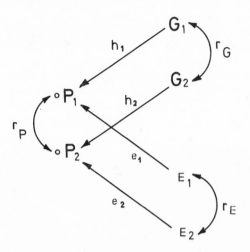

$$r_{p} = h_{1}\, r_{G}\, h_{2} + e_{1}\, r_{E}\, e_{2}$$

FIGURE 7.1. Causes of phenotypic correlations.

equilibrium proportions are attained. Assume that locus A influences trait X (gene frequencies p, q) and that locus B influences trait Y (gene frequencies r, s). Genic variances, expressed in gene substitution effects, are $2pq$ and $2rs$, and the genic covariance is

$$\text{cov } G_{XY} = \Sigma\, fxy - \Sigma\, fx\, \Sigma\, fy = 2d$$

where d denotes the linkage disequilibrium (Section 1.6). The genetic correlation is then

$$r_{G} = \frac{2d}{(2pq \times 2rs)^{1/2}}$$

The correlation is proportional to the linkage disequilibrium and will disappear if d becomes zero. Disequilibria can exist between independent loci after a recent cross, for example, and will disappear only gradually under random mating (Chapter 1). Linkage will retard the dissolution of the disequilibrium. Therefore, when linkage is close, a correlation caused by linkage disequilibrium may persist for some time.

Finally, genetic correlations may be caused by different breeding goals within a population. For example, if in one segment of a population breeders select for dairy merit and against fleshiness of cows, while in another segment selection favors muscularity and is directed against too great a milk yield, in cows used as beef dams the two traits will become negatively correlated. An analysis of the whole population without regard to the subdivision will suggest a negative genetic correlation. The disappearance of subdivisions with different breed ideals should lead to a disappearance of the genetic correlation, but this, too, will require time. We have then two causes of transient genetic correlations which are basically similar insofar as they owe their existence to subdivisions of a population and therefore they will disappear if the population becomes homogeneous.

Two different traits may be assumed to be influenced by a set A of genes in a synergistic fashion and by a different set B in an antagonistic fashion. Continued selection to improve both traits will eventually lead to fixation of the A genes, while B genes will continue to segregate. Therefore in populations subjected to long-term selection most of the genetic variability should be caused by genes with antagonistic effects on the different traits, and with continuing selection genetic correlations should become more negative.

Rendel (1967) suggested a similar model, which, however, is based more on physiological arguments. Both traits have a common resource (C). This is partitioned between trait 1, which receives V, and trait 2, receiving $1 - V$ of the resource. One set of genes influences C, another set V. If the gene frequencies and genetic effects cause mainly variability in C, the genetic correlation will be positive. If genetic variance is larger for V, selection, as shown by James (Sheridan and Barker, 1974), can in certain situations (gene frequencies) cause a transient increase in genetic correlations. This is to be expected in particular if selection is directed against two traits. Sheridan and Barker observed such a pattern of response in *Drosophila* experiments of medium duration, but not in long-term experiments. A similar prediction would have been made by the traditional model, but Rendel's hypothesis specifically emphasizes the possibility of an increase in the genetic correlation due to selection.

Two traits of an individual may be correlated via the common environment. For example, very liberal feeding of concentrates will decrease milk fat-% but increase milk protein-%. Also, it will increase both milk yield and weight of the animal. Therefore neither size nor even sign of a phenotypic correlation permit inferences about those of the genetic and/or environmental correlation. Assume a statistical model as discussed in Chapter 4 and independence of genotype and environment. The phenotypic correlation between traits 1 and 2 has the composition

$$r_P = \frac{\text{cov } G_1 G_2 + \text{cov } E_1 E_2}{[(V_{G_1} + V_{E_1})(V_{G_2} + V_{E_2})]^{1/2}}$$

where cov $G_1 G_2$ and cov $E_1 E_2$ denote genetic and environmental covariances and V_{G_1} and V_{E_1} denote genetic and environmental variances. The expression can be rewritten as

$$r_P = h_1 h_2 r_G + e_1 e_2 r_E$$

where $e_1 = (1 - h_1^2)^{1/2}$ and r_G and r_E are the correlations between genetic and nongenetic effects. It is evident from these expressions that the phenotypic correlation arises not only from the genetic and the nongenetic correlations, but also from the heritabilities of each trait. If these are small, the correlation will be influenced predominantly by the nongenetic correlation. If $r_G h_1 h_2$ is larger than r_P, the genetic and environmental correlations will differ in sign.

Analogous to heritability, genetic correlations can be taken in the "broad" and "narrow" senses. Genic effects, dominance, and epistatic deviations that affect both traits may be correlated. If the one-locus model is assumed, i.e., epistatic effects are ignored, the models for the two traits are

$$A = \alpha_1 + \alpha_2 + \delta_{A12} + e_A$$

$$B = \beta_1 + \beta_2 + \delta_{B12} + e_B$$

Gene 1 has the average effect α_1 on trait A and β_1 on trait B. The covariance between the two traits is

$$\text{cov } AB = 2 \sum p_i \alpha_i \beta_i + \sum p_i q_j \delta_{Aij} \delta_{Bij} + \sum p_i q_j e_{Aij} e_{Bij}$$

The first quantity on the right-hand side represents the genic or additive-genetic covariance cov_A, and the second quantity represents the covariance due to dominance effects cov_D. The observed covariance between trait A in relative X and trait B in relative Y is

$$\text{cov } AB = 2C_{XY} \text{cov}_A + U_{XY} \text{cov}_D$$

In pleiotropy genetic effects arise from the genes at a locus. If the covariance is due to linkage, genetic effects refer to the chromosome segment that remains intact in the course of the selection experiment or breeding effort.

The correlation between additive gene effects will be denoted henceforth as genetic correlation, while genotypic correlation refers to the correlation between all genetic effects—genic, dominance, epistatic—that influence the two traits. In terms of average gene effects the genetic correlation is

$$r_G = \frac{2 \sum p\alpha_i\beta_i}{(2 \sum p\alpha_i^2 \times 2 \sum p\beta_i^2)^{1/2}} = \frac{\text{cov}_A}{(V_{G_A}V_{G_B})^{1/2}}$$

The genetic correlations and the genetic covariance can be estimated by analogous methods to that for heritabilities and genetic variances, i.e., from correlations (respectively, covariances) between relatives, or from the selection response relative to selection differential. In each case, of course, different traits are involved.

The ratio of the response of trait 2, called the realized response, to the selection differential of trait 1 permits one to estimate the realized genetic correlation (g_i and p_i denote, respectively, genic and phenotypic standard deviation):

$$CR(2.1) = \frac{\Delta G_{2.1}}{\Delta P_{1.1}} = \frac{i \text{ cov } 2.1}{p_1} \frac{1}{ip_1} = \frac{r_G g_1 g_2}{p_1^2}$$

where CR(2.1) denotes the correlated response in trait 2 when selection is for trait 1; the direct response in trait 1 is

$$DR(1.1) = g_1^2/p_1^2$$

Therefore, the ratio of correlated to direct response gives

$$\frac{CR(2.1)}{DR(1.1)} = r_G \frac{g_2}{g_1}$$

and the product of the symmetric ratios yields the square of the genetic correlation,

$$\frac{CR(2.1)}{DR(1.1)} \times \frac{CR(1.2)}{DR(2.2)} = r_G^2$$

The approach is illustrated in Table 7.1 with data from the first generation of a two-way selection experiment on boar taint.

Hazel (1943) introduced the concept of genetic correlation into animal breeding theory. For the estimation, covariances and regression coefficients of offspring traits with or on other traits in parents were

TABLE 7.1
Realized Heritability and Genetic Correlation[a]

Sires		Sons		
O	W	n	O	W
58	101	12	66	98.9
135	109	14	63	107.5
3	92	8	34	90.3
10	104	11	34	94.5

Parameters estimated from selection response[b]

$$h^2_{R(O)} = \frac{2(64.5 - 34)}{96.5 - 6.5} = 0.68 \qquad \frac{CR(W.O)}{DR(O)} = \frac{103.2 - 92.4}{64.5 - 34} = 0.354$$

$$h^2_{R(W)} = \frac{2(101 - 94.6)}{106.5 - 96.5} = 1.28 \qquad \frac{CR(O.W)}{DR(W)} = \frac{48.5 - 50}{101 - 94.6} = -0.234$$

Parameters estimated from regression coefficient[c]

	$O(S){\cdot}O(P)$	$W(S){\cdot}W(P)$	$W(S){\cdot}O(P)$	$O(S){\cdot}W(P)$
b	0.237	0.884	0.119	1.317
s_b	0.118	0.375	0.016	1.479

$$h^2_{b(O)} = 0.47 \pm 0.24, \; h^2_{b(W)} = 1.77 \pm 0.75, \; r_{b(OW)} = 0.86 \pm 0.23$$

---

[a] O, Androstenone level, μg. W, kg weight at 190 days.
[b] CR($W{\cdot}O$), Correlated selection response in W when selecting for O. DR(O), Direct selection response.
[c] $O(S){\cdot}W(P)$, Covariance of son's odor with sire's weight.

---

used. The genetic correlation results from the ratio of cross-covariances, i.e., covariance between trait A in sire and trait B in son, to the covariances between the same traits in parent and offspring. When assumptions of no environmental correlations and of random mating are justified, the parent–offspring cross-covariance yields half of the genic covariance between traits A and B:

$$\text{cov } A_P B_O = \tfrac{1}{2} \text{cov}_G AB$$

Consequently, the genetic correlation is given by

$$r_G = \frac{(\tfrac{1}{2} \text{cov}_G AB \times \tfrac{1}{2} \text{cov}_G BA)^{1/2}}{(\tfrac{1}{2} V_{G(A)} \times \tfrac{1}{2} V_{G(B)})^{1/2}}$$

If covariances in the numerator (or denominator) differ in sign, the arithmethic instead of the geometric mean is used:

$$\tfrac{1}{2} \left(\tfrac{1}{2} \, \text{cov}_G \, AB + \tfrac{1}{2} \, \text{cov}_G \, BA \right)$$

The genetic covariance between two traits can be estimated from the covariance components of an analysis of covariance in analogy to the estimation of genetic variances from variance components (Hazel *et al.*, 1943). The ratio of "between-families" covariance component to "between-families" variance yields the genetic correlation; for example, for half-sibs

$$r_G = \frac{\text{cov}_S}{\sigma_{S(A)} \sigma_{S(B)}}$$

The data given in Table 6.4 permit us to estimate the genetic correlation between milk yield and fat-%:

$$r_G = \frac{-1.4468}{(12{,}769 \times 0.0097)^{1/2}} = -0.13$$

The sire covariance component can be considered as a covariance between the average of two traits of very large progeny groups, in analogy to the sire variance component. The expectation of such a covariance component is analogous to the expectation of a variance component, but it contains the contributions of genetic covariances instead of genetic variances. The half-sib covariance component will have the expectation

$$\text{cov}_{HS} = \tfrac{1}{4} G_A G_B + \tfrac{1}{16} I_{AA} I_{BB}$$

and full-sib covariance component

$$\text{cov}_{FS} = \tfrac{1}{2} G_A G_B + \tfrac{1}{4} D_A D_B + \tfrac{1}{4} I_{AA} I_{BB} + M_A M_B$$

where $G_A G_B$, $D_A D_B$, and $I_{AA} I_{BB}$ denote, respectively, covariances between genic effects and dominance and epistatic deviations that influence traits A and B, and $M_A M_B$ denotes a covariance due to common maternal environment.

In analogy to heritabilities, the genetic correlation estimated from half-sib components of variance and covariance will approximate closely a genetic correlation in the "narrow sense." A genetic correlation computed from similarity of full-sibs may contain a correlation between dominance effects and maternal influences, and therefore may not be as relevant as the correlation derived from half-sib data for prediction of correlated selection response.

A great many estimates of genetic correlations have been published; most of them derive from half-sib analyses, but many also derive from parent–offspring covariances. The errors attached to the estimates are rather large. Therefore individual estimates deserve less confidence than heritability estimates, other things being equal. Genetic correlations between some important and/or interesting traits are given in Table 7.2. The size of the interval gives a hint of the lack of precision of many estimates. In spite of this, genetic correlations seem to be more susceptible to differences between environments and between the genetic structures of populations on which they are estimated. A few examples follow which seem to support this contention.

The genetic correlation between milk yield and milk fat content varies between -0.07 as reported from Danish progeny test stations and -0.67 for U.S. Jerseys. The strongly negative correlation in Jerseys could by explained by the above hypothesis of continued segregation of genes with antagonistic effects on two traits. At any rate it appears as if real differences exist for this genetic correlation between individual breeds. However, they are also affected by environment. Touchberry (1963) reports that the genetic correlation is less negative at high production levels than in herds with lower production. It may be assumed that a high level of feeding permits higher average gene effects in both traits than more restricted feeding, where competition for resources between the yield and quality compartments may enforce a stronger negative relationship, the untoward effect of very high grain rations on fat-% not withstanding. Genetic correlations between milk fat-% and milk protein-

TABLE 7.2
Genetic Correlations between Economic Traits ($\times 100$)

Cattle					
Milk yield	First lactation–adult lactation	70	to	85
	Fat–%	-7	to	-70
	Percent fat-free solids	0	to	20
	Percent protein	-10	to	-50
	Feed efficiency	80	to	95
	Milk flow	20	to	30
	Milk in fore udder	-20	to	90
	Stayability	10	to	20
	Wither height	30	to	70

(Continued)

TABLE 7.2 *(Continued)*

Cattle				
Milk yield *(continued)*	Heart circumference	0	to	40
	Daily gain on feed	0	to	20
	Percent carcass fat	− 5	to	15
	Percent carcass meat	− 15	to	10
	Percent carcass bone	10	to	40
	Meat/bone ratio	− 10	to −	40
Fat-%	Percent fat-free solids	30	to	70
	Percent protein	40	to	70
Body weight	Feed efficiency milk	− 10		
Gain on feed	Killing out %	10	to	40
	Percent carcass fat	0	to	10
	Percent carcass meat	− 5	to −	50
	Dystocia	20	to	35
Swine				
Gain on feed	Feed efficiency	50	to	100
	Back-fat thickness	− 25	to	30
	Body length	− 50	to	10
	Area of longissimus dorsi muscle	− 10	to	40
	Meat color	− 20	to −	40
Back-fat thickness	Body length	− 25	to −	50
	Feed efficiency	− 5	to −	40
	Area of longissimus dorsi muscle	− 15	to −	40
	Meat color	70	to	90
Area m. longiss. dorsi	Meat color	− 20	to −	40
Sheep				
Fleece weight	Clean fleece weight	65	to	75
	Staple length	0	to	20
	Body weight	− 10	to	0
Clean fleece weight	Staple length	30	to	40
	Body weight	− 10	to	25
Poultry				
Egg number	Egg weight	− 25	to −	50
	Body weight	− 20	to −	60
	Age at sexual maturity	− 15	to −	50
	Percent hatchability	− 20	to	30
	Resistance to Marek's disease	10	to	30
	Feed efficiency	50	to	100
Early egg production	Late egg production	0	to −	10
Egg weight	Body weight	20	to	60
	Shell thickness	10	to −	40
	Feed efficiency	20	to −	60
	Percent hatchability	− 20	to −	40

% are very high and appear to leave little room for changing one of them independently of the other. Yet breeds differ considerably in this respect, thus indicating considerable genetic variability at the breed level. The genetic correlation between milk and heart circumference is positive before freshening, but becomes negative during lactation, as high-yielding cows lose more weight. Therefore the correlations before and after freshening pertain to somewhat different traits. Rather extreme differences between the genetic correlations of different breeds are reported from Danish progeny test stations. The genetic correlation between milk yield and fore-udder proportion was estimated as 0.9 in Danish Jerseys but as − 0.5 in Danish Friesians, both managed at stations under identical routines and conditions.

The correlation between feed conversion ratio (feed/gain) and gain is a correlation between a variable and a ratio in which the variable is the denominator. Turner (1959) and Sutherland (1965) have described how to estimate this correlation from the correlation between the numerator and denominator variables and from their coefficients of variation. This pertains also to the genetic correlation. The correlation between feed consumption and gain is positive and the coefficient of variation of feed consumption (numerator) is smaller than that of gain (denominator). This causes a negative correlation between feed conversion ratio and gain which becomes increasingly negative as the coefficient of variation of the numerator diminishes. Therefore the strongly negative correlations as reported, for example, by Danish test stations are partly an automatic consequence of restricted feeding where variation of feed intake is severely curtailed. Data from populations under restricted feeding also yield a negative correlation between gain and back-fat thickness, while this correlation is zero, or even slightly positive, in populations under *ad libitum* feeding regimes.

Genetic correlations estimated from full-sibs, breed, or strain averages contain, in addition to genic correlations, contributions of dominance and epistatic influences. Jerome et al. (1956) utilized a diallel mating design to analyze correlations between performance traits of poultry. The phenotypic correlation between egg number and egg size was − 0.1 and the genic correlation − 0.51, which implies that selection for egg number will impair egg weight and *vice versa*. However, the correlation between nongenic effects, mainly dominance deviations, was 0.32 and between environmental effects nearly zero (0.02). The estimates are uncertain, but they do demonstrate, first, the composition of the phenotypic correlations and, second, the not infrequent observation that means of two traits in strains, crosses, or breeds may not be correlated, or may even show a positive relation, despite a negative genetic correlation.

In contrast to all the apparent variability and uncertainly of genetic correlations, Taylor (1965) discerned in an analysis of linear body measurements of cattle twins a regular pattern. Genetic correlations between the same measurements at different ages seemed to be related to the differences in degree of maturity Δu of each measure at the two ages by way of the equation $\ln(r_G) = -0.77 \, \Delta u$. Taylor reasons that in view of the similarity of growth processes in mammalian species, this proportionality should pertain to species other than cattle.

7.1. Variance of Genetic Correlations

The estimation of the error of genetic correlations is more difficult and formulas are more complicated than for estimation of the error of heritability. In general, heritabilities and genetic correlations are estimated from the same data, but the errors of the latter are considerably greater than errors of heritability estimates. However, a data structure optimal for one is also optimal for the other.

Errors of genetic correlations estimated from hierarchic variance–covariance analyses with equal subclass numbers have been developed by Mode and Robinson (1959), Tallis (1959), and Robertson (1959a). The formula derived by Tallis is

$$
\begin{aligned}
V_{r_G} = \frac{1}{v t_1 t_2} &\left[(1 + g^2)(1 + z^2)V_1 V_2 - 2gz(t_1 t_2 V_1 V_2)^{1/2}\left(\frac{V_1}{t_1} + \frac{V_2}{t_2}\right) \right. \\
&\left. + g^2 \frac{(t_1 - t_2)^2}{2 t_1 t_2} \right] + \frac{1}{w k^2 t_1 t_2} \left[(1 + g^2)(1 + i^2)(1 - t_1)(1 - t_2) \right. \\
&\left. - 2g_i[t_1 t_2(1 - t_1)(1 - t_2)]^{1/2}\left(\frac{1 - t_1}{t_2} + \frac{1 - t_2}{t_2}\right) + g^2 \frac{(t_1 - t_2)^2}{2 t_1 t_2} \right]
\end{aligned}
$$

in which v and w ($=N - v$) denote degrees of freedom, t_1 is the intraclass correlation of trait 1, $V_1 = [1 + (k + 1)t_1]/k$, and g, z, and i are, respectively, genetic correlation and correlation between and within families. When applied to the data of Table 6.4 this rather unwieldy formula gives 0.208 as the error of the genetic correlation (-0.13). The comparatively poor precision of estimation of r_G becomes evident when it is compared with the heritability estimates from the same data: milk, $h^2 = 0.15 \pm 0.027$; fat-%, $h^2 = 0.38 \pm 0.06$.

Robertson (1959a) suggested a simple approximation for the variance of genetic correlations from half-sib analyses:

$$\frac{V_{r_G}}{[V(h_1^2)V(h_2^2)]^{1/2}} \approx \frac{(1 - r_G^2)^2}{2h_1^2 h_2^2}$$

If the genetic correlation is small and if heritabilities are approximately equal, the expression can be simplified further:

$$s_{r_G} = (1 - r_G^2)s_{h^2}/(\sqrt{2}\, h^2)$$

i.e., the error is a bit smaller than the coefficient of variation of heritability. Robertson's approximation, however, requires optimal structure of data. Applied to the example of Table 6.4, it yields coefficients of variation of the heritabilities as 0.16 for both milk and fat-%. The error computed from the simple formula is 0.118, in contrast to 0.21 as computed from Tallis' formula. However, as already mentioned, the structure of this large body of data is not optimal, so that there is a rather large sampling error.

Reeve (1955) also developed formulas for errors of regression estimates of genetic correlations:

$$V_{r_G} = \frac{g^2}{4df}\left[\frac{1}{r_{14}^2} + \frac{1}{r_{23}^2} + \frac{1}{r_{13}^2} + \frac{1}{r_{24}^2} + 2\left(\frac{r_{21}r_{43}}{r_{41}r_{23}} + \frac{r_{31}r_{42}}{r_{41}r_{32}}\right.\right.$$

$$- \frac{r_{34}}{r_{41}r_{31}} - \frac{r_{12}}{r_{14}r_{24}} - \frac{r_{12}}{r_{23}r_{13}} - \frac{r_{34}}{r_{32}r_{43}} - \frac{r_{21}r_{34}}{r_{31}r_{24}}$$

$$\left.\left. - \frac{r_{41}r_{32}}{r_{31}r_{42}} - 2\right)\right]$$

Hammond and Nicholas (1972) somewhat generalized this formula, but did not achieve much simplification. Traits in parents are subcripted 1 and 2 (for example, O and W, respectively in Table 7.1), traits in progeny 3 and 4. Here r_{14} is the correlation between trait O in parent and trait W in offspring and r_{12} denotes the correlation between the traits O and W in parent animals. Application of this formula to the values in Table 7.1 gives $s_{r_G} = 0.229$ ($r_G = 0.86$). Van Vleck and Henderson (1961) tested Reeve's formula by computer simulation and found reasonable agreement of computed with empirical sampling errors when sample size was $N > 500$. Deviations occurred mainly where heritability of one or of both traits was small. The investigators concluded that computation of genetic correlations would only be justified on data where the heritabilities had coefficients of variation of less than 20%. The coefficient of variation can be guessed *a priori* as CV $= 100(4 + h^4)/(df \times h^4)$ provided the data are normally distributed.

All of the formulas given are large-sample approximations, and assume normal distributions and equal subclass numbers. In modern animal breeding large data sets are continuously generated and large progeny groups are common where AI is applied. Therefore such data could be divided randomly and empirical estimation made possible, circumventing the somewhat unsatisfactory assumptions necessary for estimating errors from the formulas.

8

Genotype–Environment Interaction

In a first approximation the phenotype is considered as the sum of genetic and environmental effects. In many instances such a model proves unsatisfactory even for the limited range of situations to which it is applied. It must then be extended by inclusion of a quantity for interactions between genotypes and environment. The concept of interaction is understood in terms of mathematical statistics. It describes simply the fact that the sum of genetic and environmental effects is insufficient to account for the observation, in other words, it salvages the additive model when in fact the underlying functional relationship between performance and genotype and environment is not additive. Assume two genotypes and two environments. The phenotype can fall into any one of four classes:

		G	
		1	2
E	1	$p_{11} = g + e + i$	$p_{21} = -g + e - i$
	2	$p_{12} = g - e - i$	$p_{22} = -g - e + i$

The interaction effect can be estimated in general from $i_{ij} = p_{ij} - g_i - e_j$ provided that residual errors can be excluded. The statistical analysis can follow the well-known technique of two-way cross-classification analysis of variance.

The biological causes for interactions are manifold, but can be condensed into the statement that certain genotypes fit exceptionally well into certain environmental niches and poorly into others. Interactions in a statistical sense turn up when true differences between genotypes vary in different environments even though the ranking remains alike. More serious interactions from the point of view of animal improvements involve changes in ranking between genotypes in different environments. In practical breeding an important concept concerning genotype–environment interaction is adaptability, and a strain may fit very well into a particular environment even though its general performance may not be outstanding.

The recent discussion of genotype–environment interaction goes back to work by Haldane (1946) and Hammond (1947), who held that genotypes will be expressed fully only under a high level of management and that therefore it would be economical to confine selection to such environments.

The phenotypes of the four genotype–environment combinations given above can be described by the three quantities g, e, i. Haldane showed that six possible relations may exist, four of which lead to interactions. In all four situations the interaction constant is greater than either the genetic or the environmental effect or than both. If desirability decreases from 1 to 4, the relationship $g > i > e$ will lead to the following sequence of phenotypic values:

		G	
		I	II
E	I	1	4
	II	2	3

In each environment the ranking of genotypes will remain the same, i.e., in either environment genotype I is the more desirable. However, the reverse is not true, for genotype I produces more in environment I, but genotype II produces more in environment II. The reverse sequence of effects $e > i > g$ gives the following order among phenotypes:

$$1 \qquad 2$$

$$4 \qquad 3$$

The ranking of genotypes changes between environments I and II. In the third and fourth settings, $i > g > e$ and $i > e > g$, respectively, the ranking of environments and genotypes is changed:

$$\begin{array}{cc} 1 & 3 \\ 4 & 2 \end{array}$$

Interaction may be caused by different scales in each environment, i.e., as Hammond pointed out, genotypic differences may be greater in one environment than in others. This type of interaction can be removed by suitable transformation of data or, in practice, for example, simply by increasing the size of progeny groups in the weakly discriminating environments. Heritability of milk yield seems to be smaller in low-yielding herds (Gravert, 1958; Mason and Robertson, 1956). Similar observations have been made on broilers, where Bowman and Powell (1962) reported a decrease of the within-strain variance while the between-strain variance remained constant or increased.

The second form of genotype–environment interaction, where ranking varies between environments, is difficult to handle. Falconer's suggestion (1955) of considering performance in each environment as different traits represents a considerable conceptual advance and aid in prediction of the consequences of selection. Following Falconer, the logical measure for genotype–environment interaction is the genetic correlation between performance in environment I and that in II. A missing interaction implies the same gene effects in both environments and therefore a genetic correlation of 1. If genetic effects present in one environment are only partly repeated in another, and if the genotypes active in two environments are only partly identical, the genetic correlation will be less than 1. The genetic correlation between egg production on floors and that in cages has been found to be considerably less than 1 ($r_G < 0.7$). This may be explained if, on floors, maximal egg production requires, in addition to egg laying potential, also a certain aggressiveness at the feed trough, which is clearly of less advantage in single cages. In turn, good production in cages may require genes conferring resistance to perosis which are not required on floors, etc. Correlations between performance in the two environments arise from elements "common" to both, i.e., by genes influencing the egg laying potential *per se* (provided that the genes for specific adaptation to the two environments are not antagonistic).

The genetic correlation can be formulated as the ratio of the variance caused by gene effects common to all environments to the sum of this variance plus the interaction variance:

$$r_G = \frac{V_G}{V_G + V_{GE}}$$

In any one environment the set of genes common to all environments and a set specific to the particular environment are active. The genetic variance caused by both sets is given by the denominator: $V_{G'} = V_G + V_{GE}$. Increasing homogeneity of an environment will increase V_G. In a comparison between genotypes across environments, the variance caused by interaction disappears and the variance between genotypes will be reduced to V_G. Breeders know that breed or strain comparisons in a variety of environments will reduce the differences between them or, said differently, there are no genotypes that are outstanding under many different conditions. The genetic correlation can be taken as the ratio of the general genetic variance to the average genetic variance within environments. If environmental conditions are very variable, the general genetic variance may vanish and $r_G = 0$.

If genetic variances in the formula are replaced by covariances, negative genetic correlations between performance in various environments can be taken into account. Robertson (1959a) and Dickerson (1962) have shown how to consider various genetic scales in the different environments. The existence of different scales becomes manifest in different genetic standard deviations which cause, or magnify, the interaction between genotypes and environment. Their influence can be removed:

$$r_G = \frac{V_G}{V_G + V_{GE} - V_{\sigma G}}$$

where $V_{\sigma G}$ denotes the variance between genetic standard deviations in the various environments.

No correlation exists between G and GE if all environments and all genotypes are considered simultaneously. However, in any one environment the covariance between G and GE need not be zero. Therefore, for testing the genotypes an environment should be chosen where the covariance $G \times GE$ is positive and where therefore the utilizable genetic variance in increased (Comstock and Moll, 1963).

The factorial analysis can be based on models where environments are considered to be fixed or random. In the first model it is hypothesized that the tested environments represent all environments in which the genotypes will have to perform, such as in poultry production both floor and cage management, and the genotypes to be tested are distributed to both. In contrast, under the random model the test environments are considered, but only a random sample of all environments is taken. For example, with respect to possible diseases, the maladies encountered in a test run can surely be considered only as a sample of all possible, or even of all likely to occur in the next few years. The two models have different expectations of the mean squares (Table 8.1). The genetic variance is greater if environments are fixed than if they represent a random sample of all possible environments. As elaborated above, under the fixed model the inferences drawn refer to the limited range of environments tested, while under the random model they refer to a much greater range of conditions.

In Table 8.1 variance components for sires and sire × environment, computed from progeny test data, illustrate the different consequences of the model. Under the hypothesis of fixed environments the sire variance components for body weight and egg production are almost double those under the hypothesis of random environments. The agreement between performances of individual progeny groups in the three test environments is rather limited. Therefore an even poorer agreement is to be expected if the range of environments is extended. It must be emphasized that this example has general significance insofar as limitation of resources requires testing of strains, progeny groups, etc., in

TABLE 8.1
Genotype–Environment Interaction and Random and Fixed Models[a]

Cause of variance	Variance			
	Hypothesis I		Hypothesis II	
Genotype	$R + k''G'$		$R + kI + k''G$	
Environment	$R + kI + k'E'$		$R + kI + k'E$	
Interaction	$R + kI$		$R + kI$	
Residual	R		R	
	G'	I	G	I
Percent egg production	2.2	2.8	1.2	2.8
Body weight	3.8	5.6	2.0	5.6

[a] Source: Pirchner (1968). R, I, G', G, E, Variance components. E', Square of environmental effects.

a limited number of environments while their progeny have to produce in much more varied situations. If $G \times E$ interaction exists, it must be assumed that the superiority of a selected genotype will be only incompletely repeated in another environment. In other words, even though one must apply the hypothesis of fixed environments in testing, in reality the hypothesis of random environments is true. Dickerson (1963) introduced the notion "genetic slippage" to signify the experience of incomplete repeatability of selection gains in different environments.

Mather and his school (Bucio-Alanis, 1966; Freeman and Perkins, 1971) use a somewhat different model to handle genotype–environment interactions, which Moav and Wohlfarth (1974) applied to fish breeding:

$$y = \mu + g_j + (1 + \beta_j)e_i + w_{ij} + r_{ijk}$$

Improvement of environment e_i by one unit will increase performance by $b_j = 1 + \beta_j$ units. The $G \times E$ interaction effect is split into two parts: one part is linearly related to the genotype, $\beta_j e_i$, and the second part w_{ij} is linearly independent of the genotype. The quantity β_j can be subdivided further: part sg_j is due to the average genotypic effect g_j (s denotes the regression coefficient of b_j on g_j) and part b_j' is independent of the genotypic potential:

or
$$\beta_j = sg_j + b_j'$$
$$b_j = 1 + sg_j + b_j' = \bar{b}_j + b_j'$$

Moav and Wohlfarth (1974) find for the weight of a European carp strain $b_j = 1.22$, $s = 0.0021$, and $g_j = -42.86$. Therefore, one would expect $\bar{b}_j = 0.91$. The difference $1.22 - 0.91 = 0.31$ represents the specific reaction of genotype j to environmental improvement that is not caused by its general level of performance. In contrast, a Chinese strain had the following parameters: $b_j = 0.48$, $\bar{b}_j = 0.62$, $b_j' = -0.14$. Its general reaction to environmental improvement is weaker, due to its poorer genotype, $\bar{b}_j = 0.62$. In addition, it exhibits a specific insensitivity, linearly independent of its genotype, to the environment, $b_j' = -0.14$, which makes it tolerant of environmental deterioration.

The genetic correlation between performance in different environments is

$$r_G = \frac{V_G + e_1 e_2 [s^2 V_G + (b_j')^2]}{V_G + e_1 e_2 [s^2 V_G + (b_j')^2] + V_R}$$

where V_R comprises the residual variance and the variance caused by differences in the w_{ij}. The correlation is 1 when interaction is solely caused by scale differences. This model permits the prediction of the direction of $G \times E$ effects. It facilitates extrapolation to a heretofore untested range of environments. For example, the estimates given predict that European carp would do well in good environments, while for poor environments Chinese strains should have advantages.

Genotype–environment interaction is expected where considerable variation exists within either or both terms. Bowman and Powell (1962), for example, found highly significant interactions and changes in ranking when poultry strains were tested on several farms, while Oakes (1967) found no such interactions when sire progenies were compared over the same range of environments. Dunlop (1962) has arranged the possible combinations as follows:

	Diff — G	Diff — E	
	Differences		
	G	E	
1	small	small	animals in one herd
2	large	small	breeds in one herd
3	small	large	animals in different areas
4	large	large	breeds in different areas

Interactions are expected under 4, which covers the classic example of *Bos taurus* and *Bos indicus* cattle in temperate and in tropical areas. An example for 3 could be provided by sire progenies on floor or in cages where correlations are as low as 0.7 (von Krosigk and Pirchner, 1974). Examples for 2 or even 1 would be difficult to find.

Genotype–environment interactions are important where the environment cannot, or can only insufficiently, be controlled, such as in plant production, or in animal production with range beef cattle or range sheep, or fish production. In contrast, where environmental control is feasible, such as in milk production or poultry production, $G \times E$ interactions are of minor importance. However, different economic conditions may lead to $G \times E$ interactions when instead of one trait the economic value of the total production is considered. For example, grading up of European dual-purpose cattle has increased the milk yield but impaired the carcass value. The improvement in dairy performance presupposes a high level of feeding. If for some reason this is uneconomical but on the other hand beef performance remains economically attractive, dual-purpose types may be more profitable than they have been in the past. Similar examples can be drawn from related areas of beef breeding. A Simmental × Hereford cross may be generally more productive than pure Herefords, but in poor grazing regions its higher milk potential may lead to unsatisfactory fertility and make the pure beef breed superior. Obviously the breeding goal is to be determined by the economics of the particular situation.

9

Selection

Selection as a cause of gene frequency change has been dealt with in Chapter 2. Here, the effect of selection on quantitative traits will be discussed. The question of primary importance to the breeder in the past, as well as today, concerns the extent to which parental superiority in performance is repeated in the progeny. It is not surprising, therefore, that in the 19th century biometricians concerned themselves so much with the inheritance of quantitative characters.

More or less three types of selection can be distinguished: stabilizing, diversifying, and directional selection (Fig. 9.1). In the first case phenotypes from around the population mean are selected. Even though common in nature, in domestic animals it is practiced mainly in the breeding of pets and show and pleasure animals. However, in another, less common form of stabilizing selection, the extremes of both directions are selected and paired with each other so that the mean of the population remains unchanged, but variability is increased. In diversifying selection extremes are chosen but mated *inter se*. It is difficult to find examples from animal breeding; possibly it is practiced when sires are chosen for "maleness" and dams for feminine expression. In directional selection animals are chosen for which the performance excels in one direction—high milk yield, high growth rate, high racing speed, etc. Insofar as outstanding performance presupposes not only a certain harmony of body proportion but also probably a balanced functioning of physiological processes, some selection that looks outwardly directional may be, as far as underlying physiological processes are concerned, in reality stabilizing selection. In the following paragraphs we shall deal mainly with directional selection.

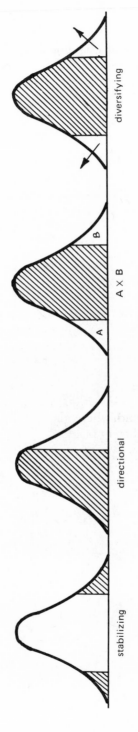

FIGURE 9.1. Types of selection.

The theoretical problems of directional selection were treated by Cochran (1951). Phenotypes are assumed to show continuous distribution. Progeny are assumed to receive genes only, and maternal effects, "social inheritance," etc., are not considered. Progeny are also exposed to random environments, so that the expectation $E(\Sigma e) = 0$. Therefore,

$$E\left(\frac{\Sigma P}{n}\right) = E\left(\frac{\Sigma G}{n} + \frac{\Sigma e}{n}\right) = G$$

and thus the expected value of progeny phenotypes equals the parental genotype or the breeding value of the parents.

Selection of extreme phenotypes will pick up all those individuals that were exposed to above average environments and/or possess very favorable dominance and epistatic combinations. All these effects are not repeated in the progeny (provided parents were randomly mated, and no correlation exists between parental and progeny environments). Also, average genotypes are more common than extreme genotypes. Therefore, extreme phenotypes will comprise a proportionally large share of individuals with average or even mediocre breeding values modified in a positive direction by favorable environmental effects and by nonadditive genetic effects. Of course the reverse also occurs, but the modification in a negative direction of rare superior genotypes is far outnumbered by the positive modification of average genotypes. The same occurs at the negative end, where many of the poorest phenotypes are genetically better than they appear but have met unfavorable conditions. Offspring receive only genes and are, theoretically at least, not affected by nonheritable modifications affecting the parents. Therefore, offspring will in general move toward the mean of the parental population, i.e., show a regression, a term coined by Galton to describe this phenomenon and now widely used, mostly in a somewhat different sense, in statistics.

The goal of selection is in general the improvement of the progeny performance. Elementary considerations lead to selection of the highest performing animals. If 10% of the animals are required to guarantee reproduction of the population, one would choose the 10% best animals. However, as will be discussed, other selection goals than maximum possible progress can be pursued.

The performance level that is surpassed by the chosen 10% best animals, and that is missed by the 90% culled animals, is a threshold which separates selected and culled animals and thus truncates the population. Truncation selection, even though possibly not very realistic, guarantees maximal progress.

The superiority of the selected parents, the animals above the truncation point relative to the population average, is the selection differential ΔP. The selection gain or selection response ΔG is the product of the selection differential times the regression coefficient of the breeding value, or additive genotype, on the phenotype, in other words, the heritability:

$$\Delta G = b_{GP} \, \Delta P = h^2 \, \Delta P$$

Assumptions necessary for this relation to hold are continuous distribution and a monotonic distribution function of genotypes, but neither normality nor linearity are necessary conditions. The relation can be standardized by dividing the quantities by their respective standard deviations:

$$\frac{\Delta G}{\sigma_G} = \frac{\Delta P}{\sigma_P} \times \frac{\sigma_G}{\sigma_P} = \frac{\Delta P}{\sigma_P} \times r_{GP}$$

The genetic response expressed in units of σ_G equals the selection differential in units of σ_P times r_{GP}. The relation $r_{GP} = \sigma_G/\sigma_P$ results from $\text{cov} \, PG = \text{cov}[(G + E)G] = V_G$ if $\text{cov} \, EG = 0$, and $r_{GP} = V_G/\sigma_G\sigma_P$.

If the population is to remain stable numerically, the percentage of animals to be retained as breeders p depends on the reproductive rate and varies between species and sex. On the average each parent must have one offspring. If n is the number offspring of the same sex per parent, $p = 1/n$. The percentages of animals to be saved as breeding animals are given in Table 9.1 for each sex for different species. This percentage must be large if the population is to increase, i.e., if it is to be successful and expand. This happened, for example, to the Pietrian pig in the 1950s and 1960s, and to some small imported groups of purebred animals where further imports are difficult (most European continental cattle breeds in North America). On the other hand, populations that decline numerically require proportionally fewer animals to be retained for breeding, permitting more intensive selection. Improvements and inventions in reproductive techniques, such as incubation of eggs, artificial insemination, embryo transplantation, induction of twinning in monoparous animals, etc., likewise permit reductions, often drastic, in the fraction of animals to be saved for reproduction. Conversely, infertility and disease reduce reproductive potential and may impede selection or even exclude it, in particular among females. In such situations resumption of selection requires an improvement of health and hygiene.

TABLE 9.1
Fraction to Be Retained for Reproduction

	Fraction, %	
Cattle		
♀	50–80	
♂ AI	5–20	5–40[a]
NS	5–10	
Horse		
♀	40–50	
♂	1–5	
Pig		
♀	10–20	
♂	$\frac{1}{2}$–3	
Sheep		
♀	30–60	
♂	1–5	
Poultry		
♀	3–15	
♂	$\frac{1}{2}$–2	

[a] Progeny testing.

If performance is normally distributed and if selection is by truncation, the selection differential expressed in standard deviations i equals the ratio between the ordinate of the truncation point z and the fraction saved p: $i = z/p$. The standardized selection differential can be computed from tables of normal distributions. These values, however, presuppose large populations. For smaller populations the i values represent the superiority in standard deviations of the best individuals in ordered sequences (Fisher and Yates, 1963). Some values are given in Table 9.2. For large populations Smith (1968) suggested the approximation

$$i \approx 0.8 + 0.41 \ln(n - 1)$$

where n is the number offspring available for breeding. Five progeny of the same sex require 20% to be retained and the above formula gives $i = 1.37$ instead of the 1.40 of Table 9.2.

In populations where a trait is distributed normally, truncation selection will have the response

$$\Delta G = i\sigma_P h^2 = i \, \sigma_G r_{GP}$$

Selection response is proportional to the product of selection intensity measured as standardized selection differential, genetic variability, measured in genic standard deviations, and the accuracy of selection, measured as the correlation between phenotype and breeding value. The expression has been derived for mass (or individual) selection where individuals are chosen on their phenotypic merit. However, it is equally valid for situations where selection is based on different types of information, such as performance of relatives, or on combinations of this and individual performance. Genetic progress depends on the product of the three quantities and it will be zero when any one of the three equals zero.

The most important way to improve genetic progress aims at increasing the accuracy of selection r_{GP}. In mass selection, this correlation is the square root of heritability. Use of additional criteria may improve accuracy and thus can speed genetic progress. This will be discussed in greater detail in Chapters 10 and 11.

The genetic advance of the offspring is the average of the genetic superiority of the parents. Since in polygamous domestic animals sires can be much fewer than dams, the former may be selected much more intensely and accurately than the dams and thus contribute more to the genetic progress.

The performance of the progeny generation after one generation of mass selection can be described as

$$P_O = h^2 (P_P - \overline{P}) + \overline{P}$$

where $\overline{P}$ is the average performance of the parental generation before selection, P_P the performance of the selected parents, and P_O the offspring performance. Three averages are involved: offspring average, parental average, and average of the parent generation, and provided $h^2 < 1$, the inequality $P_O - \overline{P} < P_P - \overline{P}$ holds. The fact that progeny performance falls short of the performance of the selected parents, the phenomenon of Galton's regression, has led animal breeders sometimes to consider selection as ineffective and occasionally even to talk of degeneration. However, this erroneous conclusion derives from overlooking the performance of the unselected parental generation. Incomplete heritability, which is the rule, implies that only part of the selection differential $P_P - \overline{P}$ will be repeated in the progeny, but these will be superior to the parental generation average by this fraction of the selection differential (except when $h^2 = 0$).

Standardized selection differentials i pertaining to various culling fractions are given in Table 9.2. These i values are the maximum that can be achieved by strict culling. In reality the selection differentials will be smaller. A breeder must take into account other traits in addition to

TABLE 9.2
Selection Intensity[a]

Fraction of population retained, %	Selection intensity $i = (P - \overline{P})/\sigma_p$			c	x
	$n = 10$	$n = 20$	$n = \infty$		
1			2.67	0.9078	2.33
2			2.42	0.8833	2.055
5		1.87	2.06	0.8546	1.645
10	1.54	1.64	1.75	0.8225	1.28
20	1.27	1.33	1.40	0.7840	0.84
30	1.06	1.11	1.16	0.7366	0.525
40	0.89	0.93	0.97	0.6935	0.253
50	0.74	0.77	0.80	0.64	0
60	0.60	0.62	0.64	0.5715	−0.253
70	0.45	0.48	0.50	0.5125	−0.525
80	0.32	0.34	0.35	0.4165	−0.84
90	0.17	0.18	0.19	0.2793	−1.28

[a] n, Size of population. $\overline{P}$, Phenotypic mean. x, Abscissa of truncation point. $c = i(i - x)$.

the trait that is being selected for. Therefore, selection of some animals whose performance is somewhat inferior to that of other animals becomes necessary. Even if a breeder intends to concentrate on only one trait, the presence of defects of health and fertility in some of the chosen animals may necessitate their exchange for poorer candidates. Rigid selection is approximated the most closely when culling is based on some aggregate economic value of the animals. However, such values may be difficult to estimate accurately. In practical breeding, selection will mostly be as indicated in Fig. 9.2. The higher the performance, the greater will be the probability that the candidate is chosen for breeding. The average of the breeding animals is k standard deviations above the population mean and $k < i$.

For example, Puff (1976) found in a German Landrace population that index values of boars and sows, respectively, were 1.27 and 0.64 standard deviations above the population mean, which correspond to 25% of boars and 60% of sows retained for breeding, considerably lower than values to be expected if strict truncation selection had been performed (Table 9.1).

Better performance increases the probability of selecting an animal. The increase in the probability can be described by a probit transformation (Cavalli-Sforza and Bodmer, 1971). Under truncation selection the mean of the selected group permits an inference about the culling

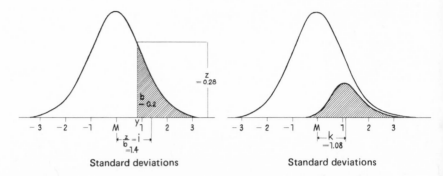

Standard deviations   Standard deviations

FIGURE 9.2. Selection differential. Shaded areas indicate the proportion selected (20%); the figure on the left corresponds to the theoretical model of selection by truncation; the figure on the right depicts the more realistic situation, in which selection is less accurate. The superiority of the selected animals in the left-hand figure is larger than that in the right-hand figure $(i > k)$.

percentage. When selection follows the probit model, mean and standard deviation of the selective value must be known. A given culling percentage under truncation selection will yield a selection differential of $i\sigma$ units. The same culling percentage will lead to $\Delta P = i\sigma/(1 + \sigma^2)^{1/2}$ under the probit model, which is smaller than the selection differential in truncation selection. It is intuitively obvious that the selection differential will be a maximum when only the best available candidates are chosen and that all other methods will lead to smaller gains.

The formulas derived so far pertain to selection gain per generation. Obviously the gain per year has more practical significance. It is influenced, in addition to the factors already discussed, by the generation interval, which can be defined as the average age of parents at the birth of their offspring. Generation intervals of domestic animals are given in Table 9.3. When the generation interval is t years, the annual genetic progress of mass selection will be

$$\Delta G/t = \frac{i\sigma_G h}{t}$$

The generation interval is to be weighted by the fertility of the age classes.

Ollivier (1974b) has investigated the relationship between the length of the generation interval and the selection differential. When the generation interval increases, the number of progeny increases and selection can be more intense. For maximum progress i/t should be maximized.

TABLE 9.3
Generation Intervals

	Interval
Cattle	
Sire–progeny, NS[a]	$2\frac{1}{2}$–4 years
Dam–progeny	5–$6\frac{1}{2}$ years
Sire–daughter	$4\frac{1}{2}$–$6\frac{1}{2}$ years
Sire–son, planned mating	$6\frac{1}{2}$–9 years
Horse	
Stallions	8–15 years
Mares	8–11 years
Sheep	3–5 years
Pig	
Pedigree animals	15–19 months
Commercial animals	2–3 years
Poultry	
Broiler	45 weeks to 1 year
Layers	13–18 months

[a] Natural service.

Increasing age at breeding may frequently also improve the accuracy of the estimate of the breeding value so that ir/t should then be maximized. When the accuracy is not improved with increasing age, it can be shown that in fertile species and in males (which are relatively fertile) early selection should be practiced and waiting to increase i does not pay. However, when the accuracy of estimating the breeding value increases with age, as in the case of traits with low repeatability, later selection and thus longer generation intervals are justified.

9.1. Goals of Selection

So far it has been assumed that directional selection is used to maximize progeny performance. This is the most important though not the only goal of selection. In addition, the probability that progeny performance is above a certain level is of interest. Cochran (1951) emphasized these alternative selection goals and Walter (1960) has categorized them into three groups: (1) maximizing the expected value, (2) maximizing the probability that progeny performance surpasses a certain minimum, and (3) ensuring that progeny performance at a given probability is as great as possible.

The procedure for the first goal has been discussed before and will be implicitly assumed in nearly all discussions to follow. In the procedure for the second goal the probability that the breeding value exceeds a certain level L is maximized. The decision criterion is $(TA - L)/\sigma_{TA}$, where TA denotes the estimated transmitting ability and σ_{TA} the standard error of the estimate. Procedure 3 applies the lower confidence interval $TA - k\sigma_{TA}$, where k denotes the standard normal deviate cutting off $\alpha\%$ of the area under the normal curve.

The various procedures can be illustrated by the following examples. Young bull A has a transmitting ability estimated from pedigree information of $+650$ kg, $r = 0.52$, $\sigma_{TA} = \sigma(1 - r^2)^{1/2} = 257$. The standard deviation of breeding values was assumed to be 300 kg. Old bull B has an estimated breeding value of $+486$, $r = 0.9$, $\sigma_{TA} = 130$. Procedure 1 selects bull A ($TA_A > TA_B$). As for procedure 2, it may be asked that the probability of breeding values exceeding 5000 kg ($=L$) is as great as possible. When the population mean is 4800 the criteria of selection are

$$\text{A:} \quad 450/257 = 1.75, \qquad \text{B:} \quad 286/130 = 2.20$$

The selection criterion for bull B is larger in spite of its lower estimated breeding value. To illustrate procedure 3, one may demand that the level that is to be surpassed by the breeding value with a probability of 90% be as high as possible. The t value for a 10% confidence interval is 1.65. Therefore the selection criteria are

$$\text{A:} \quad 5450 - 1.65 \times 258 = 5025$$
$$\text{B:} \quad 5286 - 1.65 \times 130 = 5073$$

and again bull B is to be preferred.

Procedures 2 and 3 are probably rather common even though not always made explicit.

Schneeberger et al. (1980, 1981) considered σ_{TA} as a measure of risk which the user of a breeding animal incurs and they found that, for example, semen price in the U.S. is influenced by this measure in addition to the estimated breeding value. The choice of the selection method would depend on the breeder's weighting of the expected performance versus the risk of falling short of some expected level of performance.

9.2. Influence of Selection on the Genic Variance

The total genic variance is composed of two parts (Bulmer, 1971):

$$V_G = \sum_i V_{G_i} + 2 \sum_{i<j} \text{cov } G_i G_j$$

$$= 2 \sum_i q_i \alpha_i^2 + 4 \sum_i \sum_{i<j} \alpha_i \alpha_j d_{ij}$$

The first part represents the sum of the genic variances due to individual loci, and the second part the contribution of the covariance between genic effects of different loci, which is caused by linkage disequilibrium. Since in an equilibrium population $d = 0$, the second part vanishes and the genic variance consists only of the first part, which Bulmer calls the equilibrium variance, the contribution of the covariance being conditioned on a linkage disequilibrium, which he calls the disequilibrium variance.

Selection influences the equilibrium variance via changed gene frequencies. For polygenic traits influenced by n genes, the gene frequency change at an individual locus will be proportional to $(2npq)^{-1/2}$. Therefore, where n is moderately large, changes in the equilibrium genic variance can be ignored. However, selection causes a change in the disequilibrium variance. Assume that both parents are selected. The expected value of an offspring in generation $t + 1$ is given by

$$y = \tfrac{1}{2}A_m + \tfrac{1}{2}A_f + E$$

where A_m and A_f are the breeding values of sire and dam, respectively, with variance V_A, and E incorporates all nongenic effects, with zero mean and variance $V_{A/2} + V_E$, which are both unaffected by selection in the previous or earlier generations. The variance among individuals in generation $t + 1$ will be reduced due to selected parents by $\tfrac{1}{2}h^2 i(i - x)$. Therefore,

$$V_{t+1} = \tfrac{1}{4}V_A + \tfrac{1}{4}V_A^* + V_E$$

$$V_A^* = V_A(1 - \tfrac{1}{2}h^4 c), \qquad c = i(i - x)$$

When all grandparents are selected but selection is relaxed later, the genic variance of grandprogeny will be reduced by $\tfrac{1}{4}h^4 i(i - x)$, and

$$V_{t+2} = \tfrac{1}{2}V_A + \tfrac{1}{2}V_A(1 - \tfrac{1}{4}h^4 c) + V_E$$

Relaxation of selection has reduced the disequilibrium contribution to one-half, in analogy to the reduction of the linkage equilibrium between independent loci (Chapter 1). Under continued truncation selection the variance will decrease. The same holds true for stabilizing selection, while disruptive selection will increase it. The limiting value of the genic variance V_A^* can be computed from

$$(2 - k)(V_A^*)^2 + (V_E - V_A)V_A^* - V_A(V_E) = 0$$

where k is the change in the variance caused by use of truncation selection, $k = 1 - i(i - x)$.

When the top 15% are selected, as is possible with pigs, the variance of the breeding animals will be decreased by 0.7905 of its base value. If heritability of gain on test is 1/4, continued selection of this intensity will reduce the genic variance by 3.72% to a level of 21.28%, and the heritability to 22.1%, a reduction of about one-eighth. Relaxing selection would halve the reduction of genic variance to 1.86%, provided the genes are independent, and V_A would increase to 23.96% of the base variance.

The remaining deficit would be halved again in the next generation, so that the base value of the variance would be approached quickly. Linkage will slow down the rate of approach to the equilibrium value, but Bulmer (1974) showed in simulation studies that the effect of linkage is unlikely to be important in species with a moderate chromosome number. At any rate, intensive selection, in particular of highly heritable traits, will have a considerable influence upon the genic variance as long as selection is pursued.

9.3. Changes in Gene Frequency by Selection

Gene frequencies for quantitative traits change during selection depending on the size and direction of the gene effects. The genotypic values of the gene combination A_iA_j are denoted by d_{ij} (these are equivalent to the genotypic deviations of Table 4.3; for example, in the notation used there, $d_{11} = a_{11}$). It can be shown (Griffing, 1960) that individuals of this gene combination at locus A survive selection in proportion to $w_{ij} = 1 + (i/\sigma_P)d_{ij}$. Griffing, following Kimura, termed this quantity the selection value of genotype A_iA_j. The necessary conditions for validity of this relation are that d_{ij} is small relative to σ_P, that is, that the locus has a small effect on the selected trait, and that the variance caused by the alleles at this locus is small relative to the total genotypic variance. The frequency of the genotype after selection is, approximately,

$$p_i p_j w_{ij} = p_i p_j + \frac{i}{\sigma_P} p_i p_j d_{ij}$$

Summation of this expression over loci gives the distribution of the genotypes after selection. The mean value of the selected parents is given by

$$m_s = \sum_{ij} \left(p_i p_j + \frac{i}{\sigma_P} p_i p_j d_{ij} \right) d_{ij}$$

The d_{ij} are deviations from the mean. Therefore, the expectation of the weighted sum $\sum_{ij} p_i p_j d_{ij}$ equals zero. The second quantity within the parentheses will, after multiplication by d_{ij}, equal the weighted sum of the squared genotypic deviations, that is, the total genotypic variance excluding epistasis. It follows that

$$m_s = \text{SD} \frac{\sigma_G^2}{\sigma_P^2}$$

which is equal to the selection differential times heritability in the broad sense.

The frequency of the allele A_i in the selected population will be

$$p_i = \sum_j \left(p_i p_j + \frac{i}{\sigma_P} p_i p_j d_{ij} \right) = p_i + \frac{i}{\sigma_P} p_i \alpha_i$$

The gene frequency change in the progeny generation is thus

$$\Delta p_i = i p_i \frac{\alpha_i}{\sigma_P}$$

At a given selection differential and gene frequency, the change will be proportional to the ratio between the average effect of gene A_i and the phenotypic standard deviation. Consequently, genes with large effects will be affected more by selection than genes with small effects.

The regression of gene frequency on the phenotype (Falconer, 1981) leads to similar results. Table 9.4 gives the genotypic values and fre-

TABLE 9.4
Gene Frequency Change Caused by Selection for Quantitative Trait^a

	p_ip_j	a_{ij}	w_{ij}	$p_i'p_j'$
A_1A_1	0.36	3.52	1.117	0.4022
A_1A_2	0.48	5.52	1.184	0.5683
A_2A_2	0.16	−24.48	0.184	0.0294

$m_s = \sum_{ij} \hat{p}_i\hat{p}_j,\, a_{ij} = \dfrac{i}{\sigma} \sum_{ij} p_ip_j,\, a_{ij}^2 = 3.8323$

$m_1 \simeq \dfrac{i}{\sigma} \sum p\alpha^2 = 1.866$

$\text{cov } pa = p_1(p_1a_{11} + p_2a_{12}) = p_1p_2\alpha = 0.6 \times 4.32 = 2.592$
$\Delta p = i\sigma_p b_{pa} = 2.592/30 = 0.864$
$p' = 0.402 + 0.568/2 = 0.686$

$^a a_{ij}$, Genotypic effect as deviation from mean. w_{ij}, Selective value of genotype $_{ij}$.

quencies. The frequency of the A_1 gene in the three genotypes is 1, 1/2, 0. The covariance between gene frequency and genotypic value is $pq[a + d(q - p)] = pq\alpha$, where α is the average effect of the gene substitution. The regression of gene frequency on the genotype is $pq\alpha/\sigma_P^2$. The change in gene frequency when the selection differential is $i\sigma_P$ units is $ipq\alpha/\sigma_P$. Therefore, the effect of gene substitution and the average gene effect are related as $\alpha_i = p_j\alpha$, and similar to Griffing's method, $\Delta p = ip_i\alpha_i/\sigma_P$.

The change in gene frequency due to selection of a certain intensity can be compared to the change caused by selection directed against a certain genotype, as discussed in Chapter 3. Under semidominant inheritance the approximate formula for the change in gene frequency is

$$\Delta p = \tfrac{1}{2}spq$$

Assuming equal change in gene frequency, this gives

$$\tfrac{1}{2}spq = ipq\alpha/\sigma_P$$

Furthermore,

$$s = 2i\alpha/\sigma_P$$

The selection coefficient is proportional to the product of the selection differential and the quantity $2\alpha/\sigma_P$, which Falconer calls the "proportionate gene effect." This equals the difference between the breeding values of the two opposite homozygotes relative to the phenotypic standard deviation. Taking an example from Table 9.4, where $i = 1, s = 2 \times 10.8/30 = 0.72$. The genotype of the progeny generation is found from $(\Sigma_i p_i A_i)'$. The mean of the progeny generation is given by (in deviations, $m = 0$)

$$m_1 = \frac{i}{\sigma_P} \sigma_A^2 + i^2 \sum_{ij} p_i q_j \frac{\alpha_i}{\sigma_P} \frac{\alpha_j}{\sigma_P} \delta_{ij}$$

where δ_{ij} denotes the dominance deviation (Table 4.3). We assume, as before, that α is small relative to σ_P. Therefore, the second term on the right can be neglected. The results agree with those derived at the beginning of this chapter:

$$m_1 = \text{SD} \times \frac{\sigma_A^2}{\sigma_P^2}$$

which is equal to the selection differential times heritability in the narrow sense. From Table 9.4, $h^2 = 0.0622$ ($\sigma_A^2 = 55.9872$, $\sigma_P^2 = 900$) and $m_1 = 1.866$. The second quantity on the right equals 0.1543. This rather large quantity results from the fact that the proportionate gene effect is large, contrary to the assumption in arriving at the expression.

This derivation shows clearly that the expression commonly used for genetic progress is an approximation. The neglected quantities will be unimportant for only a few generations of selection. However, under extended selection these quantities will accumulate. This might partly account for estimated progress not equaling real progress in selection experiments of long duration.

9.4. Indirect Selection

It must be assumed that most genes have pleiotropic effects. Therefore, selection for one trait will cause changes in the genotypes and phenotypes of other traits. These correlated selection effects are interesting for two reasons: first, they result automatically from selection for one or for several traits, and, second, they allow change of a trait by indirect selection.

What is the change, expected in trait B, for example, butterfat content of milk, when selection is for trait A, say milk yield? This change is estimated by the regression of breeding value for trait B on the phenotype of A. The following notation will be used: $G(A)$ and $G(B)$ are the breeding values, $U(A)$ and $U(B)$ are the environmental effects, and $P(A) = G(A) + U(A)$, where G and U are uncorrelated, and similar relations hold for $P(B)$. The covariance between breeding value for B and phenotype for A is

$$\text{cov}[G(B)P(A)] = \text{cov}\{G(B).[G(A) + U(A)]\}$$

and $$\text{cov}[G(B)U(A)] = 0$$

Therefore $$\text{cov}[G(B)P(A)] = \text{cov}[G(A)G(B)] = r_G\sigma_{G(A)}\sigma_{G(B)}$$

with the resulting regression coefficient

$$b_{G(B)P(A)} = \frac{r_G\sigma_{G(A)}\sigma_{G(B)}}{\sigma_A^2} = r_G h_A \frac{\sigma_{G(B)}}{\sigma_A}$$

This estimates the change in trait B when selection operates on the phenotypes for trait A. As given before, the selection differential equals the product $i\sigma_P$ for a normally distributed trait and truncation selection. In our example this is $i\sigma_A$. The resulting change in trait B is

$$\Delta_{G(B)A} = i\sigma_A r_G h_A \frac{\sigma_{G(B)}}{\sigma_A} = i h_A r_G \sigma_{G(B)}$$

As under direct selection, the change is the product of selection intensity i, the genetic variability in the trait to be improved $\sigma_{G(B)}$, and the accuracy of estimating genotype B from phenotype A (given by the product $h_A r_G$). The indirect change can be compared with the change expected under direct selection. Equally intense direct selection results in

$$\Delta_{G(B)} = i h_B \sigma_{G(B)}$$

Consequently, at equal selection intensities the ratio of the indirect or correlated change to the change by direct selection is

$$\frac{\Delta_{G(B)A}}{\Delta_{G(B)}} = \frac{h_A r_G}{h_B}$$

that is, the ratio of the respective correlations between the selection criterion and the breeding value. For example, we can estimate the correlated change in feed efficiency (feed/gain) caused by direct selection for growth rate in pigs. Fredeen and Johnsson (1957) report the following parameter estimates for these traits:

	Daily gain, g	Feed efficiency
h	0.60	0.67
σ_G	19.5	0.10

The genetic correlation estimate between the two traits is -0.87. Assume that the selection intensity for boars is 2.06 (the best 5% are selected) and that for sows 1.40 (the best 20%). The genetic progress in feed efficiency under direct selection is a decrease of 0.116 feed units (FU) per unit gain ($= 1.73 \times 0.67 \times 0.10$). The correlated decrease in feed required per unit gain when selection is only for growth rate is estimated to be 0.090 FU, about 78% of the change caused by direct selection. This corresponds to the ratio $r_G h_A / h_B = 0.60 \times 0.87 / 0.67 = 0.78$. The result can be looked at from two somewhat different viewpoints. First, it is an unavoidable consequence of selection for growth rate. In this example the correlated change is in the desired direction; in others it might be in the opposite direction. For example, selection for milk yield causes some decrease in butterfat content. Second, an estimate of correlated change also helps the breeder decide whether selection should be direct or indirect. In the example, selection for daily gain also changes feed efficiency in the desired direction, but only by three-fourths as much as would direct selection for feed efficiency. However, measuring feed efficiency is expensive. Therefore, a breeder using such information might decide not to measure feed eaten but to concentrate efforts and resources on improving growth rate and other traits.

When selection is indirect, say for trait A, the ratio of genetic progress to genetic progress with direct selection on trait B is $r_G h_A / h_B$. Indirect selection will be better (that is, $\Delta_{G(B)A}$ larger) than direct selection if $r_G h_A > h_B$. Searle (1965) pointed out that indirect selection can be more successful only if r_G is larger than h_B, that is, $r_G > h_B$. This condition must be fulfilled even if the heritability of the auxiliary traits is unity. When heritability is less then complete, r_G must be proportionally larger. The square root of the heritability of the auxiliary trait will be larger than h_B / r_G if correlated genetic progress exceeds progress by direct selection. However, as mentioned before, indirect selection may be much cheaper in some situations, so one might be satisfied with less actual progress in one trait if this is compensated for by more total progress per unit of expenditure. The ratio of indirect to direct selection response is composed of three parameter estimates, each with rather larger sampling errors. Therefore, in most data, the ratio is subject to considerable uncertainty.

9.5. Selection for Several Traits

The total economic value of an animal depends almost always on several traits. Consequently, the breeder is forced to consider several traits in selection. This necessitates a lower selection intensity for each trait. Although this is the only possible way to increase the total economic value, it decreases progress in individual traits. The size of the decrease depends on many factors, such as the closeness and direction of the correlations between traits, the number of traits considered, the weight given to an individual trait, and the selection method used. Basically, three methods of selecting for several traits exist (Hazel and Lush, 1943): tandem selection, independent culling, and use of the selection index.

In tandem selection the individual traits are improved successively. Trait A is improved first, then trait B, and so on. While selection is for trait A only, progress will equal the selection response for the trait alone. However, during this time no progress is expected in the other traits (assuming zero genetic correlations). Assume that the total economic value depends on k uncorrelated traits. Each trait is to be improved for one generation. After k generations the genetic improvement in the economic value is

$$h_1^2(P_1 - \overline{P}_1) + h_2^2(P_2 - \overline{P}_2) + \cdots + h_k^2(P_k - \overline{P}_k) = \sum_{j}^{k} h_j^2(P_j - \overline{P}_j)$$

Further, assume equal heritabilities, variances, and selection intensities for the k traits. The average change in economic value per generation then equals $ih\sigma_G$ or $ih^2\sigma_P$.

Independent culling means that selected individuals must surpass a certain minimum value in each character. This minimum performance is independent between traits. For example, a breeder could demand that a heifer surpass 4000 kg milk yield and 4% butterfat content if she is to be a breeding animal. If her butterfat percentage is less than 4% she will be rejected regardless of her yield. This selection method immediately decreases the selection intensity of the individual traits.

Simultaneous selection of equal intensity for k traits reduces the selection differential from $i = z/b$ to $i' = z'/b^{1/k}$, where z' is the ordinate of the point on the abscissa separating the best $b^{1/k}$ percent of the population. The expected increase in breeding value per generation is

$$\sum_j i' h_j \sigma_{G_j}$$

If the heritabilities and variances are equal in all k traits, this is reduced to

$$ni' h \sigma_G$$

Genetic progress relative to tandem selection equals ni'/i under our assumptions (equal importance of the traits, equal variances and heritabilities).

The ratio is always greater than 1, and it increases with increasing n. At higher intensity of selection it is somewhat larger than at lower intensity, i.e., independent culling levels are relatively more advantageous when selection is very intense. In Table 9.5 percentages of retained breeding animals and selection differentials are given for 1–4 traits. Percentages of breeding animals eventually retained are either 3% (e.g., bull dams) or 20% (e.g., sows). The selection differential for any one trait decreases as the number of traits under selection increases. When n equally important traits are selected, only the best $b^{1/n}$ percent with respect to a particular trait can be retained.

The decrease in selection intensity occurring as the number of traits

TABLE 9.5
Selection Differentials for Several Traits[a]

	$k = 1$	$k = 2$	$k = 3$	$k = 4$
b	0.03	0.17	0.31	0.42
SD	2.27	1.49	1.14	0.93
Σ SD	2.27	2.98	3.42	3.72
SD_k	2.27	1.61	1.31	1.14
b	0.20	0.45	0.58	0.67
SD	1.40	0.88	0.67	0.54
Σ SD	1.40	1.76	2.01	2.16
SD_k	1.40	0.99	0.81	0.70

[a] k, Number of traits considered. b, Fraction retained. SD, Selection differential. Σ SD, Selection differential for total economic value. SD_k, Selection differential for one trait if index is used for k traits.

under selection increases emphasizes the necessity of restricting selection to a minimum number of traits. The detriment to genetic progress incurred by attention to fancy points is probably caused less by any deteriorating effect these fancy points might have on economically important characters (although this may also occur) than by the necessarily decreased intensity of selection for important traits.

In the total score or selection index method, culling levels are flexible. Each trait is weighted by a score and the individual scores are summed to a total score (index value) for each animal, which is the selection criterion. By this method, superiority in some traits can make up for mediocrity in others, unlike independent culling, which discards an animal failing to qualify in one trait regardless of its performance in other traits. The selection index is a total score that, ideally, should encompass all the advantages and disadvantages of an individual. It is used intuitively by a breeder who decides to retain an animal in spite of some deficiencies because of other outstanding qualities. The total scores used in classifying type are basically selection indices.

Assume again that k uncorrelated traits are of equal economic importance. The optimal selection index weights the individual traits by their heritabilities. Thus, the selection index equals the sum of the traits, each weighted by its own heritability:

$$I = h_1^2 P_1 + h_2^2 P_2 + \cdots + h_k^2 P_k$$

Under the assumption that variances are equal, the genetic progress is

$$\Delta_{G(I)} = i\sigma_G h \sqrt{k}$$

This quantity is $\sqrt{k}$ times larger than progress under tandem selection. It is also larger than the progress expected from independent culling, but in this case the superiority of the index method depends on the selection intensity i.

Progress for a single trait using selection based on an aggregate economic value is only $k^{-1/2}$ times as large as progress under exclusive selection for that trait. Therefore, if an index is constructed for improving two equally important traits simultaneously, the progress in each amounts to 71% of the progress possible were selection based on one trait only. Using a selection index to improve three traits, one finds that the progress in any one is 58% of that of single-trait selection. These few figures demonstrate again the importance of reducing to an absolute minimum the number of traits under selection.

Abplanalp (1972) has added a fourth method, somewhat complementary to independent culling. By this method the top candidates in any one trait are chosen, regardless of their performance in other respects. When traits are independent and of equal value, selection of "extremes" is of comparable efficiency to independent culling levels, more efficient when the selection intensity is high, and less efficient when selection is weak. Selection of extremes seems to be advantageous when traits are negatively correlated.

Index selection is superior to independent culling, which, in turn, is superior to tandem selection. The advantage of index selection increases with the number of traits. In our comparisons the assumptions were equal economic importance of traits, equal heritabilities and variances, and zero correlations among traits. These assumptions are unrealistic, since traits are usually unequal in importance, have unequal variances, and are correlated. Nevertheless, the ranking of the three methods remains constant even if the assumptions are relaxed. However, if the assumptions are not fulfilled, the difference in efficiency between the methods decreases.

Young (1961) and Young and Weiler (1961) investigated the relative efficiencies of the three selection methods under more realistic assumptions. The true economic–biological importance of a trait can be expressed by the product of the economic value, the heritability, and the reciprocal of the phenotypic standard deviation: $\lambda = ah^2/\sigma_P$. Unequal importance of traits (λ) diminishes the superiority of index selection. Let Δ_I, Δ_C, and Δ_T symbolize, respectively, progress under index, under independent culling, and under tandem selection. Table 9.6 shows the ratios of the selection efficiencies when the traits are uncorrelated. Correlations likewise change the efficiencies of the various methods. If the traits are equally important, low or negative phenotypic correlations increase the relative superiority of index selection. The effect of the genetic correlation on the relative efficiency of the three methods depends on other parameters and cannot be generalized.

The advantage of independent culling appears to be its simplicity. As usual, data on performance accumulate over a period of time, and selection under independent culling can be made on some traits before data on others accumulate. Thus, time as well as facilities can be saved. If the parameters are known, the optimum culling level for each trait can be calculated.

Index selection combines the total information in an optimal way and, in principle at least, simplifies selection a great deal, since deliberations over the merits and faults of individual animals can be spared.

In reality in practical animal breeding all methods are employed side by side. A trait considered to be important in a certain area and/or

TABLE 9.6
Genetic Progress with Different Methods of Selection[a]

λ_1/λ_2	Δ_I/Δ_T	$b = 0.8$			$b = 0.1$		
		b_1	b_2	Δ_I/Δ_C	b_1	b_2	Δ_I/Δ_C
1	1.41	0.89	0.89	1.19	0.32	0.32	1.10
2	1.12	0.81	0.99	1.11	0.14	0.71	1.07
3	1.05	0.80	1.00	1.05	0.11	0.91	1.05
6	1.01	0.80	1.00	1.01	0.10	1.00	1.01

Diminished Superiority of Index Selection

r_P	r_G	τ_1/τ_2	Δ_I/Δ_T	b	Δ_I/Δ_C
0.3	-0.2	2	1.00	0.5	1.00
0.5	-0.4	2	1.09	0.5	1.09
$\lambda_1:\lambda_2:\lambda_3$					
1:1:1	1.73				
5:3:1	1.18				

[a] b, Fraction retained. b_1, Fraction retained with respect to trait 1. τ_1, Economic–biological weight of trait 1 is given by $a_1 h_1^2/\sigma_1$. Here Δ_I, Δ_T, Δ_C denote genetic progress with index, tandem selection, and independent culling, respectively. r_P, denotes respective phenotypes, and r_G is the genetic correlation between traits.

time will be emphasized at the expense of other traits, thus mimicking tandem selection. Examples are milk yield in U.S. dairy cows, or meatiness of hogs in Europe in the 1970s, or resistance to Marek's disease in poultry in the late 1960s. When traits are considered to be of great importance other traits may be nearly totally neglected. This is true of course in the case of lethals, but it is, or it was, practiced also when an animal deviates from the breed standard. Independent culling levels are inherent in certain licensing practices, where performance has to reach certain standards. Index selection—selection for a total score, the sum of advantages and disadvantages—is practiced by nearly all breeders, who, however, often intuitively weight the traits. These practices would tend to decrease the differences between the efficiencies of the selection methods.

All three methods are sensitive to selection mistakes, such as overemphasizing one trait (too high a culling level, too much weight in the index, too much time devoted to selection in the tandem method). The construction of selection indexes when traits differ in importance, heritability, and variance and are correlated, and how these factors influence the relative efficiency of the three selection methods, will be discussed later.

The three selection methods have been compared in experiments with *Drosophila* (Rasmuson, 1964; Sen and Robertson, 1964) and mice (Doolittle *et al.*, 1973). McCarthy and Doolittle (1977) used independent culling levels to change 5- and 10-week mouse weights in opposite directions and an index to change 5-week weight but to maintain the 10-week weight. Some of these selection experiments showed independent culling to be superior to index selection. In other experiments, such as in the comparison between independent culling level and restricted selection index, results followed the theoretically expected order. Large variation among the results of repeated experiments on bristle characters of *Drosophila* and litter size of mice precludes a definite final judgment on the agreement of these experiments with theoretical expectations.

10

Aids to Selection

Genetic progress can be manipulated by changing its four components: selection intensity, genetic variability, generation interval, accuracy. Altering the first three of these components frequently is difficult or even impossible in the short run; it is improvement in accuracy on which most attention is focused. In general, aids to selection become necessary if (a) they permit early selection, e.g., by the use of pedigree information; (b) the accuracy of selection is improved, e.g., by repeated measurement or by progeny testing; and (c) mass selection is not feasible or not possible, e.g., of slaughter traits.

The improvement in estimation of breeding values is achieved in two ways: by reduction or removal of nongenetic influences and by considering relatives, which share genes with the candidate. Nongenetic influences are taken care of basically in two ways: (a) standardization of the environmental influences, and (b) reduction of random errors by repeated observations.

The first can be accomplished either by physically removing environmental differences between animals or statistically correcting for them. An example of the physical exclusion of environmental differences is the *contemporary comparison,* where only contemporary records of first calvers in the same herd are compared, and therefore herd, year, and age influences are absent. A further refinement is *analogous comparison,* where records of the same season only are compared (Barker and Robertson, 1966). In horse racing, competitions are for yearlings, two-year olds, mares, etc., thus removing age and sex influences. An example of statistical correction is the use of *mature equivalent* records, which eliminate age influences, and an example which combines both comparison within physical categories and statistical correction is the *herd-mate comparison*.

The reduction of random errors is accomplished by repeating records either of the animal itself or of single (or multiple) records of relatives. If the estimate is unbiased and errors are independently distributed, repetition can be very effective in increasing the accuracy.

The rules are followed intuitively every time an object is to be measured, weighed, etc., accurately. One tries to exclude disturbing factors, one may make an intuitive correction to account for circumstances which are not sufficiently standard, and one may measure, weigh, score, etc., repeatedly to minimize random errors.

Most aids to selection are based on linear prediction of the breeding value g_i:

$$\hat{g}_i = a_i + b_i[x_{ij} - E(x_{ij})]$$

where the b_i are chosen to minimize the square of differences between estimated and actual breeding value. As will be elaborated in more detail in Section 10.3, this selection index procedure demands knowledge of variances and covariances and also of the population mean. While the former change comparatively little over a few generations or between populations, the population mean responds to selection and immigration. Therefore, a population may not be homogeneous. Furthermore, the population mean is frequently corrected for various environmental influences and it is assumed that the correction factors are accurately known, which is reasonably true in some cases, e.g., when correcting for age influences on dairy performance. In other cases, correction factors are estimated from the data. For example, in contemporary comparison the herd influence on milk yield is accounted for by simply subtracting the mean yield of contemporaries.

Henderson (1973, 1975) has categorized the various approaches to the estimation of future performance of an animal itself or of its relatives, and has emphasized the deficiency of the selection index, due to the not infrequent lack of knowledge of the population mean and to frequently poor estimates of correction factors for various environmental influences. Henderson (1973) proposed instead the best linear unbiased prediction (BLUP) of future records, in which $E(x_{ij})$ is estimated by generalized least squares. This method has received much attention in recent years and since the advent of large-scale computing facilities, and is being applied increasingly in animal improvement schemes. However, in many circumstances animals can be assumed to belong to one population and environmental influences are well known, at least to a first approximation. Breeders outside organized breeding schemes and without access to adequate computing facilities will continue to apply the more traditional methods of estimating breeding values (and probable producing abilities) and when cost–benefit ratios are taken into account, organized improvement schemes may also continue to use the well-proven methods.

10.1. Repeated Records

Repetition of measurements improves the accuracy of estimating the real producing ability (Chapter 5) and, as a consequence, the breeding value. The model used in Chapter 5 is modified,

$$P = A + E$$

where A represents the genic value and E all nongenic influences, encompassing both permanent and temporary environments and dominance and epistatic effects. The variance of n repeated records is decreased only in proportion to the variance contribution by temporary influences, V_T/n. The regression coefficient of breeding value on an average of n repeated records, or the heritability of an average, is

$$b_{AM} = \frac{V_A}{V_A + V_{EP} + V_{ET}/n} = \frac{nh^2}{1 + (n - 1)t} = h_M^2$$

The second expression is derived from the first by dividing by V_P and multiplying by n.

Recall the following quantities:

$$t = \frac{V_A + V_{EP}}{V_P}, \qquad 1 - t = \frac{V_{ET}}{V_P}, \qquad h_M^2 = \frac{V_A}{V_{\bar{P}}}$$

Repetition is the more effective the larger is V_{ET}, since it is this quantity that is decreased to $1/n$ of its size in the variance of single observations. In Table 10.1 the increase of h_M^2 with increase in the number of observations is demonstrated for traits of varying repeatability t. The heritability of a fourfold average is about three times its value based on single observations in the case of the poorly repeatable trait I (0.28 vs. 0.10), but the ratio is much less for trait III (0.71 vs. 0.50). The heritability of many repeated records cannot become 1 as long as $V_{EP} > 0$. The upper limit of the heritability of an average is h^2/t.

TABLE 10.1
Estimation of Breeding Value from Averages

	Trait I				Trait III				Trait II									
									$r_{ij} = 1.0$				$r_{ij} = 0.8$				$r_{ij} = 0.8$ $j \neq 1$	
V_A	10				50								25					
V_{E_p}	5				10								15					
V_{E_T}	85				40								60					
n	w	h_M^2	$r_{AM} = h_M$	h_M/h_1	w	h_M^2	$r_{AM} = h_M$	h_M/h_1	w	h_M^2	$r_{AM} = h_M$	h_M/h_1	h_M^2	$r_{AM} = h_M$	h_M/h_1	h^2	$r_{AM} = h_M$	h_M/h_1
1	15	10	31	—	60	50	71	—	40	25	50	—	21	46	—	20	43	—
2	23	17	41	1.32	75	62	79	1.12	57	36	60	1.20	31	56	1.22	31	57	1.29
3	35	23	47	1.52	82	68	83	1.17	67	42	64	1.08	38	61	1.33	38	63	1.43
4	41	28	51	1.66	86	71	85	1.20	73	45	68	1.35	41	64	1.39	41	67	1.52

Averages have a higher heritability but a smaller variance, which diminishes the impact of the higher accuracy of repeated records on the genetic progress per generation:

$$\Delta G = i\sigma_M h_M^2 = i \left[\frac{1 + (n - 1)t}{n} \right]^{1/2} \sigma h^2 \frac{n}{1 + (n - 1)t}$$

$$= ih^2 \left[\frac{n}{1 + (n - 1)t} \right]^{1/2} \sigma = i\sigma_A h \left[\frac{n}{1 + (n - 1)t} \right]^{1/2}$$

$$= i\sigma_A r_{AM}$$

Genetic progress is proportional to the correlation between record averages and breeding value. The ratio between this correlation and that of a single observation h is given by

$$\frac{r_{AM}}{h} = \left[\frac{n}{1 + (n - 1)t} \right]^{1/2}$$

It does not depend on the heritability but only on t, and on n of course. The correlation is given in Table 10.1, as to the ratio r_{AM}/h. The correlations increase with n, but not nearly as rapidly as the regressions. However, the ranking is the same, i.e., repeated observations increase accuracy most when t is low.

The increase in accuracy by repetition of records is partly neutralized by lengthened time intervals. The quantity ir_{AM}/t must be maximized and whether repetition of records leads to an increase in the rate of genetic advance depends on the relation between the three factors. However, repetition leads not only to increased accuracy, but frequently it also permits an increased intensity of selection. Therefore, repeated records are frequently justified for traits of low repeatability.

Even though repetition leads to higher accuracy mainly for traits with low t, it should not be overlooked that single records of highly repeatable and higher heritable traits are frequently more accurate than averages for traits with low t and h. For example, the r_{AM} of an average of four records of trait I equals 0.51, but single records of traits II and III have r_{AM} of 0.50 and 0.71, respectively. The application of the above formula is valid only if the repeated records refer to the same genotype, in other words, where the genetic correlation between repeated records equals 1.

When performance is repeated in successive periods the genetic correlation between records in various periods is frequently less than one, a manifestation of the fact that different sets of genes are active in the different performance periods.

If heritabilities of records in and genetic correlations among different periods are equal, the covariance between breeding value of n performances and the average of m observations $(m < n)$ equals

$$\frac{1 + (n - 1)r_G}{n} h^2$$

The regression coefficient of breeding value on the average is

$$b = \frac{m[1 + (n - 1)r_G]}{n[1 + (m - 1)t]} h^2$$

and $r_{AM} = \sqrt{b}$. Correlations are given in Table 10.1 which result when the breeding value comprising 4 $(= n)$ performances is to be estimated and either $r_G = 0.8$ among all performances or $r_G = 0.8$ between the first performance and later ones and $r_G = 1$ among these. When the selection goal includes many performances with identical genetic correlations, the limiting value of the correlation r_{AM} equals $r_G/t^{1/2}$ and if $r_G = 1$, $r_{AM} = 1/t^{1/2}$. The formula indicates that repetition of records will be most profitable when repeatability is small and r_G is high.

Genetic correlations among repeated manifestations of a trait may be unequal, for example, $r_{1j(j>1)} < 1$, $r_{ij(i,j \neq 1)} = 1$, which could pertain to yield traits of dairy cattle. If the breeding value is to be estimated for n performances and m records have been measured, the covariance is

$$h^2[(mn - m - n + 2) + (n + m - 2)r_G]/mn$$

The regression coefficient of breeding value on the average of m records is

$$b = \frac{m \text{ cov}}{1 + (m - 1)t}$$

and the corresponding correlation is

$$h \frac{(mn - m - n + 2) + (m + n - 2)r_G}{\{m [1 + (m - 1)(m - s)]t[n^2 - 2(n - 1)(1 - r_G)]\}^{1/2}}$$

TABLE 10.2
Apparent Improvement by Selection of n Best Out of k Records of a Cow[a]

Number n of reported best records	Number of lactations			
	$k = 2$	$k = 3$	$k = 4$	$k = 5$
1	350	530	640	720
2		270	420	520
3			220	340

[a]Residual standard deviation 360 kg.

The formulas are illustrated in Table 10.1 with values which could pertain to dairy cattle breeding where the genetic correlation between first and later lactation yields is incomplete. It is not surprising that the gain of accuracy when records are repeated is greater in such a situation than in the previous examples, as later records receive more weight in the breeding value to be estimated. Of course, the use of averages requires age correction of records if these are age dependent.

Occasionally only the best records of an animal are reported—an example is race horses, but auction catalogs of other animals offer another. Animals with more records have a greater chance of yielding an outstanding record than animals with few. In one of the many performance periods conditions may be just right for a record performance. In Table 10.2 apparent improvements of averages are given which arise from deletion of the worst records. The examples pertain to dairy performance and the standard deviation among records of a cow is assumed to be 360 kg. The values illustrate the bias accruing from such a practice. Unbiased comparison of selected top records is possible without correction only when all animals had the same number of records to begin with.

10.2. Pedigree Evaluation

An evaluation of pedigrees has long been used by breeders to judge the breeding value of their animals. The herd-book system was first introduced to provide information on ancestors. The frequently encountered opinion of breeders that an animal can only inherit that which its ancestors possessed and the assumed corollary that an accurate pedigree yields complete information on an animal's genotype are only partly true. The genotypes of the ancestors are frequently unknown. Exceptions can be provided by completely heritable characters, such as coat color in Shorthorns or blood antigens in cattle, and also by animals with complete progeny tests. Furthermore, even if the genotype is known with nearly complete certainty, it remains impossible to predict which allele a heterozygous individual will transmit to the progeny. An offspring receives a random sample of the parental genes and chance decides which of the two genes at each locus will be transmitted. Consequently, the correlation between the genotypes of parent and offspring is 0.5 and the correlation between the average of the parental genotypes and the common offspring's genotype is 0.71. In other words, the variance among full-sibs is only 50% less than that among unrelated animals. This assumes completely heritable characters. Actually, the difference in variance is even less when characters have incomplete heritability, as do most traits of economic significance.

The weight conceded to a relative in breeding value estimation depends on three considerations: the accuracy of the estimation of the performance potential of the related animal, the fraction of genes that the relative shares with the candidate, and the pedigree position of the relative in comparison to other relatives also providing information. When the breeding value of a parent is well known, data on a grandparent can contribute very little information.

In his studies on inheritance of human size Galton observed that information contributed by grandparents was half as great as parental information. Proceeding from this he and in particular Pearson formulated the "Law of Ancestral Inheritance," which neglects Mendelian laws. Fisher's and Wright's development of quantitative genetics permits correct weighting of relatives either by multiple regression or path coefficients.

Czekanowski had shown already in 1922 that Mendelian inheritance (more specifically, intermediate inheritance) leads to Galton's heritability coefficient (under random mating equal to Wright's coefficients of relationship between offspring and ancestor). In 1933, in response to a query by a Polish breeder, Czekanowski (1933) developed the principles of estimating breeding values from ancestor information. The Mendelian segregation involves a random process at each step from one generation to the next, which implies loss of information. The genic variance among individuals, explainable by complete knowledge of all ancestors in a generation, is halved at each adjacent and receding generation:

$$\text{parents (2)} \quad \tfrac{1}{2}V_A$$

$$\text{grandparents (4)} \quad \tfrac{1}{4}V_A$$

$$\text{great grandparents (8)} \quad \tfrac{1}{8}V_A$$

The residual genic variance is $\tfrac{7}{8}V_A$ when all eight great-grandparents are known and the residual genic standard deviation is $\sigma_A\sqrt{7/8} = 0.94\sigma_A$, for all practical purposes indistinguishable from the original variability. This makes it evident that remote ancestors scarcely contribute information. Therefore, "blood lines," if these comprise descendants of famous remote founder animals, have little or practically no significance as aids to selection.

Searle (1963) has derived the correlation between breeding value and information from all possible ancestors as

$$r = [h^2 + 1/2 - (h^2 + 1/2 - 2h^4)^{1/2}]2h^2$$

Of course this is only of theoretical interest, but it does indicate the limit of accuracy attainable by pedigree selection. Searle made it evident that remote ancestors barely contribute any information to this correlation. When heritability is very low ($h^2 < 0.05$) pedigree information can be more reliable than performance of an individual. The cause of this somewhat surprising result is the reduction of nongenic variability by the averaging of the many ancestor's records. This mitigates the large random element in Mendelian processes. However, this has little significance in practical breeding since the overall accuracy remains very low.

Czekanowski has shown that as a result of the Mendelian process the connection between offspring and ancestor can best be described by a linear model and can be analyzed by multiple regression. Standardized regression equations are given as

$$
\begin{array}{cccc}
1 & r_{01} & r_{02} & \cdots \\
r_{10} & 1 & r_{12} & \cdots \\
r_{20} & r_{21} & 1 & \cdots \\
\cdot & \cdot & \cdot & \\
\cdot & \cdot & \cdot & \\
\cdot & \cdot & \cdot & \\
\end{array}
$$

The first column and the first row concern the quantity to be estimated, the second row and column, the information 1. The r_{ij} are correlations; the standard partial regression coefficients to be estimated are denoted by β_i. The solution is

$$\beta_1 = d_{10}/d_{00}$$

where d_{00} is the determinant of the matrix when the first column and row are deleted, and d_{10} the determinant when the second row and the first column are left out. If in the above system of linear equations information $1(x_1)$ is provided by a parent and information $2(x_2)$ by a grandparent, the correlations are

$$r_{01} = r_{21} = h^2/2, \qquad r_{02} = h^2/4$$

and the solutions are

$$\beta_1 = \frac{\beta_{10} - \beta_{12}\beta_{20}}{1 - \beta_{12}^2}, \qquad \beta_2 = \frac{\beta_{20} - \beta_{10}\beta_{12}}{1 - \beta_{12}^2}$$

Complete heritability ($h^2 = 1$) results in $\beta_1 = 1/2$, $\beta_2 = 0$, i.e., all the information is provided by parents. When heritability is incomplete, generations more distant than parents can contribute information, the more, the smaller is the heritability. From Fig. 10.1 it can be seen that at higher heritabilities full pedigree information (Ped) barely surpasses information contributed by parents and grandparents. Table 10.3 shows that the maternal granddam contributes relatively less information when the heritability is higher (averages of three records) than when it is moderate (single records). When the dam is only inaccurately known, the maternal granddam may contribute information about her. When no dam information is available, the granddam information is half as accurate as dam information.

Ancestor information has limited accuracy. It nevertheless permits early selection and the cost is only that of compiling the pedigree. Selection for the same goal over many generations causes the individuals in a pedigree line to be strongly selected and finally it leaves very little variation for selection to operate or.. Such a tendency can be discerned in Table 5.1, where performance in the third lactations is considerably less variable than in single records. More repeatable traits could display such a tendency even more clearly and it can easily be visualized that in such a situation little room is left for pedigree selection. In contrast, traits newly recorded should leave much freedom for selection. Milk fat

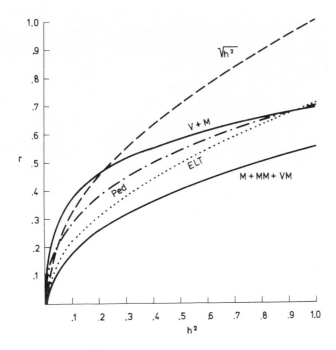

FIGURE 10.1. Accuracy of pedigree appraisal at different heritabilities. Correlation of breeding value r with own performance $\sqrt{h^2}$, full pedigree Ped, parental performance ELT, fully progeny-tested sire and performance of dam V + M, performance of dam and both granddams M + MM + VM.

and milk protein content provide examples with comparatively little variation being left among pedigrees in the case of milk fat content, while milk protein has been barely considered and therefore much variability should be extant (however, the innate variability of milk protein is much less than that of milk fat).

Pedigree evaluation is generally impossible for traits whose measurement demands slaughtering the animals. An exception to this occurs if the ancestors are slaughtered before selection is made among their offspring. Using deep-frozen semen may provide another kind of exception by making it feasible to slaughter a bull before his semen is used extensively for reproduction. Pedigree evaluation frequently is the only means possible for estimating the breeding value for a sex-limited trait in males. Furthermore, for such traits as longevity it is very nearly the only available criterion.

Environmental similarities between generations can exist. Such similarities are expected particularly on the maternal side. In populations

TABLE 10.3
Evaluation of Ancestors' Performance: Regression Coefficients and Efficiency of Various Combinations[a]

Information	k_X	k_M	R	Information	k_M	k_{MM}	k_{PM}	R
X1	0.25	—	0.50	PM1	—	—	0.06	0.13
X3	0.42	—	0.65	PM3	—	—	0.10	0.16
X1, M1	0.24	0.10	0.53	M1, MM1, PM	0.13	0.05	0.06	0.30
X3, M3	0.39	0.13	0.67	M3, MM3, PM3	0.18	0.06	0.10	0.35
M1	—	0.13	0.25	PM1, MM1	—	0.06	0.06	0.18
M3	—	0.21	0.32	PM3, MM3	—	0.10	0.10	0.23
				All great-granddams				
				One record each				0.13
				Three records each				0.16

[a] X, Candidate. M, Dam. MM, Maternal granddam. PM, Paternal granddam. Numeral following symbol, number of records. k_X, Partial regression coefficient of breeding value on candidates performance. R, Multiple correlation coefficient between breeding value and index. $h^2 = 0.25$. $r = 0.40$.

of large domestic animals the dam and granddam frequently live in the same herd. Above average care and feeding could elicit above average performance from both, and could thus simulate superior genotypes. Such environmental influences should be taken into account in the evaluation of a pedigree.

Pedigree evaluation can aid in the recognition of superior performance brought about by an epistatic gene combination. An animal of superior performance with ancestors of only moderate performance quite possibly has combinations of nonadditively acting genes causing this superiority and these combinations largely disintegrate at meiosis. Therefore, notwithstanding the superior performance, the individual may have only a few genes with good additive effects and thus its actual breeding value may be far less than the superior performance would suggest. The same analysis could be used to argue that an animal with moderate performance but an outstanding pedigree might have a breeding value far higher than its own performance would indicate.

10.3. Progeny Testing

Progeny testing has been proposed increasingly in recent decades as a method for estimating the breeding value exactly. Not infrequently, breeders think of it as selection among "genotypes" in contrast to mass selection, which is among phenotypes. However, selection based upon progeny testing also differentiates between groups of phenotypes, though not of the breeding animals themselves, but of their offspring. Therefore, in progeny testing errors caused by the environment, genetic chance deviations, and so on also can influence the results. The term progeny testing sometimes is used to mean that selection is not among the tested animals but among their progeny, but this is not progeny testing as understood here. Progeny testing means taking the performance of offspring as the criterion for selection among the parents.

An individual's offspring and his parent are equally far removed from him, and the reliability of estimating his breeding value from either a parent's or an offspring's phenotype is half as accurate as estimating it from his own phenotype. But when only the individual's own phenotype and those of his two parents are used in estimating his breeding value, environment and chance may cause greater deviations of the estimate from the true breeding value than they do when data on many offspring are used. Random environmental influences and chance deviations tend to balance out in an offspring average. Consequently, the offspring average provides in increasingly reliable measure of the breeding value as the number of progeny increases.

The covariance between the breeding value of a parent and his offspring's phenotype equals $V_A/2$. If the common genes are the only source of similarity among half-sibs, the variance among progeny groups consisting of n half-sibs equals $V_A/4 + V_R/n$. From these quantities the regression coefficient of the parent's breeding value on the progeny average is derived:

$$B_{BV.PA} = \frac{nh^2/2}{1 + (n-1)h^2/4} = \frac{2nh^2}{4 + (n-1)h^2} = \frac{2n}{n+k}$$

with

$$k = \frac{4 - h^2}{h^2} = \frac{V_R}{V_S}$$

where V_R is the residual variance within sire progeny groups and V_S is the variance between sire progeny groups.

The correlation is

$$\frac{h}{2}\left[\frac{n}{1 + (n-1)h^2/4}\right]^{1/2} = \left(\frac{n}{n+k}\right)^{1/2}$$

The regression approaches the value of 2 with large n. When n is very large, differences between progeny averages will equal half of the differences between the breeding values of any two parent individuals compared. This is because only half of the progeny genes come from the test parent. The other parent's contribution is assumed to equal the population average. The correlation between the progeny average and breeding value approaches unity with increasing n and at unity the breeding value of the parent is completely known. Figure 10.2 illustrates the increase in accuracy of the progeny test with increasing number of progeny. If heritability is high, few offspring permit a fairly accurate estimate, and a further increase in progeny numbers increases accuracy relatively little. However, for traits with low heritability the accuracy of an estimate based on few offspring is low, and increasing numbers improves it. Both cases show diminishing increases in accuracy as n increases. Correlations

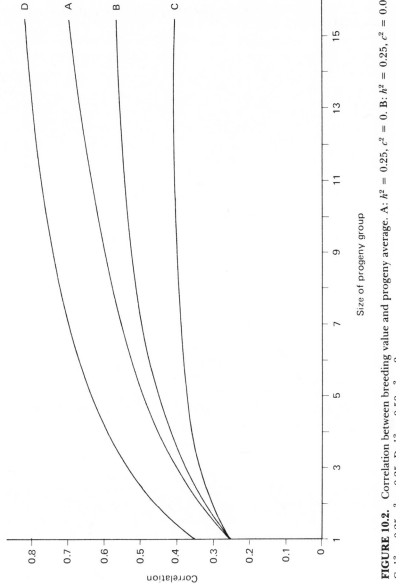

FIGURE 10.2. Correlation between breeding value and progeny average. A: $h^2 = 0.25$, $c^2 = 0$. B: $h^2 = 0.25$, $c^2 = 0.06$. C: $h^2 = 0.25$, $c^2 = 0.25$. D: $h^2 = 0.50$, $c^2 = 0$.

permit comparison of the accuracy of progeny testing and of individual performance testing:

$$h = \left(\frac{n}{n + k}\right)^{1/2}, \qquad n = \frac{4 - h^2}{1 - h^2}$$

For heritabilities of 1/4 and 1/2, respectively, $k = 15$ and 7, and five and seven progeny, respectively, are needed to achieve accuracy equal to individual performance. Consequently, for a trait such as milk yield ($r = h = 0.5$) it will very rarely be possible to progeny-test a cow with an accuracy equaling that of her own performance as a predictor of her breeding value. As heritability increases, even more progeny are necessary to achieve accuracy equal to the phenotype.

The increased accuracy of breeding-value estimates with increasing numbers of progeny shown in Fig. 10.2 is valid only when all the similarity among the progeny is caused by the genes transmitted by the common parent. When other causes for similarity (frequently called C effects; Lerner, 1950) are acting and the correlation produced by them equals c^2, the regression equals

$$b_{\text{BV.PA}} = \frac{2nh^2}{4 + (n - 1)(h^2 + 4c^2)}$$

Curve B of Fig. 10.2 was constructed under the supposition that C effects cause correlations of importance equal to that among the additive values of a half-sib progeny, that is, $c^2 = h^2/4$. This situation exists, for example, when all progeny are full-sibs and maternal effects are unimportant. Environmental influences of this order can be caused by seasonal differences in lactation milk yield or, in pigs, by persistent carryover effects of breeding-herd environment influencing performance in central test stations. Even seemingly small effects can considerably diminish the increase in accuracy resulting from increasing the size of the progeny group. A C effect of the size just mentioned holds the correlation between progeny average and breeding value at 0.71 as an upper limit for very large progeny groups. In the example depicted by curve C, breeding value and C effects were assumed to be equally important, that is, $c^2 = h^2$. The curve is still flatter than curve B and the correlation does not exceed 0.45 as an upper limit. C effects of this scale can occur when, for example, milk yields of cows from different herds are compared without considering herd effects. In general, when C effects are present, the limiting value (that is, $n = \infty$) of the correlations between progeny average and breeding value is

$$\left(\frac{h^2}{h^2 + 4c^2} \right)^{1/2}$$

So far we have assumed only one offspring per dam, as is general in cattle. In swine and poultry, as a rule, there are several full-sib groups per sire. The variance component between full-sib groups within sire progenies has the composition $h^2/4 + c^2$. The regression coefficient of breeding value on average of progeny groups consisting of l full-sib families of k individuals each is

$$b_{\mathrm{BV.PA}} = \frac{2klh^2}{4 + h^2 (kl + k - 2) + 4c^2 (k - 1)}$$

The total progeny number per sire is kl and is comparable to n of the formula given on p. 212. Even if c^2 is zero, this regression will be smaller than the one derived assuming a single offspring per mate. An increase either in k or in l, with the other quantity constant, increases $b_{\mathrm{BV.PA}}$ since the total number kl becomes larger. Figure 10.3 illustrates the change of the regression coefficients with increasing number of offspring within full-sib families. Obviously, for a given total number, the accuracy is highest with the smallest number of full-sibs per family and many families per sire. For example, eight half-sibs give almost as much information as do three families of four full-sibs each. These relations are pertinent when progeny testing of sires only is considered. Naturally, when information is used concomitantly to judge the dams the situation is different. Particularly at low heritabilities, larger full-sib families give more accurate information on the breeding value of the dam and can balance the loss in accuracy for sires caused by the less than optimal distribution of families within sires.

Progeny testing can be completely accurate. Naturally this presupposes that all assumptions are justified—random mating, and no residual correlations among offspring.

Progeny testing is particularly valuable for traits of low heritability, sex-limited traits, and slaughter traits. For practical use of the progeny test the reproductive rate must be high enough to ensure that the required number of progeny will be produced within a fairly short period.

It must be emphasized that considerable residual variability may remain between progeny-tested animals. The estimated breeding values

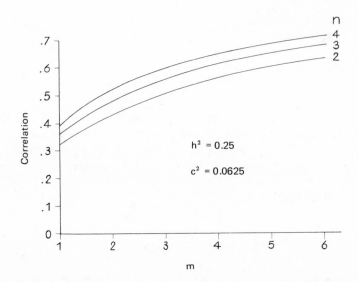

FIGURE 10.3. Correlation between breeding value and progeny average comprising a number of full-sib groups. m, Number of full-sib groups per sire. n, Size of full-sib groups.

give the expected genotype. If estimates have a correlation of r with the true value, the residual standard deviation among animals with equal estimates will be $\sigma_G(1 - r^2)^{1/2}$. Let us assume $h^2 = 1/4$ and $n = 20, 50, 100$ progeny. The residual standard deviations are, respectively, 65, 53, 36% of the original value. If σ_G of milk yield is taken as 300 kg, two-thirds of unselected bulls will have a breeding value of population mean ± 300 kg. The true breeding values of bulls progeny-tested with 50 daughters will vary around their estimate with a standard error of 0.53 $\times$ 300 = 159 kg of the estimate. If the estimated breeding value is $+ 159$ kg, one-sixth of such bulls will be below the population mean.

10.3.1. Methods of Progeny Testing

In progeny testing, problems arise due to residual correlations between offspring, due to preferential or otherwise nonrandom mating, and for other reasons. In the U.S. and in Europe daughter–dam comparisons were popular at first. The sire index was derived from the expected production of progeny (P) when dam (D) and sire (S) genotypes are given:

$$P = \frac{S + D}{2}$$

From this it follows that

$$S = 2P - D$$

This requires, of course, that every progeny has a dam with a record, and other assumptions, such as equal environmental trends in different progeny groups.

The advent of AI made the requirement for more accurate testing acute. Robertson and Rendel (1950) developed, as an approximation of the theoretically desirable least squares estimation, the contemporary comparison method, which basically consists of the sum of differences d between records of daughters of a sire and records of contemporary herd mates, i.e., cows of the same age and calving in the same year, in the case of dairy cattle. The varying reliability of these differences due to varying numbers in individual herds is taken into account by the inverse of the variance of the individual differences:

$$w = \frac{n_D n_M}{n_D + n_M}$$

where n_D and n_M refer to number of daughters and number of mates (M) in the herd, respectively. Daughters of a sire i in k herds have the weight $w_i = \sum_{j=1}^{k} w_{ij}$. The contemporary comparison of sire i is

$$CC_i = \frac{\Sigma_j w_{ij} d_{ij}}{\Sigma_j w_{ij}}$$

and the breeding value is

$$\hat{g}_i = 2b CC_i, \qquad b = \frac{w_i}{w_i + k}, \qquad k = \frac{4 - h^2}{h^2}$$

This procedure is illustrated in Table 10.4 with assumed data. The method ignores genetic herd differences and differences between mating partners of sires. In small herds where herd mates are missing no comparison is possible and the information is lost. In some countries farms are grouped (on performance) into herd classes and the contemporary comparison is computed within such herd classes. In large herds and/or large herd classes the method can be refined and the comparison made within herd–year–season class.

TABLE 10.4
Progeny Testing Methods[a]

Sires		1	2	3	4
	1	2	2	5	1
		120	170	130	180
Herds					
	2	7	6	—	5
		160	170	—	200
CC		-6.8	1.5	-3.1	8.5
MCC		-3.6	0.5	-3.5	7.5
PD		-5.9	1.1	-4.7	8.5
MQ		-6.0	1.3	-3.2	7.0
BLUP		-20.4	-13.2	-4.9	4.9

$$\text{CC}_1' = \frac{w_1}{w_1 + k} \sum_k \frac{w_{1k} D_{1k}}{w_i} = -0.28 \frac{1.6(-26.25) + 4.3(-23.6)}{5.9}$$

$$= 0.28(-24.3) = -6.8$$

$$\text{MMC}_1' = ZG_1' + \frac{1}{w_1} \sum_k w_{1k}\left(\frac{1}{n_k^*} \sum_{i=1} n_{ij} ZG_i\right)$$

$$= -6.8 + \frac{1}{5.9}\frac{1}{1.6}\left[\frac{1}{8}(2 \times 1.5 + 5 \times 3.1 + 8.5)\right]$$

$$+ \frac{1}{4.3}\left[\frac{1}{11}(6 \times 1.5 + 5 \times 8.5)\right] = -3.6$$

$$\text{PD}_1 = \frac{N_1 h^2}{4 + (N_1 - 1)h^2 + \sum_k [n_{1k}(n_{1k} - 1)/N_1]4c^2} [\bar{T}_1 - H_1' + 0.1 (H_1' - P)]$$

$$= \frac{9 \times 1/4}{4 + 8 \times 1/4 + [(2 + 42)/9] \times 0.4}[151 - 173 + 0.1(173 - 160)]$$

$$= 0.283(-20.7) = -5.9$$

$$173: \frac{1}{9}\left[2 \times \frac{8}{10}(146 - 160) + 7 \times \frac{11}{13}(184 - 160)\right] + 160$$

$$= \text{weighted adjusted mate average}$$

$$160: \text{assumed breed average}$$

$$\text{BLUP}_1 = g_1 + s_{11} = -16.8 + (-3.6) = -20.4$$

[a] CC, Contemporary comparison. MCC, Modified contemporary comparison. PD, Predicted difference. MQ, Minimum square estimate [$n/(n + k$ added to sire diagonal]. BLUP, Best linear unbiased prediction. H', Adjusted mate average, $[n/(n + 2)] (H - P) + P$.

The herd-mate comparison used in some countries employs the average of all herd mates instead of only like-aged herdmates as is practiced with CC. This presupposes age correction of records. In the herd-mate comparison, the daughter average is corrected for the herd influence by subtracting the average of all age-corrected herd-mate records:

$$\text{HM} = D - b\,\frac{n}{n + a}\,(H - \overline{H})$$

The observed deviation of the herd average $H - \overline{H}$ is corrected for its limited repeatability by multiplication with the regression coefficient of true on observed herd average:

$$\frac{V_H}{V_H + V_R/n} = \frac{n}{n + a}, \qquad a = \frac{V_R}{V_H}$$

where V_H and V_R are, respectively, the variance component between herds and the within-herd variance. For dairy records, a varies between 1 and 3. The performance of an AI heifer reflects the herd influence only to a limited extent, since half of her genes derive from outside: $b = 1 - h_H^2/2$, where h_H^2 is the herd heritability, i.e., the fraction of true herd differences that is caused by genetic differences. The covariance between a record of a daughter by a foreign (to the herd) sire and the herd average is:

$$\text{cov } DH = (H_G/2 + H_E + R)(H_G + H_E) = h_H^2/2 + e_H^2$$

and the regression coefficient is

$$\frac{h_H^2/2 + e_H^2}{h_H^2 + e_H^2} = 1 - \frac{h_H^2}{2}$$

where H_G and H_E represent the genetic and the environmental herd effects, R is the total of all within-herd effects, and e_H^2 is the fraction of the herd variance that is caused by environmental differences. The herd heritability is about 10%, and a equals 2; therefore

$$\text{HM} = D - 0.95\,\frac{n}{n + 2}\,(H - \overline{H}) - \overline{P}$$

and the estimated breeding value is given by

$$\hat{g} = 2b\overline{\text{HM}}$$

This method utilizes all records of a herd, but it requires age correction. When herds are large enough, comparisons within seasons become possible. In a U.S. investigation (Bereskin and Lush, 1965) the method fell somewhat short of expectation insofar as correlations among random daughter groups of the same sire were lower than expected. In European data, in contrast, the observed correlations agreed with expectations (Pirchner, 1970).

In the U.S. the herd-mate comparison has been modified repeatedly. In an attempt to neutralize residual correlations caused by unequal daughter numbers in individual herds the following formula for estimating the breeding value was developed:

$$\text{PD} = \frac{Nh^2/4}{1 + (N - 1)h^2/4 + \Sigma[n_i(n_i - 1)/N]c^2} \, [\overline{D} - H + 0.1(H - \overline{P})]$$

where PD is the predicted difference ($= g/2$), N is the total number of daughters, n_i is the number of daughters in herd i, $H = [n/(n + a)] \, (\overline{H} - \overline{P})$, and c^2 is the residual correlation between daughters.

None of these methods takes differences between mating partners of individual sires into account. An early suggestion to this end was made by Johansson and Robertson (1952):

$$\hat{g} = 2b(O - H_Y) + \frac{1}{2}h^2(D - H_X) + h_H^2(H_y - P) + P$$

where O, D, H, and P represent, respectively, offspring, dam, herd, and population averages and the subscripts Y and X denote herd averages pertaining to daughters and to dams, respectively. Bar-Anan and Sacks (1974) have added directly the CC value of the sires of mating partners (CC_M):

$$g = b(\text{CC}_D + \text{CC}_M)$$

Dempfle (1978) has modified this method and added directly the breeding value of sires of mating partners (g_M):

$$\hat{g} = bCC_D + CC_M = b \frac{1}{w_i} \sum_k w_{ik} (Y_{ik} + Y_{i'k}) + \bar{g}_M$$

$$\bar{g}_M = \frac{1}{w_i} \sum_k w_{ik} \frac{1}{n_k} \sum_j n_M \hat{g}_M$$

The method approaches a regressed least squares. The computation is illustrated in Table 10.4.

In a seminal paper Henderson (1973) outlined problems and possible solutions of breeding value estimation. Cochran (1951) had generalized the problem of estimating breeding values. It is desirable to minimize the square of the deviation between the estimate and the true value. Cochran had shown that the best prediction is accomplished by taking the conditional mean $f(y) = f(u)|y$, where $f(y)$ is a function of y and $f(u)$ is the definition of merit. Cochran further proved that given certain conditions, truncation selection of $f(y)$ maximizes the expectation of selected $f(u)$. However, the problem is that the distribution function may be unknown or, if it is known, the parameters needed for estimating the conditional mean may not be known, or it may be too difficult to derive the conditional mean. However, if it can be assumed that the distribution is multivariate normal, the conditional mean is linear in y and only first and second moments need to be known. If these conditions hold, best prediction is identical to the selection index approach introduced by Wright and Lush to animal breeding. The breeding value is estimated from the regression equation

$$g_i = a_i + b_i x$$

Let G denote the covariance vector between breeding value and information, e.g., daughter records, and P the phenotypic variance–covariance matrix. This can be written

$$g = E(g) + G'P^{-1}[X - E(X)]$$

As already mentioned, the means as well as the variance and covariances should be known. As for the latter, they are fairly robust and change little between similar populations, so that they can be taken from the literature. However, in the case of the mean of the population $E(X)$ it is frequently assumed that it is well known, or that it can be accurately estimated, e.g., by arithmetic means of levels of factors or by regular

LSQ (least square). The assumption is also made that the population is uniform.

However, progress in selection has introduced a time trend such that animals from different years or time periods can be considered to belong to different subpopulations, as is also true of population segments that were created by various degrees of crossing with more or less related strains, e.g., Holstein-Friesians and European Friesians. Henderson (1963) modified the prediction equation by substituting for $E(X)$ its estimator derived with generalized LSQ $X\beta$, where $\beta = (X'P^{-1}X)^-(X'P^{-1}Y)$, where X is the incidence matrix of fixed effects. This is "best linear unbiased prediction."

A common problem involves the estimation of breeding values of bulls that belong to several groups and whose daughters are distributed over several herds. The statistical model to describe the situation is

$$y = \mu + g_i + s_{ij} + h_K + e_{ijKe}$$

which differs from previous models by containing the group constant g_i. The sire influence and the residual are assumed to be random with zero mean and variances σ_S^2 and σ_e^2, respectively. The quantities μ, g_i, and h_K are fixed. Therefore, it is a mixed model of fixed and random quantities. In matrix notation the model is

$$y = X\beta + Zs + e$$

y is the vector of records, β is the vector of fixed factors (herds h_K and subpopulations g_i in our case), s is the vector of nonobservable sire effects,

TABLE 10.5
Equations for Best Linear Unbiased Prediction (BLUP)[a]

										$\hat{w}$
10	0	4	6	2	2	5	1	h_1	1410	149.7
0	18	13	5	7	6	0	5	h_2	3140	185.4
4	13	17	0	9	8	0	0	g_1	2720	−16.8
6	5	0	11	0	0	5	6	g_2	1830	0
2	7	9	0	9 + 15	0	0	0	s_{11}	1360	−3.6
2	6	8	0	0	8 + 15	0	0	s_{12}	1360	3.6
5	0	0	5	0	0	5 + 15	0	s_{21}	650	−4.9
1	5	0	6	0	0	0	6 + 15	s_{22}	1180	4.9

[a] $\hat{w}$, Parameter estimates.

and e is the vector of residuals, also nonobservable. Here X is the known incidence matrix of fixed effects and Z is the known incidence matrix of sire effects. The random variables have variance–covariance matrices of G for sire influence and E for residual errors. When sires are unrelated, G is a diagonal matrix and E is an identity matrix multiplied by the residual variance.

The estimation of the β's and the prediction of s is accomplished by solution of the equations given in Table 10.5. The equations β can be absorbed in equations s:

$$b = (X'X)^{-1}(X'y - X'Zs)$$

and

$$(X'Z)(X'X)^{-1}(X'y - X'Zs) + (Z'Z + RG^{-1})s = Z'y$$

$$s = \frac{Z'X - (Z'y(X'X)^{-1} X'y)}{(Z'Z + RG^{-1}) - Z'X(X'X)^{-1}X'Z}$$

The first expression within brackets is the diagonal of the original sire matrix $[n_{ij} + (4 - h^2)/h^2]$, which is modified after absorption by subtraction of $(Z'X(X'X)^{-1}X'Z)$. If n_{ijk} denotes the number of daughters in herd k by sire ij, n_{ij} the number of progeny of sire ij, and n_h the size of herd h, then the diagonal element of this matrix is $\Sigma_k^{nh}(n_{ij}n_{ijk}/n_{.k})$. The nondiagonal element will be $-\Sigma_k^{nh}(n_{ijk}n_{ijk}/n_{.k})$. The right term of the original matrix pertaining to sire ij is $\Sigma_k^{nh}Y_{ijk} = Y_{ij}$. which is modified by subtracting $\Sigma_k^{nh}(n_{ijk}Y_{.k}/n_{.k})$, where $Y_{.k}$ is the sum of all records in herd k. The error of the difference between estimated effects of two sires ij and ij' is given by $\sigma_e^2(c^{ij} + c^{i'j'} - 2c^{ij'})$, where the c's are the elements of the inverse sire matrix.

When nondiagonal elements of the left-hand matrix are ignored, BLUP corresponds to CC. We have

$$n_{ij.} + k - \sum_k^{nh} \frac{n_{ijk}}{n_{ij}} = k + \sum_k \frac{n_{ijk}n_{ijk}}{n_{i.k}}$$

Now

$$n_{..k} = n_{ijk} + n_{*j*k}$$

i.e., the size of the herd is the sum of the number of daughters and of contemporaries; then

$$Y_{ij.} - \sum_k \frac{n_{ijk}Y_{..k}}{n_h} = \sum_k \frac{n_h Y_{ijk} - n_{ijk}Y_{..k}}{n_h}$$

$$= \sum_k \frac{n_{ijk}n_{*j*k}(\overline{Y}_{ijk} - \overline{Y}_{i*j*l})}{n_h}$$

The difference between sire's daughter's average and the average of contemporaries is weighted by

$$w = \frac{n_{ij'k}n_{i*j*k}}{n_{ijk} + n_{i*j*k}}$$

and multiplied by $1/(k + w)$ as in CC.

In the example of Table 10.5 four sires belonging to two genetic groups and with daughters in two herds are assumed. Heritability for yield is 0.25 and $(4 - h^2)/h^2 = 15$. The predicted difference of future daughters of bull ij is $\hat{g}_i + \hat{s}_{ij}$ for bull 1: $PD_1 = \hat{g}_1 + \hat{s}_{11} = -16.8 - 3.6 = -20.4$.

The result is given, together with the estimates derived from other methods, in Table 10.4. It is evident that the ranking by BLUP differs considerably from the ranking by the other methods. Sire 3 belongs to subpopulation 2, which is superior, so that he is probably better than sire 2 in spite of the inferiority of his daughter's performance relative to those of sire 2. The change in ranking is caused by the use of *a priori* information, i.e., the membership of various groups. BLUP represents a weighted mean of *a priori* information about the origin of sires and average performance of daughters. This procedure is pursued intuitively every time a breeding value or a producing ability is to be estimated, as pointed out in Chapter 5.

The method can be used for predicting various types of performance—producing ability, i.e., future records, breeding value of multiple traits, and merits of single crosses (Henderson, 1973).

Henderson was able to show that BLUP gives the highest correlation between estimates and real values of all linear unbiased predictors and, provided that the distribution is multivariate normal, BLUP maximizes the probability of correct ranking of animals. At present, the method appears to be the best practicable from a theoretical point of view. It should be advantageous in situations where the population is heterogeneous, i.e., where several subgroups and subpopulations exist. These may be different birth cohorts or different strains and crosses as in the case in European dairy cattle which experience various degrees of crossing with U.S. dairy strains. If we assume for the example in Table 10.5 that bulls 1 and 2 are European Brown cattle and bulls 3 and 4 Brown Swiss, a breeder would intuitively use this information together with the daughter average to arrive at an estimate of the breeding value of the bulls.

Investigations of the efficiency of BLUP are numerous. In general, correlations between BLUP and other estimates, mostly from herd-mate comparison or CC, are high. However, this is expected between two estimates of the same quantity, even if one of them is considered to be better. Correlations between split samples of progeny of bulls would seem to be more pertinent for judging the practical performance of the method. Dempfle and Hagger (1980) split progenies of Bavarian Brown bulls and computed correlations between the moieties. The BLUP values agreed more closely ($r = 0.8$) than estimates from a modified CC ($r = 0.6$). However, the data were rather heterogeneous, i.e., bulls fell into several groups differing in the fraction Brown Swiss genes, and progeny groups were fairly small. One would expect that the advantage of BLUP would be smaller or even nonexistent when conditions were more favorable—more homogeneous populations, and large, randomly distributed progeny groups. For example, Schneeberger et al. (1980) found little difference in the ranking of Simmental sires for beef performance between different methods of computing progeny tests where the progeny were randomly distributed over farms. Similar conclusions were drawn by Powell and Freeman (1974) from investigations of proofs of dairy sires. Van Vleck (1967) found that even nonparametric methods of ranking bulls gave the same result as (then) conventional methods. Although BLUP was not compared, it is not improbable that under such conditions—large, randomly distributed progeny groups, largely homogeneous population—the conclusion would have been similar.

10.3.2. Accuracy of Progeny Testing

The accuracy of progeny testing increases with increasing numbers and approaches one provided no c^2 effects exist. Naturally one may ask how accurately a sire should be tested. The question as such cannot be answered in a sensible way. Let testing ratio K be defined as the ratio of testing capacity (for example, number of stalls) to number of animals eventually needed.

The interdependence of accuracy and selection intensity is illustrated in Table 10.7. and in Figure 10.4 The following quantities are

TABLE 10.6
Optimal Progeny Group Size[a]

Testing ratio	Group size
50	6–11
100	10–26
200	20–40
1000	60–90

[a] Source: Robertson (1957). $h^2 = 0.01$–0.25.

assumed: $K = 200$, $h^2 = 1/4$, $k = 15$. Most progress accrues from comparatively strong selection combined with modest accuracy when the sires of the 10% best groups of ten daughters each are to be chosen. Table 10.6 shows also that optimum group sizes are comparatively modest. For a test ratio of 200, optimal group size is between 20 and 40 and for a trait with $h^2 = 1/4$ the lower figure is relevant. However, these results are not realistic, since the economic aspects are not considered and testing costs will be higher when more animals are to be tested per candidate. Nevertheless it is apparent that optimal group size is relatively small.

Brascamp (1973) considered the time dimension. The selection of sires of bulls that generates most genetic progress does so with considerable time lag since the superiority becomes manifest only in the grand-progeny. Therefore, when this time lag between selection and manifestation of the genetic superiority is taken into consideration, by discounting selection gains, for example, the contribution of the parents of cows becomes relatively more important.

TABLE 10.7
Interdependence of Accuracy and Selection Intensity[a]

b	n	$\dfrac{b}{(b + 0.075)}$	r	i	ri
0.5	100	0.87	0.93	0.80	0.75
0.4	80	0.84	0.92	1.00	0.92
0.3	60	0.80	0.89	1.16	1.04
0.1	20	0.63	0.79	1.75	1.38
0.05	10	0.40	0.63	2.06	1.29

[a] b, Fraction selected. n, Progeny group size. r, Correlation between progeny average and breeding value. i, Standardized selection differential. Testing ratio $K = 200$.

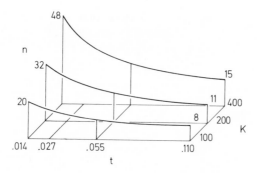

FIGURE 10.4. Optimum progeny group size with varying correlations among family members t and varying testing ratios K.

10.4. Sib and Family Testing

The progress possible with half-sib testing can be quite large and in cattle breeding it need not be much smaller than progress by progeny testing. Possible variants of the two systems are given in Table 10.8. It is assumed that a small population of 20,000 cows under test is to be improved. Thirty percent ($1 - p = 0.3$), which is 6000 cows, are available for testing 30 bulls with 40 progeny each. Six of the bulls are destined for general use and the two best of these are elite bulls. Production of 30 test bulls requires insemination of some 200 elite cows, which is 1% of the population. Such a system should yield a genetic advance of $0.176\sigma_G$ provided selection were for milk yield only.

Half-sib testing leads to 30 sire progenies, i.e., half-sib families, from which the six best ones are chosen to provide bulls for general use, five bulls from each family. Young bulls of the two top families would be used as sires of future bulls. No test insemination is necessary; therefore, the progeny group can be larger, say 100, and accuracy can be better than half of the accuracy of progeny testing ($r_{HS} = 0.47 > 0.84/2$; Table 10.8). The bulls are used for 1 year and from each family five young bulls are raised, requiring some 1000 elite cows, considerably more than with progeny testing ($i_{HS} = 2.1 < i_{PT} = 2.7$). The genetic progress per year is a little less than with progeny testing, but the costs should be reduced, too, since no long-term conservation of sperm or the laying off of bulls is necessary. Owen (1974) and Dempfle (1975) have compared progeny testing and sib testing and confirmed that only small differences exist between the two methods.

In the early stages, sib testing can even be advantageous, since progeny testing entails a longer time lag before bearing results (Owen, 1974).

TABLE 10.8
Genetic Progress under Various Testing Schemes[a]

	Progeny testing[b]						Half-sib testing					Progeny testing for elite sires only				
	b	i	r	p	irp	t	b	i	r	ir	t	b	i	r	ir	t
SS	1/15	2	0.85	—	1.70	7	1/15	2	0.47	0.94	4	1/15	2	0.85	1.70	7
	1/5	1.4	0.85	0.7	0.83	7	1/5	1.4	0.47	0.66	4	—	—	—	—	2¼
S	0	—	—	0.3	0	2¼	—	—	—	—	—	0	—	—	—	5
DS	1/100	2.7	0.5	—	1.35	5	1/20	2.1	0.5	1.05	5	1/100	2.7	0.5	1.35	5
D	9/10	0.2	0.5	—	0.10	5	9/10	0.2	0.5	0.10	5	1/10	0.2	0.5	0.10	5
Σ	—	—	—	—	3.983	22.575	—	—	—	2.75	18	—	—	—	3.15	19¼
O/G/t	—	—	—	—	0.176	—	—	—	—	0.153	—	—	—	—	0.164	—

[a] b, i, r, t, Fraction retained, standardized selection differential, accuracy, generation interval. SS, S, DS, D, bull sires, bulls for general use, bull dams, cows.

[b] p, Fraction of proven sires in AI station.

It should be pointed out that in the traditional progeny testing schemes young bulls are selected on the basis of sib performance. Skjervold and Langholz (1964) found heavy use of young bulls to be advantageous since it permits strong selection of bull sires. Taking this point further would imply that all the testing capacity was used for testing young bulls from which only bull sires are to be selected. Though similar to sib testing, in the sense that nearly the whole population is bred by young bulls, it differs in a basic way: Bull sires are selected on the basis of progeny testing, and therefore many fewer planned matings and stronger selection of bull dams are possible, with the consequence of greater genetic progress. A system such as this could be used even in a natural mating population. In such a population it can be assured that of all sires that promise reasonably accurate tests some sperm can be stored. These sperm should suffice for planned matings when progeny tests become available.

In poultry breeding the ratio of efficiencies of sib testing to progeny testing is more favorable than in cattle, since there is a greater reduction of the generation interval. When short test periods are used the generation interval with half-sib testing can be as little as 12–13 months in comparison to 20–22 months under progeny testing.

Full-sibs resemble each other in that one-half of their genes are common and one-fourth of their dominance effects are shared. Also, they share more epistatic effects than parent and offspring. When nongenic effects are unimportant, genetic progress can be greater by full-sib testing than even with progeny testing since the generation interval

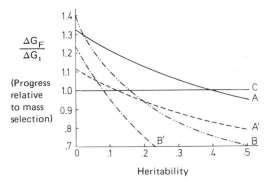

FIGURE 10.5. Efficiency of family selection ΔG_F relative to mass selection ΔG_I at various degrees of heritability. A: Full-sib family, $n = 5$. A': Candidate not tested. B: Half-sib family, $n = 25$. B': Candidate not tested. C: Mass selection.

need not be longer than for mass selection. If nongenic correlations are zero, the efficiency of full-sib testing is double that of half-sib testing at comparable family sizes.

Accuracy of estimating the breeding value is increased when the candidate itself contributes to the average. This is particularly noticeable when families are small. Figure 10.5 shows that the accuracy of estimation from family averages including the candidate's record is above the accuracy of family averages without it, and the difference is larger for small n.

11

Selection Index

A combination of information on traits of the same or of different individuals to be used as a selection criterion is called a selection index. It is a multiple regression equation with the breeding value of the candidate as dependent variable and the information on the relatives, or performance traits, as independent variables. It requires, as discussed in the previous chapter, knowledge of variances and means and it corresponds to best linear prediction (Henderson, 1973). Depending on the information available and the variable to be predicted, the following four types of indices can be discerned. Estimates are desired of (1) breeding value of single traits from records of the animal and/or relatives for the same trait, (2) breeding value of single traits from information on several traits of the animal and/or relatives, (3) breeding value of multiple traits from the performance of the animal, (4) as in (3), but information from relatives.

Alternative 3 represents the classic selection index, which was introduced to animal breeding by Hazel (1943) and to plant breeding by H. F. Smith (1936).

The breeding value A is to be estimated from information $X_1, \ldots, X_k$. The phenotypic variance–covariance matrix is denoted by V and the vector of covariances between information and breeding value g, or in the multiple trait case G, represents the genetic variance–covariance matrix. The normal equations in matrix notation are given by

$$Vb = G$$

and the solution

$$b = V^{-1}G$$

The index, or the prediction equation, is

$$\hat{g} = b_1X_1 + b_2X_2 + \cdots + b_kX_k$$

and in matrix notation

$$\hat{g} = b'X = GV^{-1}(X - \overline{X})$$

which requires knowledge of $\overline{X}$. If these are poorly known, BLUP instead of a selection index should be used and $\overline{X}$ replaced by $X\beta$, the generalized least square estimate of $\overline{X}$.

In multiple regression theory the variance of the index is the variance accounted for by regression, which is equal to the covariance between index (i.e., predictor) and value to be estimated:

$$V_I = b_1 \text{ cov } X_1A + b_2 \text{ cov } X_2A + \cdots + b_k \text{ cov } X_kA$$

$$= \Sigma \, b_k \text{ cov } X_kA = \text{cov } IA$$

In matrix notation

$$V_I = V_{bX} = b'Vb$$

$$\text{cov } IA = \text{cov}(bX')'(bX' + e) = b'Vb$$

since cov $(bX)(e) = 0$.

The regression coefficient of breeding value on index equals 1:

$$b_{AI} = \frac{\text{cov}(IA)}{V_I} = 1$$

Therefore, the concept of heritability, which implies the repeating of part of the parental superiority in the progeny, is not applicable here. This becomes obvious if a selection index is constructed from individual performance: $I = h^2(X - \overline{X})$. For example, cows with records of 1200 kg milk above average should be indexed as $\frac{1}{4} \times 1200 = 300$ if the heritability equals 1/4. The breeding value of such a cow is expected to be equal to her index. However, accuracy and reliability of an index are given by the square of the correlation:

$$R^2 = \frac{(\text{cov } IA)^2}{V_I V_A} = \frac{\text{cov } IA}{V_A}$$

since $V_I = \text{cov } IA$. Heritability, too, is a squared correlation and R^2_{IA} may be considered as such. At first sight it may be surprising that the genic variance V_A is the denominator, but the variance of the index is that part of the total variance of the breeding value that can be explained by the predictor.

11.1. Information from Different Individuals or Different Traits for Estimating Breeding Values of Single Traits

Performance values of various relatives are to be combined in a way which permits estimation of the breeding value of a candidate with maximal accuracy. The problem first became evident in evaluation of ancestor information, and, as mentioned in Chapter 10, Czekanowski (1933) used the multiple regression technique, which has been applied to such problems ever since. Searle (1963) treated the use of ancestor information in a very general manner and Leroy (1958) has published an extensive investigation of problems of pedigree evaluation.

Assume that the breeding values of a sex-limited trait such as egg laying or milk yield is to be estimated. Information is available from the dam and the two granddams as shown in Fig. 11.1, where the left half illustrates the biological relations and the right half shows the estimation problem. First we must evaluate the relative weights of the performances of the dam (M), the paternal granddam (PM), and the maternal granddam (MM), and a prediction equation (an index) can be constructed:

$$G_X = b_M M + b_{PM} PM + b_{MM} MM$$

The regression coefficients can be estimated from three simultaneous equations:

$$b_M V_M + b_{PM} \text{ cov M.PM} + b_{MM} \text{ cov M.MM} = \text{cov X.M}$$

$$b_M \text{ cov M.PM} + b_{PM} V_{PM} + b_{MM} \text{ cov PM.MM} = \text{cov X.PM}$$

$$b_M \text{ cov M.MM} + b_{PM} \text{ cov PM.MM} + b_{MM} V_{MM} = \text{cov X.MM}$$

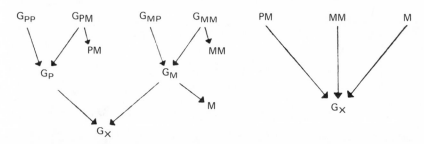

FIGURE 11.1. Pedigree evaluation. Left: Biological connections. Right: Sources of information from which a candidate's breeding value must be estimated when only phenotypes of female ancestors are known. G_{MM} and G_X represent the breeding values of maternal granddam (mother of dam) and of the candidate; PM and M represent the performances of paternal granddam and of dam.

V_M denotes the variance of the dam's performance, cov M.PM the covariance between performance of dam and of paternal granddam, and cov X.MM the covariance between the genotype of the candidate and the performance of the maternal granddam. The efficiency of such an index can be compared with other indexes or with selection based on performance of certain relatives by using the multiple correlation coefficients (right-hand column of Table 10.3).

For estimating the regression coefficients there are three equations with three unknowns. Their solution is particularly simple because cov M.PM = cov MM.PM = 0, as is evident from Fig. 11.1A. The other quantities assume the following values when expressed in standard units:

$$V_M = V_{PM} = V_{MM} = 1$$

$$\text{cov M.MM} = h^2/2, \qquad \text{cov X.M} = h^2/2, \qquad \text{cov X.MM} = h^2/4$$

Using these values, one finds the relative weight of the paternal granddam's record to be $b_{PM} = h^2/4$. The other two regression coefficients must be computed from the two remaining equations:

$$b_M + b_{MM}h^2/2 = h^2/2$$

$$h^2/2b_M + b_{MM} = h^2/4$$

which yield

$$b_{/MM} = h^2 \frac{1 - h^2}{4 - h^4}$$

$$b_M = \frac{h^2}{2}\left(1 - h^2\frac{1 - h^2}{4 - h^4}\right)$$

The coefficient for the maternal granddam is smaller than b_{PM}, and b_M is less than $h^2/2$, that is, $h^2/2$ equals the regression coefficient of offspring's breeding value on the dam's performance when the dam's record is the only information on the maternal side of the pedigree.

Table 10.3 gives the relative weights when heritability is either 0.25 or 0.42. The table also contains the partial regression coefficients for other combinations of the animal's own and his ancestor's performance. Heritability was assumed to be 0.25 and repeatability 0.40, as would pertain to milk yield of cattle. Therefore, the heritability of an average of three records is 0.42. The values of the regression coefficients in the various combinations illustrate the generalities of relationship and predictive value. For example, increasing the reliability of a dam's performance record decreases the importance of that of the maternal granddam as a predictor of the offspring's genotype. If both dam and granddam have one record ($h^2 = 0.25$), b_{MM} is weighted about 38% as much as b_M. If both dam and granddam have three records each ($h^2 = 0.42$), b_{MM} decreases to 33% of b_M. The correlation between an individual's genotype and a relatively complete pedigree—three records each of dam and both granddams—is considerably lower than that between the individual's breeding value and his own performance. Estimating the genotype solely from grandparental performance is rather inaccurate and is even more so if only information on great-grandparents is available. When an animal has its own performance records, accuracy is increased very little by considering the pedigree. The gain in accuracy from considering ancestor's performance decreases as the heritability increases. It becomes effectively zero when heritability is complete ($h^2 = 1$), since the genotype is known completely, making the addition of any information an impossibility. Breeding progress can actually be retarded if pedigrees are overemphasized, particularly when distant ancestors are overvalued. The examples in Table 10.3 show the relatively small importance of grandparents for estimating the breeding value. Obviously the more remote the ancestor, the less weight should be given its records.

Other examples in applying index theory are provided by various cow indexes, among others. In the case of cow indexes, various pieces of information must be combined. For example, one cow may have one or two records of her own and in addition an average of some 20 half-sisters, while another cow may have five records and an average of 50 half-sisters. Of course, computers can handle such situations easily, but with comparatively much computing time. Heidhues *et al.* (1961) listed regression coefficients for frequently occurring combinations of information. Likewise Flock *et al.* (1971) have published tables which permit the estimation of breeding value for a variety of traits with different heritabilities, repeatabilities, and numbers and kinds of records, i.e., from relatives of various degrees and numbers. Most of the published regression coefficients were computed on the assumption that genetic correlations between repeated records of one animal are one. However, at least for milk yield and swine litter size, this is not so, nor may it be the case for other kinds of repeatable performance. Of course this situation can be taken into consideration when setting up an estimation program.

Dempfle (1977) has used matrix methods to simplify estimation procedures where comparatively few combinations of individual records on one side are to be put together with potentially very many half-sib averages of varying reliability. Table 11.1 illustrates the procedure. The

TABLE 11.1

Combining an Individual's Performance and Half-Sib Average for Estimating Breeding Values[a]

$$
\left[\frac{A \;\;|\;\; B}{B' \;\;|\;\; C} \right] [b] = Ga
$$

$$
\begin{bmatrix} 36 & 0 & 2.25 \\ 0 & 49 & 1.80 \\ 2.25 & 1.80 & 2.925 \end{bmatrix}
\begin{bmatrix} b_1 \\ b_2 \\ b_{HS} \end{bmatrix}
=
\begin{bmatrix} 9 & 7.2 \\ 7.2 & 9 \\ 2.25 & 1.8 \end{bmatrix}
\begin{bmatrix} 1/4 \\ 3/4 \end{bmatrix}
=
\begin{bmatrix} 7.65 \\ 8.55 \\ 1.9125 \end{bmatrix}
$$

$$
b = \begin{bmatrix} b^* \\ 0 \end{bmatrix} + \frac{n}{\sigma_P^2 - \frac{1}{4}\sigma_A^2 + n\,(\frac{1}{4}\sigma_A^2 - B'A^{-1}B)} \begin{bmatrix} A^{-1}B'B'A^{-1} & -A^{-1}B \\ -B'A^{-1} & a \end{bmatrix} Ga
$$

$$
= \begin{bmatrix} 0.25 \\ 0.174 \\ 0 \end{bmatrix} + \frac{n}{33.75 + n(2.25 - 0.2067)} \begin{bmatrix} -0.0696 \\ -0.0412 \\ 1.1203 \end{bmatrix}
$$

$$
b_{n=50} = \begin{bmatrix} 0.25 \\ 0.174 \\ 0 \end{bmatrix} + 0.3679 \begin{bmatrix} -0.0696 \\ -0.0412 \\ 1.1203 \end{bmatrix} = \begin{matrix} 0.224 \\ 0.159 \\ 0.412 \end{matrix}
$$

[a] *n*, Number of half-sibs.

phenotypic variance–covariance matrix is composed of four parts: A represents the variance–covariance matrix of individual records, C the matrix of variances of half-sib averages, and B, in the case of only first lactations, reduces to the covariance matrix between the individual performance and half-sib average. The genetic variance–covariance matrix is at the right-hand side of the system of equations and the vector a gives the values attributed to individual lactations. In the example shown these values are 1 for the first lactation genotype and 3 for the later lactation genotype and the genic correlation between the two was taken as 0.8. This approach permits storage of partial regression coefficients for a limited number of individual records (e.g., 16 when only five consecutive records plus no record are considered), which are modified by the inclusion of half-sib information. However, it must be mentioned that the gain in information from including half-sib performance is relatively modest, e.g., it increases the correlation from 0.64 when only two records are used to 0.66 when in addition the average of 50 half-sisters is included in the estimation, an increase of some 5% in accuracy. Such matrix operations should be useful in particular in those situations where relatively small numbers of combinations, e.g., of an individual's records or full-sib records, are to be used with a much larger possible number of other combinations, e.g., progeny averages of different numbers of offspring.

The selection index can be applied for combining individual performance and family average (Lush, 1947):

$$\text{BV} = h^2 \frac{1-r}{1-t} P + h^2 \frac{r-t}{1-t} \frac{n}{1+(n-1)t} \text{FA} + \overline{X}$$

where P is the candidate's performance and FA is the family average, both expressed as deviations from the population average $\overline{X}$. The first quantity in this expression, $h^2 (1 - r)/(1 - t)$, corresponds to the heritability of phenotypic differences within families, and the coefficient of the family average represents the partial regression coefficient of breeding value on the family average containing the candidate's performance. The ratio of the correlation between this criterion and the breeding value to the correlation between individual performance and breeding value is

$$r_{IG} = \left[1 + \frac{(r-t)^2}{1-t} \frac{n-1}{1+(n-1)t} \right]^{1/2}$$

which is always larger than one [except when t is a large enough negative number to make $(n - 1)t$ less than one]. The size of this quantity depends mainly on the difference $r - t$. The larger this difference, regardless of sign, the larger the gain from using the index. The family average always provides additional information on the genetic value of the animal. If $r - t$ is positive, an animal from a superior family receives additional credit for this beyond what it gets for its own phenotype. If $r - t$ is negative, the situation is reversed and individual performance is corrected for the family average. When family similarity is caused predominantly by the environment, belonging to a good family thus indicates that the animal has experienced an above average environment. For example, a dairy herd can be considered a family in this context. With natural service, r is about 0.1 and t is in the range from 0.2 to 0.3. A high herd average implies that the environment was good, and to estimate the breeding value a certain amount has to be deducted from the performance of the cow. Assume the following parameters: $h^2 = 0.25$, $t = 0.3$, $r = 0.1$, $n = 30$. The intraherd heritability then is 0.32 and the regression coefficient of breeding value on herd average is 0.22. The estimate of a cow's breeding value will be increased by 32 kg for each 100 kg that her performance is above the population average, and it will be decreased by 22 kg for each 100 kg that the average of her particular herd exceeds that of the population.

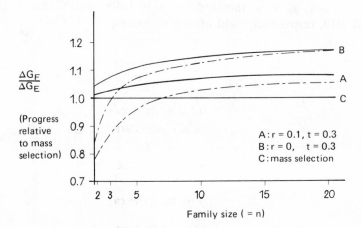

FIGURE 11.2.   Efficiency of within-family selection (broken lines) and of index selection (solid lines). A, full-sib family with n = 5, A', full-sib family, candidate not included, B, half-sib family with n = 25, B', half-sib family, candidate not included, and C, mass selection. ΔG_F represents the genetic progress under family; ΔG_E represents selection relative to mass selection.

Figure 11.2 also demonstrates that selection using an index combining individual performance and family average will always be superior to both individual and family selection. However, in many situations this advantage will be small. Since it requires individual recording, the possible additional cost over that of family selection may not justify the gain in efficiency. Selection on such an index will be justified, however, if individual records are collected anyway and thus no additional costs are incurred.

An interesting approximation of a selection index which weighs information by "progeny equivalents" instead of regression coefficients was proposed by Robertson (1959b) and developed further by Mostageer (1969). The information is converted to progeny equivalents, i.e., the number of progeny that contribute the same quantity of information. In the case of individual performance

$$h^2 = r^2 = \frac{n}{n + k} \quad , \quad k = \frac{4 - h^2}{h^2}$$

and the number of progeny equivalents is given by

$$n_P = \frac{4 - h^2}{1 - h^2}$$

When $h^2 = 1/4$, $n_P = 5$. Individual records with heritabilities of 1/4, 1/2, and 1/10, respectively, add information equal to that of 5, 7, and 4.3 offspring, which are the progeny equivalents. For m half-sibs

$$r^2 = \frac{\frac{1}{4}m}{m + k} = \frac{n}{n + k} \quad \text{and} \quad n = \frac{km}{3m + 4k}$$

which reaches a maximum of $n_P = k/3$. When $h^2 = 1/4$ the maximal possible quantity of information contributed by half-sibs is equal to that of five progeny, and 50 half-sibs contribute 3.57 progeny equivalents. When $m = \infty$, the sire is completely known. For traits with $h^2 = 1$, $n_P = 1$. Since he is related to the same degree as one progeny, the same information is provided. The information contributed by various relatives is weighted by the progeny equivalents. When passing from one generation to the previous one, the deviation is doubled, as is the case when the breeding value is to be estimated from the progeny performance. In Table 11.2 the performance of dam B corresponds to one progeny equivalent. Her deviation is doubled when computing the breeding value of daughter C. The progeny deviation of sire A is quad-

TABLE 11.2
Estimation of Breeding Value by Progeny Equivalents w

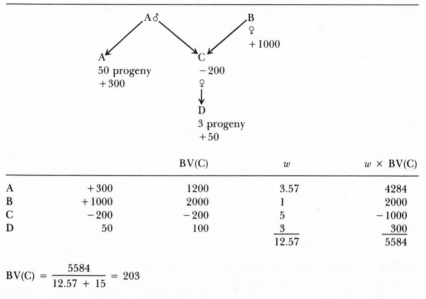

	BV(C)	w	$w \times$ BV(C)
A	+300	1200	3.57	4284
B	+1000	2000	1	2000
C	−200	−200	5	−1000
D	50	100	3	300
			12.57	5584

$$BV(C) = \frac{5584}{12.57 + 15} = 203$$

rupled, once doubled when going from the progeny to sire A and a second time when passing from sire A to C. The breeding value is the weighted sum divided by $(n_P + k)$, similarly to the estimation from progeny averages, $b = 2n/(n + k)$. The method is an approximation where covariances between sources of information are neglected. Apart from its heuristic merit, it should be useful where breeders lack access to computer facilities.

11.2. Estimation of Aggregate Genotype

As already pointed out, the selection index in the classic sense is the optimal method for considering multiple traits in selection. Hazel (1943) has expressed the aggregate genotype as a linear combination of individual genotypes weighted by their respective economic values. In matrix notation the estimation equations are given by

$$Vb = Ga$$

which are similar to the equations given above except that here G denotes
a variance–covariance matrix instead of a vector of covariances, and **a**
represents the vector of economic weights as used in Table 11.1.

The economic value of a trait is obviously important for deciding
how much consideration the trait deserves in selection. When traits are
independent of each other and have approximately equal heritability (or
when their heritabilities are unknown) the best weights are their relative
economic values.

Economic value is defined as the increment in profit occurring from
increasing the particular trait by one unit, independent of other traits.
Therefore, the additive genotype of the total economic value is the sum
of the breeding values of the single traits weighted by their respective
economic values. When the individual breeding values are denoted by
G_1, G_2, etc., the additive genotype of the aggregate economic values
corresponds to $a_1G_1 + a_2G_2 + \cdots + a_nG_n$.

The next quantity deserving attention is the heritability of a trait.
It is intuitively obvious to breeders that, other things being equal, traits
with higher heritability deserve more attention simply because they re-
spond better to selection than traits with low heritability. When characters are independent or when correlations are unknown, the appropriate selection index is the sum of the product of heritability and economic
weights of the individual traits.

Since most traits are at least partially correlated, genetic change in
one trait causes changes in others. Therefore, genetic as well as phenotypic correlations must be considered when an index is constructed.
Naturally it is desired in artificial selection to improve all useful characters, but if improvement in A causes deterioration in B, and if B is
more important than A, the breeder may have to forego improvement
in A. When animals are below average in A this indicates that their
breeding value is above average in B, which may be added to the information already given by their performance in B.

We shall demonstrate the construction of an index with an example
from dairy cattle breeding. Assume that milk yield and fat content completely determine economic value. The relevant parameters are shown
in Table 11.3. It is assumed that a 100-kg increase in yield of milk with
average fat content has $2\frac{1}{2}$ times as much value as a 0.1% increase in fat
content when milk yield is average; the population averages are assumed
to be 3500 kg milk and 4% fat. The breeding value for the aggregate
economic value is expressed by $2.5G_M + G_G = G_T$, where the G's denote
the breeding values of the individual traits. The partial regression coefficients are derived by solving the two equations

TABLE 11.3
Selection Index Construction[a]

Trait	a	V_P	V_G	h^2	cov$_G$	r_G	r_P
Milk, hl	2.5	36	10.8	0.3⎱			
Fat content, %	1	9	3.6	0.4⎰	−1.87	−0.3	0
					b_M		b_F
I_1					0.7		−0.12
I_2	$h_F^2 = 0.6$				0.69		0
I_3	$a_M = a_F = 1$				0.25		0.19
I_4	$a_M = 1, a_F = 0$				0.30		−0.21
I_5	$a_M = 0, a_F = 1$				−0.052		0.40
I_6	PT, $n = 15$				2.59		0.41
I_7	PT, $n = 50$				3.89		1.17
I_8	PT, $n \rightarrow \infty$				5		2
I_9	$\Delta F = 0$				0.64		0.33
I_{10}	$\Delta F = 3$				0.54		1.11

[a] a_M, a_F, Economic weights for milk, fat content. hl, Hectoliter = 100 liters. PT, Progeny testing.

$$b_M V_M + b_F \, \text{cov} \, MF = \text{cov} \, MG_T$$

$$b_M \, \text{cov} \, MF + b_F V_F = \text{cov} \, FG_T$$

The variances and covariances of the left-hand sides of the equations can be derived from Table 11.3. The right-hand sides represent covariances between the aggregate genotype (G_T) and milk yield (M) and fat content (F):

$$\text{cov} \, MG_T = \text{cov} \, M(2.5G_M + G_F) = 2.5V_{G_M} + \text{cov} \, G_M G_F$$

$$\text{cov} \, FG_T = \text{cov} \, F(2.5G_M + G_F) = 2.5 \, \text{cov} \, G_M G_F + V_{G_F}$$

The quantities necessary for estimating these covariances can also be derived from Table 11.3. Solving the equations yields the following index:

$$I_1 = 0.7(\text{kg milk}/100) - 0.14(\text{fat content} \times 10)$$

A cow whose production is above average by 400 kg of yield and 0.1% fat will be given $4 \times 0.7 - 1 \times 0.14 = 2.66$ points. Naturally, the absolute size of the points assigned to the two components can be changed without invalidating the index as long as the relation between the scores for milk yield and fat content remains unchanged.

In multiple regression theory the variance of the index V_I is taken as equal to the variance "accounted for by regression":

$$V_I = b_1 \text{ cov } X_1 G_T + \cdots + b_n \text{ cov } X_n G_T = \Sigma \, b_i \text{ cov } X_i G_T = \text{cov } I_G T$$

From this it follows that the regression coefficient of total breeding value upon the index is unity:

$$b_{G_T \mathbf{I}} = \frac{\text{cov } I G_T}{V_I} = 1$$

Therefore, in a sense, the heritability of an index is unity, which becomes more evident if mass selection for one trait is considered, where the index would be $h^2(X - \overline{X})$. However, a better measure for the usefulness of the index is its reliability as measured by the multiple correlation between the index and the aggregate genotype:

$$R^2_{IGT} = \frac{(\text{cov } I G_T)^2}{V_I V_{G_T}} = \frac{\text{cov } I G_T}{V_{G_T}}$$

since cov $I G_T = V_I$. The square of the correlation equals the ratio between the covariance index–aggregate genotype and the variance of the aggregate genotype, which, without considering the derivation, might be puzzling at first sight. Remembering that heritability is the square of the correlation between phenotype and breeding value, one might consider R^2 as the heritability of the index. For our example

$$\text{cov } I G_T = \text{cov}[(0.7M - 0.14F)(2.5G_M + G_F)] = 17.77 = V_I$$

and

$$V_{G_T} = V_{2.5G_M + G_F} = 61.72$$

We have

$$R^2_{IGT} = 17.77/61.72 = 0.288, \qquad R_{IGT} = 0.536$$

The expected genetic gain due to selection based on such an index is maximal under the conditions specified. The gain in the two constituent traits can be estimated from the regression coefficients of the breeding values of the respective traits on the index. The regression coefficient of the breeding value of milk yield on the index is

$$b_{G_MI} = \frac{\text{cov } G_MI}{V_I} = \frac{\text{cov}[G_M(0.7M - 0.14F)]}{V_I}$$

The variance of the index is

$$V_I = V_{0.7M - 0.14F} = 0.49V_M + 0.0196V_F - 0.196 \text{ cov } MF = 17.77$$

the same as found before, and $b_{G_MI} = 0.44$. In other words, the breeding value for milk yield increases by 0.44 units, corresponding to 44 kg of milk, when the index increases by one unit. Similarly the regression of breeding value for fat content on the index is $b_{G_FI} = -0.10$. Individuals with an index higher by one unit should therefore have average breeding values for fat percentage which are 0.1 unit or 0.01% lower. When the selection differential for this index is assumed to be one standard deviation ($= 4.22$ units) the progeny should change by the following quantities:

milk yield	$4.22 \times 0.44 \times 100$ kg $= 185.7$ kg
fat percentage	$[4.22 \times (-0.1) \times 0.1]\% = -0.042\%$
economic value	$1.857 \times 2.5 \times (-0.42) \times 1 = 4.22$ units

which is equal to the index value given to the cow. This illustrates the fact that the regression of aggregate breeding value on index is one. When selection is for milk yield only and is of the same intensity (one standard deviation $= 600$ kg) the resulting changes in milk yield and fat content are given by $b_{G_MM}\sigma_M$ and $b_{G_FM}\sigma_M$, which in our case give

milk yield	$6 \times 0.3 \times 100$ kg $= 180$ kg
fat percentage	$[6 \times (-0.05) \times 0.1]\% = -0.03\%$
economic value	$2.5 \times 1.8 + (-0.3) \times 1 = 4.2$ units

When selection is for fat content only, and again of the same intensity (one standard deviation $= 0.3\%$), the resulting changes are

milk yield	-62 kg
fat percentage	$+0.12\%$
economic value	0.35 units

A comparison of these quantities shows that selection by index is most efficient under the specified conditions. There is little difference compared to selection in which milk yield is the sole criterion, but the intention here is only to demonstrate the general principles. A situation can be constructed easily where the advantage of the index is more obvious.

When this index is applied, a cow with a high fat percentage is reduced in index points relative to one with a lower fat content. At first sight this appears to be a paradox, since obviously milk with higher fat content is worth more. The causes for this apparent contradiction are the negative genetic correlation between milk and fat content and the much higher economic value of milk yield. Due to the negative genetic correlation, the high fat content indicates that the breeding value for milk yield is lower than appears from the milk yield itself. If a high-yielding cow also has high fat content, it hints that her real breeding value for milk is not all that high and that other factors such as environment have helped to make the cow a high producer. Milk yield under our assumptions has a higher economic value than fat content. The latter is used, therefore, in this index largely to indicate the breeding value for milk yield. However, since fat content has some value of its own, it is not used exclusively as an indicator trait.

The index is valid only for the specified conditions, that is, when the specified genetic and economic parameters apply. When these change, the optimal index also changes. For example, if the heritability of fat content is assumed to be 0.6 instead of 0.4 as before, the optimal index becomes

$$I_2 = 0.69 \text{ kg of milk}/100$$

The coefficient for fat content is then zero. It was mentioned earlier that higher heritability means larger genetic gains. Therefore such a trait deserves more attention in selection. In the index I_2 this is reflected in the disappearance of the negative sign for fat percentage, so that an individual is no longer penalized for a desirable trait.

If the genetic parameters given in Table 11.3 remain unchanged but milk yield and fat content are assumed to be of equal value (a 100-kg increase in annual yield has the same value as a 0.1% increase in fat content), the index becomes

$$I_3 = 0.25 \text{ kg of milk}/100 + 0.19(\text{fat } \% \times 10)$$

The increased economic importance of fat content is reflected in the index. Fat content is relieved, so to speak, of its primary role as an indicator for breeding value of milk yield, largely the position it held in I_1.

Henderson (1963) has pointed out that all traits can be used to estimate the genic value of a single trait and that the partial indexes, weighted by the economic values of the respective traits, can be summed as a selection index for total merit. In our example the breeding value for milk can be estimated from milk yield and fat-%:

$$I_M = 0.3M - 0.208F$$

and similarly the breeding value for fat-%:

$$I_F = -0.052M + 0.4F$$

A unit of milk is valued $2\frac{1}{2}$ times a unit of fat-%, and therefore

$$I_T = 2.5I_M + I_F = 2.5(0.3M - 0.208F) + (0.4F - 0.052M)$$

$$= 0.7M - 0.12\ F$$

which is identical to I_1. With this computing method varying economic weights can easily be accounted for without the need for a complete index construction. For example, when $a_F = a_M$ the index is

$$I_T = I_M + I_F(0.3 - 0.052)M + (0.4 - 0.208)F$$

$$= 0.25M + 0.19F$$

The two-part indices for milk and fat-% deserve some attention for their own sake. Due to the negative covariance between the two breeding values the regression coefficients of the auxiliary traits are negative. For example, in I_4, $b_F = -0.21$ is much more negative than in I_1, where $b_F = -0.12$; this is explicable by the fact that in I_4 fat-% is used exclusively as an auxiliary trait.

The index is a multiple regression equation and coefficients change when type and reliability of information change. This appears self-evident when dealing with regression problems. However, as regards selection index, surprise is sometimes expressed that it changes when the information changes. This may be caused by the habit of equating the index value with the estimated breeding value, unconsciously assuming that the latter is more or less stable. However, the index is optimal only when type and quality of information remain alike. For example, for progeny testing information the index is quite different from an index for individual performance testing. When 15 progeny are tested (I_6), the coefficient of fat content is positive and is one-sixth of the coefficient for milk yield. When $n = 50$ (I_7) the fat coefficient is about one-third, and when progeny groups are very large ($n \to \infty$) the ratio of milk to fat coefficients is 5:2, like the ratio of economic values. In the index for individual records fat content receives a negative weight due to the negative genetic correlation with and the relatively modest heritability of milk yield, i.e., it is used as an indirect indicator for the economically more important breeding value of yield. In progeny testing the latter is estimated comparatively accurately and less need exists for auxiliary information. Yet even when $n = 50$ the negative covariance exerts influence, since if breeding values are independent, the regression coefficients would be 4.01 [$= 2.5 \times 2 \times 50/(50 + 12)$] for milk and 1.69 [$= 2 \times 50/(50 + 9)$] for fat-%, respectively. When the genotype is completely known, the regression coefficients are equal to the economic weights (I_8).

There may be a need to improve some traits, say m out of n traits, and to keep the other $n - m$ traits unchanged. In our example it could be reasonable to improve milk yield without lowering fat content. Kempthorne and Nordskog (1959) have shown how to construct a selection index with such restrictions. Hogsett and Nordskog (1958) reported examples of such indexes from poultry breeding. For a simple case in which one character is to be improved and a second trait held constant, Abplanalp and co-workers (1964) demonstrated a simple method for constructing the index. For example, the regression coefficient of fat content on index is

$$b_{G_FI} = \frac{\text{cov } G_FI}{V_I} = \frac{\text{cov } G_F(aM + bF)}{V_I}$$

When the genetic parameters of Table 11.3 are used and it is assumed that $a = 1$ and $b_{G_FI} = 0$, since in our example fat content must not change, b can be computed from the equation

$$b = \frac{\text{cov } G_FM}{V_{G_F}} = 0.32$$

The index is

$$I_5 = \text{kg of milk}/100 + 0.32(\text{fat } \% \times 10)$$

A selection differential of one index unit should raise milk yield 28 kg with no change in fat content. A selection differential of one standard deviation should increase milk yield 171 kg, about 92% of the progress expected when the optimal index is applied without restrictions.

When more traits are involved, the original solution of Kempthorne and Nordskog (1959) must be utilized. Let G^* be the part of the genetic variance–covariance matrix that concerns the traits to be kept constant, in our example, fat-%; $G^* = -1.871 + 3.6$. New regression coefficients b^* are found from

$$b^* = [I - P^{-1}G^*(G^*P^{-1}G^{*\prime})G^{*\prime}]b = \begin{pmatrix} 0.64 \\ 0.33 \end{pmatrix}$$

This is the same ratio as that found before (1:0.52). The covariance of this index to G_T is $a'Gb^* = 15.73$ and leads to $R^2 = 0.252$, $R = 0.502$, somewhat, though not much, smaller than for I_1.

This method permits construction of selection indexes where several traits are to be kept constant. The procedure can be illustrated with results published by Nordskog and Hogsett (1958). The optimal index, given in Table 11.4, leads to genetic changes for rate of lay in percent, egg weight in grams, and body weight in kilograms of 1.24, -0.22, and -0.059, which is optimal under the given price conditions, but loss of egg weight and/or body weight may not be tolerable in the future. Therefore, the condition may be made that the two last traits should not be changed, i.e., that $\Delta E = \Delta B = 0$. The index that satisfies the condition is given in Table 11.4. As shown above for I_9, the matrix G^* consists of the rows of the genetic variance–covariance matrix that concern the traits to be held constant (E, B). The genetic changes expected from application of the index are zero for E and B and 1% for percent of lay, and the efficiency of the index is considerably reduced to $R^2 = 0.055$. Were only E to remain unchanged, the loss of efficiency would be much less ($R^2 = 0.102$). In some breeding problems an optimum performance may exist for some traits while others need to be improved as much as possible—for example, fat content may have to be improved by k units. Tallis (1962) proposed selection indexes for optimum genotypes in which, for example, r out of p traits are to be altered by a fixed amount (the distance between the present performance and optimum performance) and progress in the remaining $p - r$ traits should become maximum. If the necessary improvement equals k units the index coefficient is found from

$$b^{**} = b^* + P^{-1}G^*(G^*P^{-1}G^{*\prime})^{-1}k$$

TABLE 11.4
Restricted Index[a]

	L	E	B
σ_P	0.1695	4.27	0.277
σ_{BV}	0.0536	2.70	0.196

$$
\begin{pmatrix} P \\ 0.0287 & 0.0362 & -0.0023 \\ & 18.2329 & 0.4140 \\ & & 0.6767 \end{pmatrix} \begin{pmatrix} b \\ 6.708 \\ 0.122 \\ -3.398 \end{pmatrix} = \begin{pmatrix} G \\ 0.0029 & -0.0724 & -0.0053 \\ & 7.2932 & 0.2645 \\ & & 0.0384 \end{pmatrix} \begin{pmatrix} a \\ 100 \\ 1.14 \\ -0.063 \end{pmatrix}
$$

$V_I = 2.273, V_{BV} = 21.735, R^2 = 0.105, R = 0.323$

$\Delta G = \Delta G = 0$

$$
G^* = \begin{pmatrix} -0.0724 & -0.0053 \\ 7.2932 & 0.2645 \\ 0.2645 & 0.0384 \end{pmatrix}
$$

$$
b^* = (I - P^{-1}G^*(G^{*\prime}P^{-1}G^*)G^{*\prime})b = \begin{pmatrix} 6.235 \\ 0.041 \\ 0.571 \end{pmatrix}
$$

$V_I = a'G'b^* = 1.194, V_{BV} = a'Ga = 21.735, R^2 = 0.055, R = 0.234$

$$
P^{-1} = \begin{pmatrix} -35.0768 & 0.1071 & -1.6520 \\ & -0.0628 & 0.3424 \\ & & -14.9367 \end{pmatrix}
$$

[a] Source: Modified from Hogsett and Nordskog (1958). L, Rate of lay, %. E, Egg weight, g. B, Body weight, kg. BV, Breeding value.

where G^* represents the submatrix involving variances and covariances of the traits to be improved by k units. If $k = 0$, the index is identical to the restricted index discussed before. When fat content of milk is to be improved by 0.3% the index is

$$
b^{**} = \begin{pmatrix} 0.64 \\ 0.33 \end{pmatrix} + \begin{pmatrix} 1/36 & 0 \\ 0 & 1/9 \end{pmatrix} \begin{pmatrix} -1.871 \\ 3.6 \end{pmatrix}(0.6505)(3) = \begin{pmatrix} 0.54 \\ 1.11 \end{pmatrix}
$$

The restriction causes further loss of efficiency and R^2 decreases to 0.2. The genetic advance from the application of such an index is

$$G_{I_{10}} = \frac{\text{cov } I_{10}G}{\sigma_I} i$$

Now, cov $I_{10}G = 3$, and $\alpha = i/\sigma_I$. When $i = 1$ ($\sigma_I = 3.52$) one generation of selection would achieve $1 \times k/3.52 = 0.284k = 28\%$ of the total difference toward the optimum, and three generations of selection would be necessary to achieve it. The regression coefficients of milk and fat content on this index are 1.07 hl and 0.848% (0.28 × 3), in comparison to -0.42% and 1.82 hl in the optimal index.

Economic weights have been assumed to be known exactly. In reality this is frequently not true. Economic weights fluctuate between areas and times and they may change between selection and realization of the genetic gain brought about by selection. Rønningen (1971) investigated the problem of incorrect economic weights and found that small deviations cause comparatively little loss of efficiency, but the loss may be considerable when the weights are seriously misjudged. Vandepitte and Hazel (1977) showed that incorrect weightings lead to a suboptimal index, which results in smaller than expected selection gains. If a^* represents wrong weights, the covariance between index and aggregate genetic merit is

$$\text{cov } IG = a^*G'P^{-1}Ga$$

The variance of the "wrong" index is

$$V_I^* = a^*G'P^{-1}Ga^*$$

The genetic gain in true genetic merit upon application of a wrong index is ($i = 1$)

$$\frac{a^*G'P^{-1}Ga}{(a^*G'P^{-1}Ga^*)^{1/2}}$$

The efficiency of using a wrong index relative to the correct index will be

$$\frac{\text{cov } GI}{\sigma_I \sigma_{I^*}}$$

Let us assume that the true economic weights of milk yield and fat content are in the relation 1:1, but that when selecting the ratio applied is 2.5:1. The covariance between an index based on the latter ratio (I_2) and the true merit is 6.0237 .The standard deviation of the assumed wrong index I_1 is 4.204 and that of the correct index I_3 is 1.5969. The relative efficiency of the wrong index is 6.0237/(1.597 × 4.204) = 0.9. Vandepitte and Hazel found that small errors (about half of the value) have little impact, but that larger errors are rather serious. The effects of wrong weights are not linear and not symmetric and in general errors on the low side are more serious than those on the high side. Random deviations have similar consequences and simulations have shown that $\Delta g/I^*$ somewhat underestimates the maximal possible advance.

These index examples are for two traits only. Seven quantities had to be assumed, and these all had to be estimated from data in order to construct a usable index. The discussed examples demonstrated how much the index depends upon the quantities used in constructing it. When these quantities are estimated from data they are subject to sampling errors; this is particularly true of genetic correlations. Of course, the errors of the estimates decrease as the amount of data increases. The size of the parameters to be estimated also influences the estimation errors. Harris (1963) showed that the difference between actual progress when an index is applied and the progress possible in theory decreases, and the success increases, when (1) heritabilities increase, (2) genetic correlations increase, and (3) environmental correlations decrease. The most unfavorable situation involves low heritabilities combined with low genetic and high environmental correlations among the component characters. However, this situation permits little progress even if the genetic parameters are well known. It is worsened by errors of estimates larger than those that would be expected under more favorable conditions.

Consequently, the application of the selection index may be fully justified only when the parameters can be estimated from a large volume of reliable data. If not, the index can only indicate the extent and directions of selection efforts of individual characters.

In certain situations economic weights may not be explicit but definite opinion exists as to the desired genetic progress of different traits relative to each other.

Such a situation exists mainly when several types of produce are desired, for example, meat and milk from cattle, or meat and wool from sheep. Also, even though a selection index utilizes given information in an optimal way, kind and quantity of information may be modified by the breeder which will influence the genetic gain. Pesek and Baker (1969) have suggested that one use the desired genetic gain to arrive at the required selection index. The gain in trait 1 by selection using an index is

$$\Delta g_1 = ib' \frac{G_{1j}}{\sigma_I}$$

where G_{1j} denotes the row of the genetic variance–covariance matrix related to trait 1 and b is the vector of index weights. The ratio i/σ_I does not influence the proportionality, and the regression coefficients can be transformed to

$$\beta_j = b_j \frac{i}{\sigma_I}$$

The desired coefficients are found from

$$\beta = \Delta g G_{ij}^{-1}$$

One may also ask for the economic weights that would lead to genetic gains in the desired proportions if an unrestricted index were applied:

$$Pb = Ga$$

and

$$a^* = G^{-1}P\beta = \frac{\sigma_T}{i} \Delta g (G'P^{-1}G)^{-1}$$

The example given in Table 11.5 refers to the case where meat and milk are to be improved simultaneously. Assume that the current economic weight for a unit of each is 30 monetary units. The information available for selection of AI bulls consists of progeny average for milk yields ($n = 50$) and of meat growth rate of bulls themselves.

Selection on the optimal index at $i = 1$ will improve the genetic merit of the chosen bulls by 2.89 hl milk but impair their meat breeding value by 1.23 units. One may wish, however, that the genetic gains in both traits should be in the ratio 10:1. The index coefficients necessary

TABLE 11.5
Desired Gain Index[a]

Unrestricted index:

$$P = \begin{pmatrix} 3.98125 & -1.75 \\ -1.75 & 25 \end{pmatrix} \qquad b = \begin{pmatrix} 35.51 \\ 5.7856 \end{pmatrix} = \qquad G = \begin{pmatrix} 6.125 & -1.75 \\ -3.5 & 6.25 \end{pmatrix} \qquad a = \begin{pmatrix} 30 \\ 30 \end{pmatrix} \qquad \Delta g = \begin{pmatrix} 2.752 \\ -0.362 \end{pmatrix} \qquad \sigma_I = 71.68 \qquad R^2 = 0.496$$

Restricted index:

$$\Delta g$$

$$G^{-1} = \begin{pmatrix} 0.19436 & 0.05442 \\ 0.10884 & 0.19048 \end{pmatrix} \qquad = \qquad b* \frac{i}{\sigma_I}$$

$$(10 \quad 1) = (2.05244 \quad 0.73468) \left[\frac{i}{\sigma_I} \right]$$

$$\Delta g = \begin{matrix} 2.0 \\ 0.2 \end{matrix} \qquad \sigma_I = 4.999 \qquad R^2 = 0.421$$

$$P^{-1} = \begin{pmatrix} 0.25925 & 0.01814 \\ 0.01814 & 0.04127 \end{pmatrix}$$

[a] Standard deviations for milk, meat 7 hl, 5 dkg (1 dkg = 10 g); heritabilities ¼.

to achieve this ratio are 2.05244 (β_1^*) and 0.73468 (β_2^*) . The genetic gain from applying such an index is

$$i \frac{\Delta g}{\sigma_I} = G'\beta = \frac{i}{5} \binom{10}{1}$$

and of course the efficiency of this index relative to the current economic situation is decreased.

In the example, $b'Ga = 330, \sigma_I = (b'Pb)^{1/2} = 4.999, (a'Ga)^{1/2} = 101.8,$ and

$$R^2 = \frac{(b'Ga)^2}{(b'Pb)(a'Ga)}$$

Therefore $R^* = 0.649$, compared to $R = 0.705$ of the unrestricted index. The economic weights resulting in an unrestricted index that would generate progress in the desired ratio of 10:1 have the ratio 2.1423:3.5638 = 1:1.66 instead of 1:1 as before.

The dependence of the optimum index on the information can be illustrated by assuming that records of 50 dairy and 50 beef progeny are available. The optimum index in such a situation has the weights $b_M = 41.3$ and $b_F = 38.1$ and the superiority of selected bulls ($i = 1$) would be 2.23 hl and 5.6 g, a ratio much narrower than desired. The solution as outlined is possible only if P and G have equal rank. Frequently P has a larger dimension. For example, information about one trait may come from several types of relatives and for such situations Essl (1981) and Brascamp (1980) have developed solutions.

11.3. Stepwise Selection

The parameters that determine genetic progress are affected by previous selection (Cochran, 1951; Rønningen, 1969; Cunningham, 1975a; Niebel and Fewson, 1976; Robertson, 1977). The changes in the selected subpopulation for which the characteristics were originally normally distributed depend on the quantity $c = i(i - x)$, where i denotes the standardized selection differential and x the abscissa of the truncation point (Table 9.2). Cochran (1951) showed, following Pearson, that selection on index j changes the covariance between quantities k and l as follows:

$$\text{cov } kl' = \text{cov } kl - (\text{cov } jk)(\text{cov } jl)\frac{c}{\sigma_j^2}$$

The change in the variance of the selected trait itself ($j = k = l$) is

$$\sigma_{j'}^2 = \sigma_j^2(1 - c)$$

The covariance between the selected trait and another trait is decreased to $\text{cov } j'k = \text{cov } jk(1 - c)$. Since both covariance and variance are decreased by the same proportion, the regression is unchanged:

$$b_{kj'} = \frac{\text{cov } jk(1-c)}{\sigma_j^2(1 - c)} = \frac{\text{cov } jk}{\sigma_j^2} = b_{kj}$$

The variance of the correlated variable is decreased:

$$\sigma_{k'}^2 = \sigma_k^2 - \frac{(\text{cov } jk)^2 c^2}{\sigma_j^2} = \sigma_k^2(1 - r^2c)$$

The effects of selection on the parameters can be demonstrated by reference to a broiler example.

The 5% heaviest animals are selected and subsequently submitted to progeny testing, after which the top 20% are chosen as elite sires. Parameters in the unselected population are $\sigma_P = 100$, $h^2 = 0.36$, and progeny group size n equals 40 (all single progeny from different dams, not very realistic for broilers), $\sigma_{\bar{P}}^2 = 11.257$. Selection of the top 5% changes variances and covariances (Table 11.6). The coefficients of the index that involve both the birds' own record (I) and the average of its 40 progeny ($\bar{P}$) are $b_I = 0.1019$, $b_{\bar{P}} = 1.4338$. The variance accounted for by the selection index is $\sigma_{I'}^2 = 18.3974$ ($\sigma_{I'} = 4.289$).

Selection has reduced the genetic variance and $R^2 = 0.7383$ ($R = 0.859$). If selection had not taken place and all broiler males or a random sample of them were progeny-tested, the regression coefficients would have been the same but the variance accounted for by the index would have been larger ($\sigma_I^2 = 29.476$) and consequently the accuracy would have been greater ($R^2 = 0.8188$).

The selection index applied at the second selection step utilizes both sources of information. The reduction of the accuracy compared to the

TABLE 11.6
Stepwise Selection

$\sigma_I = 10$ (body weight, 10 g), $h^2 = 0.36$

1. Mass selection

$p_1 = 5\%, i_1 = 2.06$

2. Progeny testing

$p_2 = 20\%, i_2 = 1.40, c = i_1(i_1 - x) = 2.06 (2.06 - 1.645) = 0.8549$

$$\begin{bmatrix} (1 - c)\sigma_I^2 & (1 - c) \operatorname{cov} I.\, P \\ (1 - c) \operatorname{cov} I.\, P & (1 - r_{I.P}^2 c)\sigma_P^2 \end{bmatrix} \begin{matrix} b_{I'} \\ b_{\bar{F}} \end{matrix} = \begin{matrix} (1 - c) \operatorname{cov} I.BV \\ (1 - r_{IZ}^2 c) \operatorname{cov} P.BV \end{matrix}$$

$\sigma_{I'}^2 = \sigma_I^2(1 - c) = 100(1 - 0.8549) = 14.51$

$\sigma_{BV}^2 = \sigma_{BV}^2(1 - h^2 c) = 36(1 - 0.36 \times 0.8549) = 24.92 \sigma_Z' = 4.99$

$\operatorname{cov} IBV' = 36(1 - 0.8549) = 5.2236$

$\operatorname{cov} I.\bar{P}' = 18(1 - 0.8549) = 2.6118$

$\sigma_{\bar{P}}^2 = 11.275(1 - 0.6922)$

$\operatorname{cov} \bar{P}.BV' = 18(1 - 0.3078) = 12.4602$

$b_I' = 0.1019, b_{\bar{P}}' = 1.4338, \sigma_I^2 = 18.3974, \sigma_{I'} = 4.289, R^2 = 0.7383$

(only theoretically possible) alternative, progeny testing all animals, is about 22% ($\Delta G' = 4.289i$; $\Delta G = 5.429i$).

One alternative, and probably the rule in practical breeding, is to utilize at the second selection the new information only and to neglect the performance test. The regression coefficient of breeding value on progeny average must be computed from the reduced variances and covariances: $b_{\bar{P}}'' = 12.4602/8.5051 = 1.465$, $\sigma_I^2 = 18.255$, $R^2 = 0.738$. The correlation coefficient of the index has been reduced to 0.856, in comparison to 0.859, a negligible difference mainly because of the comparatively large group size of the progeny. The selection gain would be 74.4% of the theoretical optimum. Niebel and Fewson (1976) pointed out that the loss of accuracy can be serious (10–20%) when the first selection is accurate but relatively mild and the second selection is intense but moderately accurate. In most situations, however, the loss of accuracy should be relatively small.

Since the variance is changed, the relationship between the fraction culled and the superiority of the selected animals no longer holds. The computation of the exact selection differentials in a previously selected population (Utz, 1969; Niebel and Fewson, 1976) is relatively difficult. However, Dickerson and Hazel (1944) have pointed out that changes in selection differentials will be small. If the fractions retained at either step vary between 10 and 50%, the first not less than 10% and at the second step not more than 50%, the selection differentials change not more than 3%. This small underestimation of the selection differential is partly balanced by the overestimation of correlations when these are computed from unselected populations. The genetic superiority of the 5% heaviest broiler males is

$$\Delta g_1 = 2.06 \times 10 \text{ g} \times 0.36 = 7.42 \text{ g}$$

The genetic gain from selecting the 20% best progeny-tested males (neglecting the change of the selection differential) is

$$\Delta g_2 = 1.4 \times 4.289 \text{ g} = 6.01 \text{ g}$$

The generation interval of broilers is relatively short. It is assumed to be 45 weeks ($= T_1$) and that progeny testing is to lengthen it by 15 weeks ($= \Delta T$). Consequently, $6.01 > 7.45 \times 15/45$ and judging from genetic gain, progeny testing would be justified. However, progeny testing may involve much additional cost and therefore another conclusion may be justified.

Stepwise selection is the rule with animals that are used over several periods. For example, cows are exposed at every lactation to culling. Consequently, records of later lactations are no longer normally distributed and correlations, etc., computed from such records are reduced. This may contribute to the reduced heritability of later lactation records.

12

Empirical Tests of Selection Theory

The concepts discussed in the previous chapters were deduced from Mendelian laws and knowledge of mating structure. Obvious questions to be posed concern the validity of these concepts and if and to what extent predictions based on them hold in the real world. Testing with domestic animals would be desirable, but time and costs involved limit and even exclude such undertakings, at least with domestic mammals. Nevertheless, reference will be made to some of the few experiments that have been performed. The Institute of Animal Genetics of the University of Edinburgh pioneered the use of laboratory animals for testing the application of quantitative genetics to animal breeding. A newer development is computer simulations of selection experiments. However, comparison of actual or realized selection progress with that expected should permit inferences on the validity and power of selection theory. Differences between expected and realized progress arise from various causes. Some of the causes, which are not mutually exclusive, are given in Table 12.1.

12.1. Selection Experiments

Differences between expected and realized advances may be caused in small populations by random drift. Such phenomena can be nicely studied with simulation experiments. However, in one of the very first experiments devoted to checking the quantitative genetic theory consid-

TABLE 12.1
Causes of Discrepancy between Expected
and Realized Selection Differentials

1. Random deviations of small populations, founder effect, "bottleneck"
2. Heritability
   a) Neglect of nongenic correlations
   b) Maternal effects
   c) c^2 Effects
3. Asymmetry of selection response
   a) Scale effects
   b) Asymmetric gene frequencies
   c) Directional dominance
   d) Maternal effects
   e) Composite traits
4. Counterselection
   a) Multiple goals
   b) Natural counterselection
5. Genetic homeostasis
   a) Fitness of intermediates
   b) Superiority of heterozygotes

erable variability was observed in genetic progress between the individual selected lines (Clayton *et al.*, 1957).

The relevance of random variation to selection was studied by De Fries and Touchberry (1961) in an experiment with *Drosophila affinis*. Five replicates of each selection group were carried. Selection was of various intensities for change in body weight. Each group in each generation was reproduced from one pair of parents. Therefore, selection was among full-sibs. Since the effective population size was minimal and the selected trait had a low heritability ($h^2 = 0.06$), the variability of the results from the lines should be maximal. The experiment depicts the effect of random variation on selection. The average selection progress of all groups corresponded to expectation. After ten generations of selection, the 95% confidence intervals of the individual selection groups were as follows:

selection group	10%+	10%−	25%+	25%−	control
95% confidence interval, μg	590–1064	504–774	531–1009	406–865	475–908

(10%+, the heaviest individual selected of ten measured; 10%−, the lightest; 25%+, the heaviest of four, 25%−, the lightest of four). This experiment illustrates that random deviations can be important in comparing selection responses in experiments and in breeding programs, particularly in populations of small effective size N_e. With increasing numbers in populations under selection, the importance of these sources of variability declines and the repeatability of selection response increases.

Small effective population size makes composition of the progeny generation increasingly unpredictable (Section 2.4). However, in addition to random drift, other factors affect the result.

Selection differentials may vary from line to line, or even if these are similar, the average genic value of the few chosen individuals may vary between lines. Examples of differences between expected and realized selection differentials are provided by Kramser (1974) from pigs and by Botkin and Stratton (1967) from sheep. In the pig herd studied by Kramser the replacement sows had to be taken from the upper 60% of the herd because an accommodation of show points and teat numbers made a large number of replacement gilts ineligible. Botkin and Stratton (1967) also found barely any freedom remaining for selecting utility traits because culling on conformation and fancy points disposed of a number of otherwise eligible stock.

The absence of any freedom to compensate is a considerable disadvantage experienced in small, closed herds, i.e., the above average level of health, performance, etc., in one herd cannot be utilized to increase selection differentials in other, poorer herds. In larger herds, or in open herds where exchange of animals is customary, compensation and thus larger selection differentials are possible.

The evolution of some species is known to have sprung from one or from very few individuals. Such species evolve rapidly. The phenomenon has been called the "founder effect." The history of several better known breeds reveals that animal numbers in the founding generations were very small and inbreeding was considerable. The population, or rather its present-day genes, had to pass a bottleneck. It must be assumed that gene and genotype proportions quite different from what would have been chosen by selection escaped the bottleneck. A demonstration of such a founder effect is provided by an experiment in which selection for and against 8-week weight of mice was made (Butler, 1981). The empirical standard errors of realized heritabilities from eight lines of eight full-sib families each derived from an F_2 of a four-way cross were less than 0.01, while they were about 0.05 when similar lines were derived from an outbred mouse population. As gene frequencies in F_2-derived lines are at least 0.25, such lines experienced no bottleneck, in contrast to the lines from the outbred population.

Lack of agreement between expected and realized responses can be caused by the use of the wrong heritability values. Yamada and co-workers (1958) analyzed data from a flock of White Leghorns, which had apparently leveled off in egg production. Considering heritability estimated from sire and dam components together, some response was expected; however, heritability coefficients computed from sire components alone indicated that additive-genetic variance was practically exhausted.

Maternal effects may simulate high heritabilities and thus lead to unrealistic expectations. A negative maternal effect on sow litter size was discovered by Revelle and Robison (1973). Gilts from large litters tend to have smaller litters due to an unfavorable maternal environment and *vice versa*. Alsing *et al.* (1980) found correlations to be much smaller between the litter size of a gilt and the size of her own litter of origin compared to the correlation with other litters of her dam (Section 6.4). As a consequence of this, dam–daughter similarity is diminished, leading to underestimated heritabilities from regression and to wrong expectations of genetic progress. The c^2 effects, such as herd influences, seasonal influences, or even carryover effects from herds to test stations, if not corrected for, can influence whole progeny groups in positive or negative ways and increase correlations between group members. Failure to consider such residual correlations will cause overestimation of heritabilities and of possible genetic progress. Thomson and Freeman (1970), for example, found correlations between milk yields of half-sisters in different herds to be considerably smaller than expected from the heritability, and similar observations were made on Austrian dairy data (Pirchner, 1970). Rutzmoser (1977) and Rutzmoser and Pirchner (1979) encountered substantial residual correlations among data on half-sib beef performance in progeny test stations. As discussed in Chapter 10, such c^2 effects prevent an accurate estimation of genetic merit, even from numerous offspring. Accuracy of breeding value estimation may fall short of expectation by a considerable margin, and consequently realized genetic progress will be less than expected.

Asymmetry of genetic progress has attracted considerable attention. The initial observation was made in laboratory animals, but more recently has also been seen in some divergent selection experiments with domestic animals. For example, Hetzer and Harvey (1967) selected for or against back-fat thickness in Duroc and Yorkshire swine (Fig. 12.1). The greater advances of the plus lines than of the minus lines resulted from their greater variability and therefore from greater selection differentials. When measures are expressed in terms of standard deviations, asymmetry is reversed in the first five generations (Table 6.5).

Causes of asymmetry may be scale effects, as illustrated by the selection advance of Hetzer's lines (Fig. 12.1, left). Another cause may be asymmetric gene frequencies. As outlined in Chapter 2, selection is rather ineffective in moving genes at high or low frequencies toward fixation and this is accentuated when genes are dominant.

Effects of inbreeding can contribute to asymmetry. For instance, an increase in homozygosity, either of favored genes because of selection, or of random genes because of drift, will be a brake on the further improvement of heterotic traits, but at the same time will add to their decrease in lines selected in the opposite direction. An example is provided by Hetzer's experiments, where the association between fat thickness and degree of inbreeding was negative and correction for it (lines were more than 25% inbred) reduced asymmetry between line-specific heritabilities.

In consequence of large variability between replicated selection lines Falconer (1973) emphasized the possible chance nature of asymmetry in unreplicated experiments; yet asymmetry seems to be a real phenomenon in many selection experiments. For example, Butler (1981) observed strong asymmetry repeated in different lines. Hetzer detected a strong asymmetry of growth changes in lines selected for back-fat thickness. In Durocs the 20-week weight changed −0.08 kg in the plus line and +0.44 kg in the minus line, which is explicable by a negative relationship between weight gain and back-fat thickness in Durocs. In Yorkshires, weight changes were −0.21 kg in the plus line and −0.59 kg in

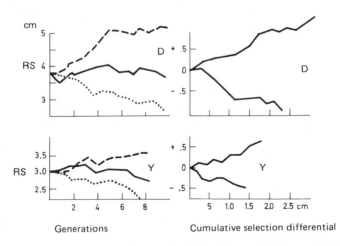

FIGURE 12.1. Selection for and against back-fat thickness. D: Durocs. Y: Yorkshires. BF: Back-fat thickness. [Modified after Hetzer and Harvey (1967), with permission.]

the minus line. Hetzer and Miller (1972) explain this by attributing to weight gain and back-fat thickness a positive relationship and by assuming that in this breed weight is a heterotic trait. Therefore inbreeding depression and a correlated response augment each other in the minus line and neutralize each other in the plus line.

Artificial, conscious or unconscious, and natural counterselection can exert a powerful influence on selection progress. In general, consideration of additional traits reduces the realized selection differential. Numerous examples have been published and some will be discussed in Section 12.4. Hickman (1971) reported from a Canadian dairy selection experiment the following expected E(SD) and realized R(SD) selection differentials (kg dry matter yield):

	E(SD)	R(SD)
progeny-tested sires	1.75	0.59
sire dams	2.06	1.75
cow dams	0.35	0.16

The actual superiority was much less than the selection differential possible if selection had been solely for dry matter yield. A large proportion of the deficit was caused by the accidental loss of a selected sire, thus illustrating the risk inherent in small populations.

The classic selection experiment of Lerner (1954), in which shank length of White Leghorns was changed, vividly demonstrates natural counterselection. The ratio of realized to expected selection differentials decreased in successive 5-year periods from 0.93 to 0.78 to 0.63. Suspension of selection improved reproductive performance even though the degree of inbreeding remained unchanged. Therefore, the cause of the lack of progress must be sought in natural counterselection. Lerner highlighted the resistance of populations to change by coining the expression "genetic homeostasis," which has found wide application in the interpretation of selection efforts. Nordskog and Wehrli (1963) have defined fitness of their poultry selection lines as a product of rate of lay, fertility, hatchability, and viability, i.e., as number of viable offspring per dam. Lines were selected for increased and decreased egg weight, and fitness thus defined declined by 39% and 16%, respectively. The latter decrement can be chiefly explained by unavoidable inbreeding, but the much greater fitness reduction in the large-egg-weight line can be attributed to above optimum egg weight in the base population, in addition to the inbreeding depression. Selection in the small-egg-weight line probably moved egg weight closer to its natural optimum and thus was less detrimental than selection in a positive direction.

One of the principal causes of genetic homeostasis is heterozygote superiority. Natural and artificial counterselection may simulate overdominance, as exemplified by the gene for halothane susceptibility. Selection for meatiness favors stress-susceptible pigs, which incur high losses. Heterozygotes will be preferentially represented among the selected breeder pigs and selection progress will be moderated.

12.2. Success of Practical Breeding

In the course of time the performance changes observed in animal populations are caused by genetic G_i and by environmental E_i influences, whose effects cannot be disentangled without some effort. Performance at time t can be described by

$$P_t = G_t + E_t + e$$

where e is the random deviation. In the subsequent period it changes to

$$P_{t+1} = P_t + \Delta P = G_t + \Delta G + E_t + \Delta E + e'$$

The apportionment of the phenotypic change ΔP to the genetic change ΔG and to the environmental change ΔE is possible only if one of these can be totally or partially eliminated. Environmental changes, which comprise all nongenetic changes, are practically unavoidable, with the exception of some defined diseases which can be controlled. Therefore, efforts to disentangle the genetic and the environmental components of genetic change concentrate on eliminating all or part of ΔG either by use of genetic controls or by repeat matings.

12.2.1. Control Populations

In these populations genetic changes are minimized. This permits measurement of environmental changes over the course of years, and an estimation of ΔG by a direct comparison of the selected population P_i with the control C_i:

$$C_{t+1} = G_t + E_t + \Delta E + e'$$

$$P_{t+1} = G_t + \Delta G + E_t + \Delta E + e$$

$$P_{t+1} - C_{t+1} = \Delta G + (e - e')$$

The residuals e and e' can be controlled by the size and structure of the lines to be compared. The success of isolating the genetic trend is illustrated in Fig. 12.2. Problems connected with the use of controls were treated by Gowe *et al.* (1959) and Hill (1972*a*). Genetic changes in controls can arise from artificial and/or natural selection. Artificial selection can be avoided and natural selection can be minimized by optimal management and hygiene. Genetic changes can be caused by drift, which increases the genic variance between generations by

$$V_d = \frac{V_A}{N_e}$$

and over t generations to

$$V_{d_t} = 2FV_A = 2V_A\left[1 - \left(1 - \frac{1}{2N_e}\right)^t\right]$$

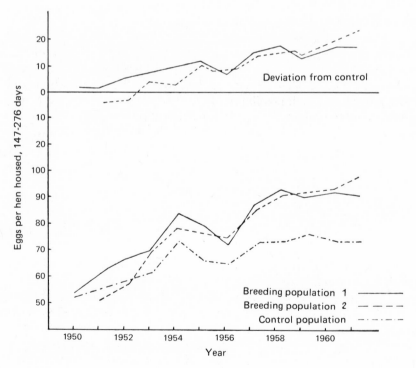

FIGURE 12.2. Selection progress in egg-laying performance. (Gowe, 1962.)

both depending on effective population size, the basic aspects of which have been discussed in Chapter 2. In connection with application to control populations, some additional considerations are desirable. Assume S parents with k_i offspring each. The variance of family size is σ_k^2 and $\Sigma_i^S k_i = T$. The effective size of the population is derived from

$$\frac{1}{N_e} = \frac{1}{T} + \frac{S\sigma_k^2}{T^2}$$

The first quantity reflects the sampling of random gametes of heterozygotes, i.e., Mendelian segregation. The second quantity is due to sampling from families with unequal size. When family sizes are randomly distributed, $\sigma_k^2 = T/S$, the variance equals the mean, and $1/N_e = 1/(2T)$, as discussed in Chapter 2. It should be emphasized that an absence of selection does not imply equal family size; rather, it is assumed that each parent has an equal chance to have progeny.

The gene flow between generations follows four pathways: sire → son (mm), sire → daughter (mf), dam → son (fm), and dam → daughter (ff). Mendelian segregation leads to the same drift along all four paths. In contrast, the variance of family size differs greatly among the four combinations (cf. the enormous variability among progeny numbers of different sires). The progeny numbers of the four combinations of parents and offspring are not entirely independent, as an elite bull will have an above average number of sons and daughters in the population. Latter (1959) considered such possibilities in deriving an expression for N_e (N_m and N_F denote numbers of sires and dams, respectively, σ_{mf}^2 the variance of number of daughters from sires, etc.):

$$\frac{1}{N_e} = \frac{1}{16N_M}\left[2 + \sigma_{mm}^2 + \frac{2N_M}{N_F}\,\text{cov(mm.mf)} + \left(\frac{M}{F}\right)^2 \sigma_{mf}^2\right]$$

$$+ \frac{1}{16N_F}\left[2 + \sigma_{ff}^2 + \frac{2N_F}{N_M}\,\text{cov(ff.fm)} + \left(\frac{F}{M}\right)^2 \sigma_{fm}^2\right]$$

A reduction of drift can be sought in two ways—by eliminating Mendelian segregation or by reduction or even elimination of differences in progeny numbers. The first approach can be realized by the use of inbred lines or of their F_1 progeny. Homozygous lines have genetically uniform progeny, apart from the small mutation-caused increment in variability. Therefore, drift from such a source should be negligible. The inbreeding depression in performance of such lines can be circumvented by use of their F_1 hybrids. Such control is possible in laboratory animals, where numerous inbred lines are available, and it may represent a feasible approach in poultry. However, the principal avenue to drift reduction in populations of domestic animals is via constancy of family size. This implies the use of one son from each sire and N_F/N_M daughters, and similarly from each dam one daughter and N_M/N_F sons. The first three combinations have constant progeny numbers and $\sigma^2_{mm} = \sigma^2_{mf} = \sigma^2_{ff} = 0$. The path dam $\rightarrow$ son leads to variable numbers of sons and $\sigma^2_{fm} = (N_M/N_F)(1 - N_M/N_F)$. If covariances are assumed to be zero, the effective population size is given by

$$\frac{1}{N_e} = \frac{3}{16N_M} + \frac{1}{16N_F}$$

Use of equal numbers of males and females in addition to a constant family size doubles the effective size compared to a random mating population:

$$\frac{1}{N_e} = \frac{1}{8N_M} + \frac{1}{8N_F}$$

In other words, the effective size of such a population is twice the number of parents.

Effective size and genic variance caused by drift in control populations are given in Table 2.11, taken from a publication by Gowe *et al.* (1959). Constant family size leads to a noticeable reduction of the genic drift variance (σ_A was assumed as 20 eggs/hen housed). However, its disadvantage would be the necessity of keeping pedigrees.

Drift-induced performance changes of controls can be decreased by reduction of the genic variance. Two approaches are possible for its achievement—null selection differential and negative assortative mating. If it is possible to create a perfect negative correlation between mating partners ($r = -1$), negative assortative mating can increase the effective population size by about 20%. However, the reduction of the drift variance will have occurred only with respect to the trait for which the correlation between mating partners was that low.

Similar considerations are relevant to the variance reduction via zero selection differentials. Hill (1980) has pointed out that by selecting the fraction b of a population the genic drift variance will be reduced to

$$[1 - h^2(1 - b)]V_A/N_e$$

When equal numbers of the two sexes are assumed, the expression for effective population size is

$$\frac{1}{N_e} = \frac{1}{4}\left(\frac{1}{N_M} + \frac{1}{N_F}\right)[1 - h^2(1 - b)]^{-1}$$

Avoidance of selection contributes little when family size is constant. On the other hand, and in particular when family size is large, b small, and h^2 large, drift variance can be halved.

So far, considerations relate to generations. When generation intervals vary, variance caused by drift should be related to years. The effective population size on a yearly basis is $N_y = N_e t$.

Large genetic differences between control and breeding populations can lead to genotype–environment interactions, which may bias comparisons. Therefore it is advisable to control selection advance by lines or strains that are genetically not too distant from the selected population.

When a control population is initiated and thus selection stopped, epistatic combinations may disintegrate and cause "genetic slippage" (Dickerson, 1963). This could lead to overestimation of genetic progress, at least in the first generations.

The variance of the difference between controls ($\overline{C}$) and the selected population ($\overline{P}$) becomes, after t generations (Hill, 1972a),

$$V_{G_t} = \frac{2}{t^2}(V_{\overline{P}} + V_{\overline{C}}) + \frac{V_A}{tN_y}$$

Drift variance is divided by t, and the variance of the means by t^2. Therefore, drift variance assumes increasing weight with longer time intervals.

Specific control populations should be distinguished from those used over extended time periods and regions. An example of specific controls created for a particular project is the control Hetzer used to monitor progress in the back-fat selection experiment. On the other hand, the Cornell Randombred Control population is a well-known example of the other type of control, used on a world-wide scale for many purposes.

12.2.2. Repeat Matings

These permit the replication of all or part of a genotype over periods of time. The expectation of the genotype of full-sibs born in various time periods remains the same. Therefore, the expected values of performance differences between such full-sibs are the environmental differences between time periods. Subtraction of such differences from differences between time period averages of the selected population (ΔP) yields the genetic differences ΔG. Goodwin *et al.* (1960) suggested a system of repeat matings for poultry which can also take care of recombination losses and maternal effects. Application of such a repeat mating scheme enabled Dickerson (1963) to quantify the effects of both natural selection and disintegration of epistatic gene complexes relative to maintenance of a selection plateau.

For large domestic mammals such mating schemes are too costly and Dickerson (1962) suggested that the performance of successive progeny groups of sires be used for disentangling genetic and environmental changes. Smith (1962) has elaborated the theory and application of such schemes. The difference between the progeny averages of a sire in successive years should correspond to the expression

$$\Delta P' = \Delta G/2 + \Delta E = E(b_{S.t})$$

and the difference between successive yearly population averages is

$$\Delta P = \Delta G + \Delta E = E(b_{P.t})$$

$E(b_{P.t})$ and $E(b_{S.t})$ represent, respectively, the expected values of regressions of the population average and of half-sib average on time. The difference between the two expressions should give half the genetic change in the time period: $\Delta P - \Delta P' = \Delta G/2$. Figure 12.2 illustrates the situation. The method picks up only the linear component of the change and its validity assumes that sires are mated to a comparable batch of females in the respective time periods and that no selection of sires occurs. Since such conditions are rarely met, various amendments to the procedure are required.

The estimation of genetic progress is usually accomplished by the determination of regression coefficients:

$$\Delta G = 2(b_{P.t} - b_{S.t}) = -2b_{(P - S)t}$$

$b_{(P - S)t}$ denotes the regression of the difference between progeny and population averages on time. When full-sibs can be reproduced in successive time periods, genetic change is estimated from

$$\Delta G = b_{P.t} - b_{FS.t}$$

$b_{\text{FS}.t}$ signifies the regression of full-sib performance on time. Mating of sires in successive time periods to different types of dams changes the expectation of the regression coefficient. Examples of such a practice are mating of older boars exclusively with sows, or, in AI, insemination of heifers exclusively with semen of tested (for calving ease) old bulls. In Fig. 12.3 an increase in age of mating partners is assumed, as was found by Puff (1976) in Bavarian pig populations, where the average age of the sow increased by 5 months for each additional year of age of the boar. Mating to partners from earlier time periods causes the regression to be reduced:

$$E(b_{S.t}) = \Delta E + \tfrac{1}{2}\Delta G(1 - b_{A.t(S)})$$

$$E(b_{P.t}) = \Delta E + \Delta G(1 - \tfrac{1}{2}b_{A.t})$$

$$\Delta G^* = \frac{2}{1 + b_{A.t(S)} - b_{A.t}} (b_{P.t} - b_{S.t})$$

where $b_{A.t(S)}$ and $b_{A.t}$ specify, respectively, the regressions of age of mating partners on time, and of age of females on time.

The regression is biased in the reverse direction by mating of a sire to selected partners. If older sires tend to be mated to above average partners, $b_{S.t}$ will be biased upward and ΔG will be underestimated. For example, if the performance of mates of old sires is SD units above

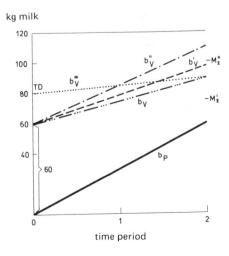

FIGURE 12.3. Estimation of genetic trends. (1) $2(b_P - b_V) = 2(30 - 15) = 30$ (unbiased). (2) $2(b_P - b_V') = 2(30 - 18.75) = 22.5$ (mating of dams of previous generation, $M_2' = 75$). (3) $2(b_P - b_V'') = 2(30 - 25) = 10$ (selection of mates of chosen sires, $M_2'' = 100$). (4) $2(b_P - b_V''') = 2(30 - 5) = 50$ (selection of sires on first progeny test, daughter difference = 40).

average, half of their genetic superiority should be subtracted (Powell and Freeman, 1974):

$$\Delta G^{**} = 2(b_{P.t} - b_{S.t} + \tfrac{1}{2}h^2 SD)$$

If sires are selected on the basis of their first progeny test, and consequently repeated progeny averages derive from selected sires, $b_{S.t}$ is biased downward (Syrstad, 1966). Assume that the superiority of n first progeny of a sire is SD units above the population average. The first $n/(n + k)$ of this difference is expected to be repeated in later offspring. Therefore, the bias of the regression coefficient of t successive half-sib groups of a sire on time will be

$$f_b = \frac{\tfrac{1}{2}SD[1 - n/(n + k)](t - 1)}{\Sigma^t (t - \bar{t})^2}$$

and

$$\Delta G' = 2(b_{P.t} - b_{S.t} - f_b)$$

Application of such corrections generally reduces the apparent genetic gain. This is particularly true of f_b, which cannot be neglected in AI cattle breeding schemes. For example, Kögel (1976) found the estimate of genetic progress in the German Brown breed to be halved to 24 kg/year when the correction for progeny test selection was applied. The same correction reduced the estimate for Gelbvieh (German Yellow cattle) by only about one-sixth (Hindemith, 1978); the difference between the extent of bias in the two populations is probably explained by different selection intensities. Puff (1976) likewise found the estimate of genetic progress in a swine population to be reduced after corrections were applied.

   Smith (1962) showed that the variance of such an estimate of genetic gain is approximated by the expression

$$V_{\Delta G} = \frac{4s^2}{\bar{n} \, \Sigma[y(y^2 - 1)/12]}$$

where y is the number of years in which the sires have progeny, $\bar{n}$ is the average number of progeny per year and sire, and s is the standard deviation of the trait.

   As the denominator contains y^3, prolonging the use of sires is the most effective way to increase the precision of the estimate. For example, when ten bulls have 20 progeny each, five of the bulls in 2 years and five in 3 years, the standard error of the yearly genetic change ΔG is about $s/8$. When five bulls are used for 3 years and the other five for 5 years each, the standard error drops to $s/17$. For this estimate it is not necessary to have progeny in intervening years; it would suffice to use bulls in the first and last years of the period.

Smith (1977) suggests that frozen semen and possibly frozen embryos could be used for estimating genetic progress. When frozen semen of control animals is rotated on daughters of other control sires and if the base generation females are average, the genetic progress can be expressed as

$$\Delta G = b_{(P - C).t}$$

i.e., the regression coefficient of the performance difference between the population and the control animals on time. This estimate has the variance

$$V_{\Delta G} = \frac{6s^2}{St(t + 1)(t + 2)}$$

where S denotes the number of sires both in the control and in the population. The procedure appears to be efficient and if t is not too small, comparatively few sires will suffice to keep the variance small.

Repeat mating procedures have the advantages, relative to control populations, of no accumulation of drift variance and of little chance that genotype–environment interactions will develop. On the other hand, many assumptions—random partners, no selection, etc.—will hardly be fulfilled in practice and corrections become necessary which impair the precision of the estimates. Short time periods lead to relatively large errors of the estimates, but use of frozen semen can help to alleviate some of these problems.

Estimation by regression yields only the linear trend. Application of mixed models (Section 10.3) permits grouping of sires by birth-year cohorts and leads to an unbiased estimate of their breeding values (Henderson, 1973). The difference between the mean breeding values of such birth-year groups contains half of the genetic change in the sires. Lederer and Averdunk (1973), Hindemith (1978), and Kögel (1976) have used the mixed model approach together with the regression method to estimate genetic change in German breeds. In general both methods yielded similar results.

12.2.3. Retrospective Selection Indices

A check on the validity and power of selection theory demands comparison between expected and realized gains. In single-trait selection experiments the expected gain can be estimated in the way outlined in Chapter 9. In practical breeding, however, several traits should be considered and any attempt to estimate genetic progress must take into account both number of traits and the weights attributed to them in the selection of animals. In addition, good estimates of variances and covariances are needed.

The problem was faced by Dickerson *et al.* (1954) when they attempted to estimate genetic change in a trait from a combination of mass and family selection. Their solution was comparable to the proposal of Magee (1965) relating to multiple-trait selection. By this proposal selection pressure on individual traits, or on individual and family performance, is estimated by means of a retrospective index. The procedure need not be adhered to if selection follows a predetermined index and if *a priori* chosen traits are solely considered in selection. Allaire and Henderson (1966a,b, 1967) have used this method to investigate the relative importance which dairy breeders attribute to yield and type traits. Application of index $I = b_1 x_1 + b_2 x_2 + \cdots$ leads to the selection of animals whose phenotype x_1 will be changed, on average, by

$$\Delta x_1 = i \frac{\text{cov } Ix_1}{\sigma_I} = \frac{i}{\sigma_I} \Sigma \ (\text{cov } x_1 x_j) b_j$$

analogous to the genetic change after index selection. If the selection differential equals i standard deviations, the covariance can be deduced:

$$\text{cov } Ix_1 = \frac{(\Delta x_1)\sigma_I}{i}$$

The weighting attributable to individual traits is given by the regression coefficients that result from solving the equations

$$b_1 V x_1 + \cdots + b_n \text{ cov } x_1 x_n = \Delta X_1 c$$

$$b_2 \text{ cov } x_1 x_2 + \cdots + b_n \text{ cov } x_2 x_n = \Delta X_2 c_1$$

.

.

.

The phenotypic variance–covariance matrix and the vector of weights are on the left-hand side. The vector of selection differentials multiplied by σ_I/i is on the right-hand side. σ_I/i $(=c)$ is constant, leaving the ratio of weights unchanged, and therefore can be neglected. The genetic gain from selection by the retrospective index can be computed:

$$\Delta g_i = \sum_j b_j \frac{\text{cov } g_i x_j}{V_I} \Delta I$$

ΔI denotes the selection differential for the index. Puff (1976) has estimated such retrospective indices for Bavarian Landrace and the indices indicated that breeder's weights differ from the recommended weights. The *a priori* index recommended by the advisory service was

$$I_A = 0.14(\text{W/A}) - 3.98\text{BF} - 47.7\text{FC} - 47.4(\text{M/F})$$

and the retrospective index

$$I_R = 0.07(\text{W/A}) - 0.83\text{BF} - 103\text{FC} - 187(\text{M/F})$$

where W/A denotes weight for age, BF ultrasonic back-fat measure, FC feed conversion, and M/F meat/fat ratio. The first two measures are from the individual, the latter two from station-tested full-sibs. The differences between the indices indicate that buyers give much less weight than the theoretical index provides to performance criteria recorded by breeders. At any rate, estimating the genetic progress from the *a priori* index will lead to wrong expectations. The genetic gains expected on the basis of I_R are juxtaposed in Table 12.2 to the realized gains. The differences between the two columns are considerable, but data volume and structure were unfavorable (brief use of sires). Therefore no conclusions are warranted as to possible asymmetry of selection response,

TABLE 12.2
Expected (G_E) and Realized (G_R) Selection Response in Bavarian Landrace[a]

	G_E	G_R
Feed conversion rate, kg feed/kg gain	−0.026	−0.065
Area of longissimus dorsi muscle	0.52	0.15
Fat area/muscle area	−0.016	0.001

[a] Source: Puff (1976).

etc. Nevertheless, an examination of the validity of quantitative genetic theory in field data requires a knowledge of the actual selection which has taken place. Retrospective selection indices should make appraisal of actual selection possible.

Knowledge of the actually applied selection index permits estimation of the selection intensity applied by breeders. The standardized selection differential can be estimated from the ratio between the index selection differential ΔI_R and the index standard deviation σ_{I_R}, $i = \Delta I_R/\sigma_{I_R}$, and the culling percentage can be deduced from i. In the investigation of realized selection in the Bavarian Landrace, the i values for boars and sows were 1.27 and 0.64, respectively. Consequently breeders selected with regard to I_R the best 25% boars and the best 60% sows. The index concerned only four traits, while breeders surely must have taken into account more traits when pondering selection—for example, legs, reproductive performance of relatives, etc. However, information on these was not available for construction of the index. Therefore, it appears that the culling intensity with regard to the four traits is near the practical limit.

Yamada (1977) has suggested that one estimate the whole set of variances and covariances from selection differentials. The standardized selection differential is $i = \Delta P_k/\sigma_k$. Therefore $\sigma_k = \Delta P_k/i$. Further, the secondary selection differential in trait j when selection was for trait k is

$$\Delta P_{jk} = \Delta P_k \frac{\text{cov } jk}{\sigma_k^2}$$

and

$$\text{cov } jk = \frac{\Delta P_{jk}}{P_k} \sigma_k^2 = \frac{\Delta P_{jk} \cdot \Delta P_k}{i^2}$$

These quantities can be inserted in the index equations to estimate the relative weights

$$b_i = P^{-1} \Delta X_i = \frac{1}{i^2} \begin{bmatrix} \Delta X_{11}^2 & \Delta X_{12}\Delta X_{22} \\ \Delta X_{21}\Delta X_{11} & \Delta X_{22}^2 \end{bmatrix}^{-1} \begin{bmatrix} \Delta X_{1.I} \\ \Delta X_{2.I} \end{bmatrix}$$

$$= \frac{1}{i^2} \left\{ \begin{bmatrix} \Delta X_{11} & \Delta X_{12} \\ \Delta X_{21} & \Delta X_{22} \end{bmatrix} \begin{bmatrix} \Delta X_{11} & 0 \\ 0 & \Delta X_{22} \end{bmatrix} \right\}^{-1} \begin{bmatrix} \Delta X_{1.I} \\ \Delta X_{2.I} \end{bmatrix}$$

In a study on realized selection for dairy performance of Bavarian Fleckvieh, Reinhardt (1981) found that 82% of the cows survived to the second lactation ($i = 0.47$). With the culling intensity employed, the following direct and indirect selection differentials result:

$$\Delta M = 350 \text{ kg}, \qquad \Delta F\% = 0.15\%$$
$$\Delta M.F\% = -10 \text{ kg}, \qquad \Delta F\%.M = -0.006\%$$

The realized selection differentials were

$$\Delta M.I = 125 \text{ kg}, \qquad \Delta F\%.I = 0.015\%$$

ΔM is the selection differential for milk (superiority of the 82% best cows), $\Delta M.F\%$ the secondary selection differential for milk when the 82% best cows in fat-% are selected, and $\Delta M.I$ the actual selection differential for milk yield. The retrospective index used for selection is

$$i^2 \begin{pmatrix} b_M \\ b_F \end{pmatrix} = \begin{bmatrix} 122{,}500 & -1.5 \\ -2.1 & 0.0225 \end{bmatrix}^{-1} \begin{pmatrix} 125 \\ 0.015 \end{pmatrix} = \begin{pmatrix} 0.00023 \\ 0.1515 \end{pmatrix}$$

Where genetic changes are used instead of phenotypic selection differentials, a modification of this approach was applied by Harvey (1972) and Berger and Harvey (1975) to estimate realized genetic parameters from multitrait selection experiments.

Let $\delta_{ij} = \Delta_{ij}/\sigma_r$ denote the standardized selection differential for trait i when selection was for trait j, and Δg_{ij} the analogous genetic change. The realized genetic parameters can be estimated from the equations

$$\begin{pmatrix} \delta_{11} & \delta_{21} \\ \delta_{21} & \delta_{22} \end{pmatrix} \begin{pmatrix} h_1^2 & h_1 r_G h_2 \\ h_1 r_G h_2 & h_2^2 \end{pmatrix} = \begin{pmatrix} \Delta g_{11} & \Delta g_{12} \\ \Delta g_{21} & \Delta g_{22} \end{pmatrix}$$

Vangen (1980) used such methods to analyze the results of a Norwegian selection experiment for growth rate and against back-fat thickness in pigs. The realized genetic parameters found in this investigation showed reasonable agreement with the parameters estimated from half-sib covariances.

12.3. Selection Limits

Long-term selection in a closed population should eventually exhaust genetic variability and approach a limit. The existence of such selection plateaus is known not only to animal geneticists but also to practical animal breeders. However, rather few examples are evident in practice. The evolution of racing speed in thoroughbreds could indicate a selection plateau. Cunningham (1975b) found that speed in the classic English events has not improved in the past 70 years, equivalent to about seven generations. Lerner (1958) concluded from a careful analysis of the Berkeley White Leghorn flock that it was on a selection plateau for egg numbers. An attempt was made to induce genetic variability by irradiating sperm (Abplanalp $et\ al.$, 1964), but the nonirradiated segment of the strain responded when selection was resumed, so that it had not been on a selection plateau after all. Selection limits can arise for various reasons: (1) exhaustion of utilizable variability (genic variability for within-population selection), (2) overdominance, (3) multiple selection goals.

The rate of exhaustion of genetic variability will depend on the number of segregating loci, as was pointed out by Castle. If gene action is strictly additive, gene effects are constant between loci, and gene frequencies are equal, the difference in performance between the two alternative homozygotes will be $2na$. The genic variance of the base population out of which the two opposite lines had been selected is $2npqa^2$. The difference between the two lines, one homozygous for all positive alleles, the other for all negative alleles, will be $(2n/pq)^{1/2}$ genic standard deviations. If the base population was an F_1 between two inbred lines, $q = p = 0.5$ and the difference between the opposite limits equals $(8n)^{1/2}$. The prediction of the limit requires knowledge of the gene number, or at least of the number of segregating factors, which is not available. However, the relationship between gene numbers and selection limits can be used in reverse to estimate numbers of loci important for the trait that segregate in the base populations. Also, a number of rather tenuous assumptions are necessary for application of the formulas. Nonvalidity of most assumptions leads to an underestimation of selection limits.

Robertson (1960) used the probability of gene fixation (Section 2.4) to develop an approach to the problem of limits which does not require knowledge of gene numbers. If the product $N_e s$ (s is the selection coefficient) is small, the limit caused by fixation at one locus will be approximately

$$u_0(p_0) - p_0 = 2N_e\, \Delta p$$

The change in the mean, which is influenced by n equally important loci, where each of the homozygotes differs by $2a$, will be $2N_e(\Delta p)2na$. Selection will change the mean in one generation by $2(\Delta p)na$. The ratio of total possible progress to progress in one generation is $2N_e$ generations when gene action is additive and $2N_e(1 + p)/3p$ generations when traits are influenced by recessive genes. The half-time of genetic progress of traits that are largely influenced by additive genes is $1.4N_e$ generations. Intermittent reduction in the size of the population, i.e., bottlenecks, reduces the response. James (1971) has shown that the intermediate response is reduced to $1 - 1/(2N_e)$ of the response of lines maintained at the original numbers. The long-term response is expected to be reduced to a proportion lying between $1 - 1/(2N_e)$ and $1 - (1 - p)^{2N_e - 1}$ of the response expected from an unrestricted line.

Selection between full-sibs can double the effective population size. In the base population the utilizable genic variance between full-sibs is half of the population variance. However, in a selected population the genic variance among selected parents can be considerably diminished, as is evident from formulas (Section 9.2) developed by Bulmer (1971) and by what has been stated in this context by Dempfle (1975). If an intensely selected trait has high heritability, the genic variance available for mass selection may not be much larger than that utilized from selection among family members. Consequently, due to a larger N_e, a higher selection limit can be expected. In the example discussed in Section 9.2 selection of the best 15% of animals reduced the fraction of genic variance from 25% to 21.3%. The reduction affects only the variance between families, while the genic variance within families remains at 12.5% of the original phenotypic variance. Extrapolating to a whole species, one may argue that its subdivision into several subpopulation— breeds, strains, etc.—ensures a higher selection limit than selection within a uniform, more or less diffuse population.

Robertson has pointed out that the selection intensity should be small to ensure sustained progress, and to maximize long-term limits 50% should be culled. James (1972) and Dempfle (1974) have also pointed out in this respect that economic considerations, reflected in interest rates, are important in animal breeding. If selection gains are permanent, one round of selection yields in k years (generation interval $t = 1$ year) an economic return of

$$\frac{\Sigma^k gv}{\Sigma (1 + z)^k} - C = \frac{gv}{z} - C$$

where v is the value of one unit improvement, z is the interest rate, and C is the cost of selection. Continuous selection for k generations accumulates financial gains to

$$\sum^k \frac{g_k v - zC}{z(1 + z)^k}$$

Estimation of selection gains in future generations requires knowledge of the number, frequencies, and effects of alleles (Section 9.3). Dempfle concluded from computer simulation that optimal selection intensities increase when interest rates rise. This is particularly relevant when desirable genes are rare and consequently long-term selection gains can be expected. The need to discount future gains focuses more emphasis on immediate and short-term selection gains and increases optimal selection intensities.

There are no examples of practical breeding which enable one to judge the validity of the concepts discussed in this section. The poultry strains analyzed by Yamada *et al.* (1958) and Jerome *et al.* (1956) which purport to show little selection progress, harbored little genic variance; but their selection history is not known, so no testing of the theoretical development is possible. The causes for the plateau in racing speed have not been analyzed in detail, but considerable genetic variance of racing speed is present in spite of selection for over 200 years for it.

Several investigations of selection limits in experimental populations have been made. Roberts (1966) analyzed selection experiments on mice and it turned out that the half-times of selection response were considerably less than $1.4N_e$ generations. Therefore, the conclusion is drawn that the base frequencies of favorable genes were high. Selection progress over very long times, about 70–100 generations, were reported by Yoo (1980) for *Drosophila* and by Enfield (1977) for *Tribolium*. Furthermore, in all the plateaued populations genic variance was still present, as revealed by successful back-selection and/or by changes upon relaxation of selection. In the *Drosophila* experiment, where the average (over replicates) total response was in excess of 36 genic standard deviations, this total response at the limit was unlikely to exceed N_e times the response of one selection generation and the variation between the selection response of different lines was large and unpredictable. Similar observations were made in the *Tribolium* experiment. Common to many long-term selection experiments is the occurrence of transient plateaus and periods of accelerated response. Events such as rare recombinations, mutations, and even quantitative changes of ribosomal RNA genes observed by Frankham (1980) may be explanations for such patterns of response.

The exhaustion of utilizable genic variability can in principle be diagnosed from analyses of correlations and covariances among relatives. However, errors of the estimates are frequently too large to permit early diagnosis of an imminent plateau. Exhaustion of genic variance naturally permits no success in back-selection nor any change when selection is relaxed.

Selection plateaus caused by exhaustion of genic variance can be broken if new genic variability can be created. This is possible through mutations or by immigration from other populations. As to mutation, the experiments on the Berkeley White Leghorn flock yielded inconclusive results. *Drosophila* experiments have indicated that very high levels of irradiation are required to induce utilizable variability and consequently too many lethal mutations would occur to be endured by animal populations (Clayton and Robertson, 1964). However, investigations on mice (Bailey, 1959) and *Drosophila* (Mukai, 1979) have shown that spontaneous mutations are a source of continuous new quantitative variability, which could marginally affect long-term selection limits, in addition to the rare events mentioned above or earlier.

If superior or at least comparable populations are accessible, genetic immigration is possible and selection plateaus can be passed. In order to break such limits Osman and Robertson (1968) and Roberts (1975) have investigated the recourse to immigrants from the base population or from inferior populations. It appeared that such measures can create utilizable variation. However, several generations of selection are necessary to bring the cross populations to the level of the plateaued line, which under continued selection can then be passed—not very encouraging to practical breeders. Harrison and Mather (1949) and Lerner (1954) have suggested that one should permit several generations of panmixis after crossing to dissolve linkage disequilibria. Osman and Robertson (1968), however, observed in their experiments that immediate selection was just as successful and concluded that sufficient alleles were combined onto individual chromosomes even without waiting. As *Drosophila* has only four chromosome pairs and no crossing-over occurs in the male, any effects due to an initial disequilibrium should be more serious than in species with many chromosomes and free recombination in both sexes.

Falconer (1971) used inbreeding to break a selection limit. Ten inbred lines were derived from a line plateaued for litter size, and one of the four that survived avoided inbreeding depression. This, incidentally, excludes overdominance as a cause of the plateau. Selection was resumed after crossing the four surviving lines and the previous plateau was passed by some 10% of its value. The success can be explained by elimination of undesirable recessive genes through the inbreeding process. This experiment should be of interest in situations where gene immigration from superior populations into plateaued strains is not possible.

A selection plateau can be caused by overdominance even if genic variance is present. In such a situation a strong depression will be observed upon inbreeding, but relaxed selection should not cause much reduction in performance. Selection for combining ability (recurrent reciprocal selection) can move the population above the plateau, as Richardson and Kojima (1965) showed in an experiment with *Drosophila*.

One, possibly the most, important cause of insufficient progress and of genetic plateaus is the need to simultaneously improve several traits. With n traits of equal heritability and with equal correlations among them, the regression coefficient of the aggregate genotype on the average performance is

$$\frac{1 + (n - 1)r_G}{1 + (n - 1)r_P} h^2$$

The coefficient and thus the genetic progress will be zero when $r_G = -1/(n - 1)$ (Dickerson, 1963). A situation like this need not be confined to multitrait selection in the usual sense, but may also occur when the population is to perform in several environmental niches and genotype–environment interactions are important.

The condition known as opposing natural selection can be considered as a special case of multitrait selection with unfavorably correlated traits. Opposing natural selection is reflected in reduced viability and fertility. Upon relaxation or backward selection both traits are improved. Natural selection and improvement of fertility subsequent to relaxing artificial selection became evident in Lerner's shank-length selection experiment (Section 12.1). In the long-term *Tribolium* selection experiment reported by Enfield (1977) the regression coefficients of percent fertile matings on late selection generations were all negative and so were the regression coefficients of selection differential on generation number. Crossing of the selected lines failed to improve fertility, which was then restored to the level of the control population when selection was relaxed. The unfavorable correlation between meatiness of pigs and their resistance to stress could imply a selection limit. The possibility of selection limits caused by a given genetic and economic situation can be visualized from the example of the restricted selection index of Table 11.2. The correlation between the unrestricted index and the aggregate genotype is 0.32. If the conditions of maintenance of egg weight and of body weight are introduced, the correlation is reduced to 0.23. It is easy to visualize further restrictions dictated by the market which reduce or even nullify genetic progress in a population with given genetic parameters.

Selection limits of this sort are usually circumvented by recourse to selection for individual traits of the aggregate genotype in different lines. Examples of this are broilers and layers in poultry, or beef cattle and dairy cattle. This solution will not be possible where components of one and the same trait are negatively correlated, such as, for example, meat quantity and stress resistance in the case of meat production from pigs.

In reality, breeding populations are rarely closed to the extent achieved in selection experiments and to the extent assumed in the theoretical treatment. Therefore, one may assume that selection limits due to the exhaustion of genetic variability are rare and that in most cases of impending plateaus an infusion of genes will have been achieved.

13

Threshold Traits

As with Mendelian traits, so threshold traits have a discontinuous distribution, that is, their phenotypes fall only into two or three classes. However, they do not follow Mendelian rules. Many familial traits and diseases in humans must belong to this class, since only some 15% seem to be caused by major genes, 2% by chromosomal abnormalities, and 3% by environmental causes. The etiology of the remaining 80% is unknown and at least some of these can be considered as threshold traits. In domestic animals examples of such traits are disease resistance if only sick and healthy are differentiated, litter size in monoparous species, where twins and, even more, triplets are rare, and digit number in guinea pigs and some mouse strains, where deviations cannot be ascribed to single genes (Roberts and Mendell, 1975). Wright faced the problem when analyzing digit number of guinea pigs, and he suggested a model which has proved very useful for the analysis of such traits. It is assumed that the observed variation on the visible scale (p scale) is caused by variation of a continuously distributed underlying substance. The manifestation of the trait is possible in two or three classes only—for example, single, twin, and triple births. The change on the p scale occurs when the underlying substance surpasses a certain quantity—the threshold— on the x scale. In the case of twin births the underlying substance may be a hormone or combination of hormones which induce polyovulation (which does not necessarily imply twin births). In the case of disease resistance the causative substance may be an antibody, in polydactyly a morphogenetic substance, etc.

The traits show discontinuous distribution, but at the same time some aspects of quantitative inheritance, and so are called quasicontinuous traits (Fig. 13.1). For the threshold analysis to be valid the x trait must be normally distributed. Parameters can refer to traits on the visible p scale or to the hypothetical substance of the cause x scale. The correlation between the proportion of individuals showing the trait p and the causal variable x is $z/[\bar{p}(1 - \bar{p})]^{1/2}$. The covariance between the variable on the x scale and the area under the normal curve cut off by the variable x equals its ordinate z. The variances of the standardized variable x and of the p variable are 1 and $\bar{p}(1 - \bar{p})$, respectively. The heritabilities on the two scales are related by

$$h_p^2 = h_x^2 r_{xp}^2 = h_x^2 \frac{z^2}{\bar{p}(1 - \bar{p})}$$

and an analogous relation exists for repeatabilities (Lush *et al.*, 1948). Values of r_{xp}^2 for various frequencies of a trait are given in Table 13.1. The denominator of r_{xp}^2 is $\bar{p}(1 - \bar{p})$, which is small beyond 0.7 and 0.3 and very small at the extremes of the distribution ($\bar{p} > 0.95$ or $\bar{p} < 0.05$). The heritabilities of the manifestation of a threshold trait depend on its frequency and are highest in the middle ranges of the distribution. Heritabilities from different populations can only be compared when binomial heritability h_p^2 has been transformed into h_x^2.

Heritability and repeatability of binomial traits are estimated, similarly to quantitative traits, from variances between and within groups. Individuals displaying the trait are coded 1, those missing the trait 0. Robertson and Lerner (1949) have pointed out that this is equivalent to $2xN\chi^2$ (N groups). If a_i denotes the number of resistant animals in a family of n_i individuals, $p_i = a_i/n_i$, the genic variance is given by

$$V_A = \frac{(\Sigma\, a_i p_i - \bar{p}\, \Sigma\, a_i) - (N - 1)\bar{p}(1 - \bar{p})}{2kC}$$

C is the kinship coefficient and

$$k = \Sigma\, n - \frac{\Sigma\, n_i^2}{\Sigma\, n} - (N - 1)$$

Division by the phenotypic variance $\bar{p}(1 - \bar{p})$ yields the heritability. The data fall into two groups—for example, resistant and nonresistant—and can be arranged in a $2 \times N$ table. The heterogeneity χ^2 is $(\Sigma\, a_i p_i - \bar{p}\, \Sigma\, a_i)/\bar{p}(1 - \bar{p})$, the first part of the numerator above divided by the phenotypic variance. The remaining part of the numerator equals the ex-

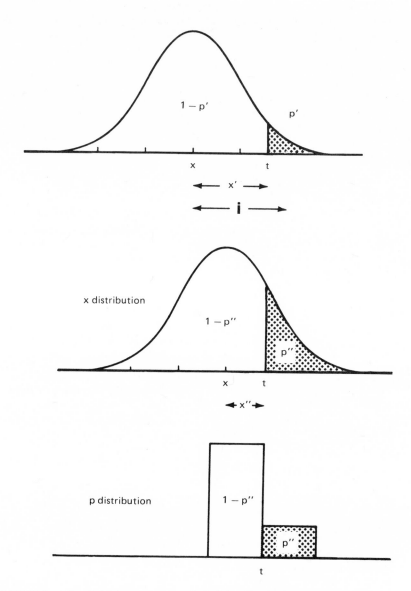

FIGURE 13.1. Threshold character. Top: Underlying distribution in general population. Center: Distribution in relatives of *propositi*. Bottom: Observable distribution.

TABLE 13.1
Comparison of Threshold Heritabilities with Expected Values[a]

	Frequency of trait $\bar{p}$								
	0.1	0.2	0.3	0.4	0.5	0.6	0.7	0.8	0.9
$z^2/[\bar{p}(1 - \bar{p})]$	0.342	0.490	0.576	0.622	0.637	0.622	0.576	0.490	0.342
h_p^2 (transformed from $h_x^2 = 0.362$)	0.12	0.18	0.21	0.22	0.23	0.22	0.21	0.18	0.12
h_p^2 (estimated directly)	0.10	0.19	0.29	0.31	0.23	0.23	0.24	0.26	0.20

[a] Source: Dempster and Lerner (1950).

pected value of the χ^2 (degrees of freedom). The sum of squares caused by true family differences is given by the difference between observed and expected χ^2. The heritabilities given in Table 13.2 were estimated by this method.

Natural selection acts against diseases, specifically against diseases affecting animals of prereproductive age, since susceptible animals do not survive to reproduce. Robertson and Lerner (1949) demonstrated how heritability can be estimated from the ratio of selection response to the selection differential. If survival is coded by 1 and death by 0, the

TABLE 13.2
Heritability of Threshold Traits

	$\bar{p}$	h^2
Layer mortality[a]	0.42	0.089
Lymphomatosis[a]	0.08	0.048
Marek's disease in pullets[b]	0.2–0.4	0.15–0.30
Twin birth in sheep[c]	0.51	0.126
Twin birth in cattle[d]	0.036	0.022
Calving traits in heifers[e]		
Dystocia	0.069	0.043,[f] 0.022[g]
Perinatal mortality	0.091	0.042,[f] 0.018[g]
Calving traits in cows		
Dystocia	0.016	0.005,[f] 0.004[g]
Perinatal mortality	0.041	0.013,[f] 0.004[g]

[a] Robertson and Lerner (1949).
[b] Von Krosigk et al. (1972).
[c] Rendel (1956).
[d] Bar-Anan and Bowman (1974).
[e] Bar-Anan et al. (1976).
[f] Direct effect.
[g] Maternal effect.

part p_0 of the population is resistant, and the susceptible individuals are culled, selection response is given by the expression

$$E(p_1 - p_0) = \frac{V_G}{p_0}$$

where p_1 is the resistant part of the population in the succeeding generation, E is the expectation, and the subscripts indicate generation numbers. The phenotypic selection differential is $1 - p_0$, since 1 is the phenotype of the survivors and the population average is p_0. The heritability equals the ratio of selection response to selection differential; thus

$$h^2 = \frac{V_G}{p_0(1 - p_0)}$$

Robertson and Lerner further demonstrated that the heritability of resistance to two diseases depends on the genetic variances and the covariance of both traits as well as on the frequency of the individual traits. The genetic variance of resistance against two diseases V_T can be expressed as

$$V_T = \bar{p}_A^2 V_B + \bar{p}_B^2 V_A + 2\bar{p}_A\bar{p}_B r_G(V_A V_B)^{1/2}$$

where all variances are genetic. From this, the heritability is

$$h_T^2 = \frac{V_T}{\bar{p}_A\bar{p}_B(1 - \bar{p}_A\bar{p}_B)}$$

In the simple case where the incidence and variance of both diseases are equal ($\bar{p}_B = \bar{p}_A = \bar{p}$, $V_A = V_B = V_G$), the heritability of resistance to both diseases is

$$h_T^2 = \frac{2V_G(1 + r_G)}{1 - \bar{p}^2} = h^2(1 + r_G^2)\frac{2\bar{p}}{1 + \bar{p}}$$

Thus, the heritability of the combined trait of resistance against two diseases depends on the individual heritabilities, the genetic correlation, and the mean incidences, and usually is lower (except at r_G close to ± 1 and high $\bar{p}$) than the heritability of a single-component trait.

In general, the hypothetical variable x cannot be measured. Falconer (1965) suggested an interesting approach to estimating heritability of threshold traits that is basically similar to the method utilized to estimate realized heritability, in which the increased performance of the progeny of selected parents is taken as the selection response and compared to the selection differential. The term "liability" is preferred by Falconer to "susceptibility," since the latter is restricted to the innate property of the individual to contract disease or to manifest the character, while liability encompasses both internal circumstances and external ones, such as degree of exposure, that influence the manifestation of the threshold character. The individual will succumb to disease or manifest the trait when "liability" surpasses the threshold. In estimating the heritability of liability to disease (or to manifestations of any other threshold character), the increased incidence of the trait in relatives of the *propositi* is considered to be the response and the incidence in the affected individuals (the *propositi*) estimates the selection differential. The threshold is assumed to be constant. Furthermore, the variance must be assumed to be the same in both the general population and in the relatives of affected individuals. As shown in Fig. 13.1, the population mean is x' units below the threshold in the general population. If the trait has a hereditary component, the incidence will increase in the relatives of the affected individuals. The average of liability among these relatives will be only x'' units below the threshold. The selection of affected individuals has caused the mean liability of their relatives to change by $x' - x''$ units, which is, so to speak, the selection response. The mean of the "selected" population, that is, the *propositi*, is i units away from the population mean. Therefore the regression of response in the relatives of the *propositi* on the difference in liability between *propositi* and general population is given by $b = (x' - x'')/i$. Then, the regression of selection response on the selection differential is $b = (\text{cov}G_RP)/V_P = 2Ch^2$.

The same approach can be used to estimate repeatabilities and genetic correlations. Here the change in the second trait $y' - y''$ is compared to the selection differential of the first: $b = 2Ch_xh_yr_G$.

The method can be illustrated with data from Bavarian Simmentals (Table 13.3). Eight hundred and eighteen daughters of cows with twin births had at their second calving 44 twin births ($= 5.38\%$), while among their 1453 herd mates of equal age, 58 twins were counted (4%). The mean for the daughters of twin-bearing cows is 1.609 ($= x'$) standard deviations below the threshold, and the herd-mate mean is 1.751. The selection differential, distance between the mean of the herd mates and the population mean, equals 2.154. Therefore, $b = (1.751 - 1.609)/2.154 = 0.066$ and $h^2 = 0.132$. Twin calvings at different ages can be considered as different traits. The regression of twinning frequency at

TABLE 13.3
Twin Frequency in Bavarian Simmentals

Twin birth in dams	Twin births in second calving		Twin births in third calving	
	Daughters	Herd mates	Daughters	Herd mates
Second calving				
n	818	1453	304	502
%	5.38	4.00	5.59	5.18
b	0.066		0.018	
Third calving				
n	1436	2323	645	981
%	5.92	3.53	6.67	4.59
b	0.112		0.087	

[a] Source: Johansson *et al.* (1974).

the third calving of daughters on the twinning frequency of the second calving of dams and the reciprocal can be combined:

$$b = 0.13 = r_G \times 0.132 \times 0.176, \qquad r_G = 0.86$$

Falconer proposed an approximate variance of these regression coefficients:

$$V_b = \left[\frac{1}{i} - b(i - x)\right]^2 W_p + \left(\frac{1}{2}\right)^{1/2} W_v$$

$$W_v = \frac{p}{i^2 a} = V_x$$

Subscripts v and p refer to relatives and to population, but in our example to daughters and herd mates. The variance of the heritability of twinning at the second calvings is

$$V_{h^2} = 4 \left\{ \left[\frac{1}{2.154} - 0.066(2.154 - 1.751)\right]^2 \frac{0.96}{2.154^2 \times 1453} \right.$$

$$\left. + \frac{1}{2.154^2} \frac{0.946}{2.032^2 \times 818} \right\} = 0.00032$$

$$s_{h^2} = 0.0178$$

The method can be used for estimating the repeatability r of individual records and of herd differences. Johansson *et al.* (1974) found in Swedish

Friesians that cows with one twin calving (population average 3.42%, x' = 1.812, i = 2.216) had in later calvings 9.87% twins (x'' = 1.287). Therefore, r = (1.822 − 1.287)/2.216 = 0.241. In Bavarian Simmentals, the twinning frequency at second calving is 3.34% (x' = 1.833, i = 2.226). Herd mates of twin-bearing second calvers had a twinning frequency of 3.63% (x'' = 1.796). The repeatability of herd differences is b = (1.833 − 1.796)/2.226 = 0.17.

The estimates of parameters for continuous variables can be transformed into estimates for binomial variables. The "binomial" repeatability of Swedish Friesians r_p is 0.046, and the repeatability on the x scale is 0.241. The population frequency is q = 0.0324; therefore $z^2/\overline{pq}$ = 0.174 and r_p = 0.174 × 0.291 = 0.042, in satisfactory agreement with the theoretically expected value.

An interesting comparison between the p scale and the continuous x scale was published by Dempster and Lerner (1950). The variable chosen for the x scale was yearly egg production of a poultry flock. Its heritability was estimated to be 0.362. The distribution was made discontinuous by coding; for example, the top 10% of the birds with 1 and the rest with 0, then the top 20% with 1 and the rest with 0, and so on. This made a threshold character out of laying performance, since only the top 10% possessed the trait. This was repeated for other thresholds between 0.1 and 0.9 and the heritabilities were estimated by χ^2 as just outlined. The results as well as the expectations arrived at by multiplying the h_x^2 of 0.362 by $z^2[\overline{p}(1 − \overline{p})]$ are given in Table 13.1. As shown in the table, both the transformed and the estimated heritabilities follow the same general trend and agree quite well. The estimated genetic response Δ_G equals the product of the selection differential $(1 − p)$ and h^2. The low heritability at $\overline{p}$ = 0.1 affects breeding progress, with the consequence that, regardless of intense selection, success remains below that of a population with less selection intensity ($\overline{p}$ = 0.2–0.4) but higher heritability.

The frequency of the trait has a direct influence on the selection differential. If the fraction retained b equals the frequency of the trait p, the selection differential is maximal. If $p > b$, the possibility of more intense selection is forfeited since differences in genic value between carriers of the trait cannot be discerned and breeding animals represent a random sample of all carriers. If the situation is reversed, $b > p$, the selection differential is reduced to

$$\frac{p}{b}\left(\frac{z}{p}\right) + \frac{b − p}{b}\frac{−z}{1 − p} = \frac{z(1 − b)}{b(1 − p)}$$

Again, genetic progress is reduced, since breeding animals that lack the trait are only a random sample of their phenotypic group and have below average breeding value. Assume $b = 30\%$, but $p = 20\%$; the selection differential is $\frac{2}{3} \times 1.4 + \frac{1}{3} \times (-0.35) = 0.82$ provided that animals without the trait can be used for breeding.

The severe reduction in heritability at extreme frequencies can induce slower genetic progress at high selection intensities than if selection is less intense. At $p = 30\%$, $h_p^2 = 0.576\ h_x^2$ and $\Delta G = 0.576h_x^2(1 - 0.3) = 0.403h_x^2$; at $p = 10\%$, $h_p^2 = 0.342h_x^2$ and $\Delta G = 0.308h_x^2$, which is considerably less than at a lower selection intensity. Van Vleck (1972) and Rønningen (1976) have investigated the agreement between progress estimated from h_p^2 and from h_x^2. The genetic progress on the binomial scale is underestimated when the trait is at a low frequency and *vice versa*. If frequency and fraction selected are equal, progress estimated on either scale is the same, but the difference increases with increasing discrepancy between the two. Danell and Rønningen (1981) investigated indexes of continuously and binomially distributed variables and found that agreement between expected and realized (in simulation experiments) genetic progress is greater the closer the frequency of the trait is to the percentage selected.

Heritabilities h_x^2 estimated by the proband method permit estimation of the frequency of a binomial trait in relatives. For example, one may ask for the twinning frequency in daughters of sons from twin-bearing cows. At $h_x^2 = 0.153$ and a population average of 4.3% ($x = 1.717$, $i = 2.125$) the solution is found from

$$(0.153/4)2.125 = 1.717 - x''$$

(regression of granddaughter on granddam equals $h^2/4$). The threshold x'' for granddaughters is 1.636, which implies 5.1% twinning, an increase of 0.8%. Smith (1974) has described equations for eugenic advice which involve more complex situations.

Heritability of threshold traits is highest at medium frequency. This poses a dilemma when selection is for disease resistance, as breeding herds obviously cannot be subjected to such a level of infection. Therefore, in poultry breeding, test flocks are used and selection is based on half-sib and progeny information (von Krosigk *et al.*, 1972).

Prospects for selection can be improved if all-or-none traits can be made to show continuous variation. Lush *et al.* (1948) used age at death as a measure of resistance to leucosis in poultry and Anderson (1960) used age at manifestation of lid cancer in cattle for the same purpose.

Buschman *et al.* (1976) have analyzed antibody production in mice and swine after experimental antigen challenge with this purpose in mind.

An increasing number of major genes have been discovered which cause or influence to a large degree resistance to diseases such as leucosis in poultry (Crittenden *et al.*, 1970) and Marek's disease. These discoveries were surprising since disease resistance was considered to be a classic example of a low-heritability trait where implicitly no major genes are expected. Also, long-term selection progress, as observed by Cole and Hutt (1973) for the two traits, would suggest that many genes contribute to resistance. However, in disease resistance maternal influences such as congenital infection, congenital immunity, degree of infection, etc., would seem to be important and may obliterate the effects of major genes.

Thresholds are important for the understanding of the canalization of phenotypes, i.e., the phenomenon that a phenotype remains relatively uniform and "normal" in spite of large genetic and environmental variability (Rendel, 1967; Roberts and Mendell, 1975).

14

Mating Systems

The mating of animals in a selected group, or in any group for that matter, may be random or there may be a tendency for the mating of like with like or with unlike. In the first instance the mating is positive assortative, and in the second negative assortative. The similarity may refer to the genotype or to the phenotype. Genetic positive assortative mating is inbreeding, and the reverse is outbreeding. If the criterion is the phenotype, the corollary is phenotypic assortative mating, positive or negative, whichever is the case. Unlike mutation, migration, and selection, the mating system does not influence gene frequency, but affects the distribution of genotypes. Phenotypic positive assortative mating for size implies mating of large animals with large partners and small with small, so that frequencies of genes affecting size, for example, are not changed. However, homozygotes will be increased somewhat.

After selection the matings can be manipulated in such a way that genes are combined in optimal genotypes. Consequences of positive genetic assortative mating—inbreeding—relative to the increase of homozygosity and to differentiation of individual lines of a base population were discussed in Chapter 3. The effects of inbreeding and crossbreeding on mean and variance will be treated here.

14.1. Influence on Average

The model for the influence of a single locus on the population mean has been discussed in Chapter 3. The mean of an inbred population with an F degree of inbreeding is given by

$$M = a(p - q) + 2pqd(1 - F)$$

which is less than the mean of a panmictic population if d is positive, i.e., if dominance deviations exist. The difference from a panmictic population is $2pqdF$. If the trait is additive, $d = 0$, and means of inbred and panmictic populations are equal. If dominance exists and if it is directed, which seems to be the rule, the effects at several loci sum to $2 \Sigma pqdF$. This is the inbreeding depression. The formula indicates that inbreeding depression is affected mainly by loci where gene frequencies are intermediate. The expression makes it evident that in the one-locus model the extent of inbreeding depression, or, more generally, inbreeding change, is linearly related to F and that this is true for dominance and overdominance.

The epistatic effects are manifold and it is impossible to describe them in a simple manner as was done in the case of dominance and overdominance. Crow and Kimura (1970) have investigated the behavior of the mean under inbreeding for a number of gene models. Their results can be summarized as follows: Interactions between additive effects of different loci cause no inbreeding depression, interactions between additive effects at one locus and dominance effects at other loci cause inbreeding depression linear in F, and interactions between dominant effects at two loci lead to inbreeding depression curvilinear in F. If epistasis is diminishing, the curve relating performance to F will be concave upwards—considerable inbreeding depression at first but comparatively little depression at higher F. When epistasis is reinforcing, inbreeding depression is again curvilinear in F, but the curve is convex upward, implying little effect of inbreeding at low F values and a marked decline in performance at higher levels of inbreedng.

The majority of observations can be satisfactorily described by the one-locus model, i.e., inbreeding depression is linear in F.

Crossbreeding increases heterozygosity and if dominance is directed, performance is improved. In the one-locus model, the performance increment is proportional to the increase in heterozygosity, which in turn depends on the difference between the gene frequencies of the parental strains. Population P_1 has gene frequencies p and q, and population P_2 has gene frequencies p' and q', where $p' = p - r$ (Table 14.1). If dominance is directed, the F_1 mean is higher than the mean of the two parental strains:

$$M_{F_1} - M_{\bar{P}} = r^2 d$$

This quantity is related to the inbreeding of the parental lines. Assume that the two strains represent a random sample of possible strains of the base population. The variance between such random observations equals

TABLE 14.1
Heterosis in Crosses between Populations

	Gene frequencies	Mean
P_1	p, q	$a(p - q) + 2pqd$
P_2	$(p - r), (q + r)$	$a(p - q - 2r) + 2d[pq + r(p - q) - r^2]$
$\bar{P}$		$a(p - q - r) + d[2pq + r(p - q) - r^2]$
F_1	$(p - r/2), (q + r/2)$	$a(p - q - r) + d[2pq + r(p - q)]$
F_2	$(p - r/2), (q + r/2)$	$a(p - q - r) + d[2pq + r(p - q) - r^2/2]$

$$H_{F_1} = M_{F_1} - M_{\bar{P}} = r^2 d = (p - p')^2 d$$
$$H_{F_2} = M_{F_2} - M_{\bar{P}} = \tfrac{1}{2}r^2 d = -(M_{F_1} - M_{F_2}) = \tfrac{1}{2}(p - p')^2 d$$

		P_1			
		A_1	A_2		
		p	q		
	$A_1 (p - r)$	$p(p - r)$	$q(p - r)$	A_1A_1	$p^2 - pr$
P_2				A_1A_2	$2pq - r(p - q)$
	$A_2 (q + r)$	$p(q + r)$	$q(q + r)$	A_2A_2	$q^2 + qr$

the expected value of half the square of the difference (r) between observations in each random sample: $r^2 = 2\sigma_q^2 = 2Fpq$, i.e., double the Wahlund variance. The heterosis effect or heterosis increment is the improvement possible by crossing and it has the expected value:

$$H_{F_1} = \Sigma\, dr^2 = 2\, \Sigma\, Fpqd$$

which corresponds to the inbreeding depression caused by homozygosity in the parent lines.

The *inter se* mating of F_1 individuals with gene frequencies $p - r/2$ and $q + r/2$ increases homozygosity and reduces heterosis. In the F_2 the heterosis increment is reduced to one-half of its value in the F_1 (Table 14.1) and $H_{F_2} = M_{F_2} - M_{\bar{P}} = \tfrac{1}{2} \Sigma\, dr^2$. Continued *inter se* mating of F_2 causes no further decline in large populations. Gene frequencies remain stable. Therefore, a synthetic population will retain half of the total heterosis increment relative to the base population.

The favorable effects of heterozygosity are present in purebred populations, where heterozygotes are also abundant ($2\overline{pq}$). Therefore, crossbreds between such populations will show heterosis increment to the extent that their gene frequencies differ; in contrast, hybrids between completely inbred lines will manifest total heterosis, which is the difference between the performance of homozygous parents and that of the completely heterozygous hybrids.

The different extent of heterosis in various crosses can be deduced from the inbreeding coefficients of the various crossbreds relative to the F_1. Figure 14.1 illustrates some possible matings between purebreds and crossbreds. It is assumed that the equally inbred base lines A and B ($F_A = C_{AA'} = F_B = C_{BB'}$) are unrelated ($C'_{AB} = 0$).

The kinship coefficient between two F_1 individuals (respectively, the inbreeding coefficient of their offspring) is given by

$$C_{X_1X_2} = \tfrac{1}{4}(2C_{AA'} + C_{AB}) = F/2$$

and the kinship coefficient between hybrids of a four-way cross is

$$C_{W_1W_2} = \tfrac{1}{4}(C_{X_1X_2} + C_{Y_1Y_2} + 2C_{X_1Y_1}) = F/4$$

The inbreeding coefficient of the F_2 offspring of a mating between two F_1 individuals is half as large as the inbreeding coefficient of the base lines. Relative to the F_1 with a maximum of heterozygosity, the F_2 will fall back half the way toward the base population, provided that effects of individual loci are independent (only additive and dominance effects). It is a loss relative to the F_1, but, equally, it is a gain relative to the parent populations. Progeny of a four-way cross have an inbreeding coefficient one-fourth as large as the base population and the decrement in performance relative to the four-way cross is expected to be one-fourth of the difference between crossbreds and parental lines and three-fourths of the heterosis increment will remain.

The crossing of inbred lines of a set that represent random samples of the base population, such that all lines are crossed in all combinations, is called a diallel. The expression was coined by the Danish geneticist Johannes Schmid to denote the mating of two males with two females. Today the expression is used for a system of crossing where k lines are mated in all k^2 combinations, which are composed of $k(k - 1)$ crossbred

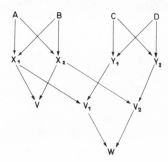

FIGURE 14.1. Inbreeding in hybrid progeny. $C_{AB} = C_{AC} = C_{CD} = 0 = C_{X_1Y_1}$. $C_{AA'} = C_{BB'} = F = C_{CC'}$. $C_{X_1X_2} = F_V = \tfrac{1}{4}(C_{AA'} + C_{BB'} + 2C_{AB}) = F/2$. $C_{V_1V_2} = F_W = \tfrac{1}{4}(C_{X_1X_2} + C_{Y_1Y_2} + C_{X_1Y_2} + C_{X_2Y_1}) = F/4$.

matings and k inbred matings. The diallel mating restores the lost heterozygosity and its average performance should equal the performance of the panmictic base population. Individual crosses possess large heterosis increments, which, however, other crosses may lack. If gene frequencies in two parental lines are nearly equal, the resulting cross will have fewer heterozygotes than the base population and, correspondingly, performance will be poorer. Inbred lines represent a random sample of lines from the base population. Therefore, the average of gene frequencies among all lines corresponds to the population gene frequency and crossing of lines in all combinations including each line with itself results in random union of gametes and consequently equilibrium genotype frequencies. In Fig. 14.2 a base population with gene frequencies of 0.5 is assumed. The difference between alternative homozygotes $2a$ equals ten units, the dominance deviation d is 20, and the performance of the poor homozygote is five units. A random sample of inbred lines from the base population will have a mean performance of ten ($M_{aa} = 5, M_{AA} = 15$) and random mating among all lines including inbreeding yields the original population average of 20. A modified diallel has purebreds excluded and embraces the $k(k - 1)$ crosses (Griffing, 1956). In the example of Fig. 14.2 the average performance of all crosses is 23.3 units, an improvement of 3.3 when compared to the means of the base population and of the complete diallel. The performance increment of the $k(k - 1)$ crosses relative to the k^2 combinations [k inbred lines, $k(k - 1)$ crosses] is $2pqdF/(k - 1)$, i.e., the average of the inbreeding depression of the lines divided by $k - 1$.

As already mentioned, some crosses will have means lower than the mean of the base population, but poor performance due to homozygosity also occurs in some individuals of the base population. Unproductive crosses result from the isolation of such combinations, which are reproduced by the repetition of the particular cross, while in the panmictic population both poor-performing homozygotes and top-producing het-

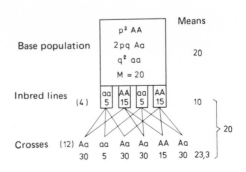

FIGURE 14.2 Diallel mating system.

erozygotes are random phenomena. These considerations underline the importance of selection among crosses. Crossbreeding *per se* does not guarantee record performance. If the one-locus model is accepted, lines with complementary gene frequencies must be chosen, which ensure maximal heterozygosity of the crossbreds and thus heterosis.

14.2. Effects on Variances

Inbreeding and crossbreeding affect the distribution of genetic variances. Inbreeding may occur in a diffuse manner as a consequence of isolated, more or less random matings between relatives, or it may lead to line formation. Individuals within lines mate at random, but lines are isolated in various degrees from each other. In such a situation the variance is affected. Exact predictions are possible only for the genic variance. While inbreeding does not affect the mean of traits whose genetic character is exclusively genic, it increases and redistributes the variance of such traits. If one visualizes lines with complete homozygosity ($F = 1$), gametes are, so to speak, doubled and differences become more abrupt and obvious than in noninbred populations, where more gradual transitions are the rule.

The genic variance of an exclusively additive trait is $2pq\alpha^2$ (α is the gene substitution effect = genotypic effect a; Chapter 4). In panmixis, genic effects of the two alleles of a zygote are independent and

$$V_{\alpha_i + \alpha_j} = V_{\alpha_i} + V_{\alpha_j} = 2V_\alpha$$

In inbred populations genic effects are correlated (the inbreeding coefficient was defined by Wright as the correlation between genic effects of gametes) and the genic variance is increased to

$$V_{AP} = 2V_\alpha(1 + F)$$

The genic variance within an inbred line decreases linearly with the degree of inbreeding and consequently it will be zero in a homozygous line:

$$V_{AW} = 2[p_0 q_0(1 - F)]\alpha^2 = V_{A0}(1 - F)$$

The subscript zero refers to the base population.

The genic variance between lines is given by the difference,

$$V_{AL} = 2FV_0$$

As discussed in Section 6.1, the covariance between observations in one group is identical to the variance component between groups and equals $2C_{XX}V_A$, where C_{XX} is Malécot's kinship coefficient. This in turn is identical to the degree of inbreeding of potential offspring from group members.

The between-line variance can be directly derived. The line mean for a purely additive trait is given by

$$M = a(p - q) = a(1 - 2q)$$

The variance of line means is

$$V_{a(1 - 2q)} = 4Fpq\alpha^2 = 2V_{A_0}F$$

The degree of inbreeding may differ between parents and offspring. If F is the degree of inbreeding of parents and f that of their offspring, the within-line variance of additive traits is given by

$$V_{Aw} = (1 + F - 2f)V_{A_0}$$

No such general solutions are possible for traits influenced by dominance. Robertson (1956) has investigated the behavior of the dominance variance in the case of recessive genes. The estimation of dominance variance requires knowledge of the gene frequencies. If recessive genes are rare, as in the case of lethals, inbreeding at first (to about $F = 1/2$) increases the within-line variance. This surprising outcome is explicable by an increase of homozygotes, which in noninbred populations are exceedingly rare. Inbreeding beyond about $F = 1/2$ leads to a decrease and finally disappearance of genetic variability within lines. The between-line variance increases progressively over the whole range of the degree of inbreeding.

The variance between crosses depends on the degree of diversity in the parent lines, i.e., their inbreeding coefficients. Cockerham (1967) has discussed these problems. Figure 14.2 illustrates kinship of two- and four-way crosses. Remember that the genic variance between groups is $2CV_A = 2FV_A$. Progeny of an F_1 of a two-way cross are inbred to the extent of $F/2$. Therefore, the genic variance between F_1 crosses of different inbred lines will be FV_{A_0} provided the parental lines were not only equally inbred but were random samples from the same base population and had many ancestors in the base population (in other words, animals in any one line should all be related to each other to the same degree). The dominance variance between such F_1 crosses is $F^2V_{D_0}$, which shows that the nongenic variance between crosses becomes increasingly important with an increasing degree of inbreeding.

The redistribution of the variance implies that crosses between fully inbred parental lines will have the total genic variance between them (Table 14.2). Consequently, the best cross will have the genic superiority of the best individual in the base population. The genic value of the best crosses between lines with $F = 1/2$ should equal the best full-sib family in the base population, and the best four-way cross between lines with $F = 1$ should be comparable to the best half-sib group, etc. The variability among crossbred individuals of a particular combination will show the reverse relation—no genetic variability between members of an F_1 cross from completely inbred lines, genic variance halved among F_1 individuals with $C_{XX} = 1/2$, etc. Genetic differences between F_1 crosses are more pronounced than those between comparable (as to C_{XX}) four-way crosses. Consequently, selection between F_1 combinations can utilize more genetic variability than selection among three-way or four-way combinations.

Inbreeding and crossbreeding may also affect the environmental variance. Inbreeding impairs physiological homeostasis, i.e., the ability of an individual to maintain its physiological norm against the impact of environmental influences, and heterozygosity improves it. Skarman (1965) found in crosses from Landrace and Yorkshire swine an improvement of average performance but also a reduction of the within-group variance (Table 14.3), which can be interpreted as an improvement in homeostasis. Klupp (1979) found in three trout strains standard deviations of 50-day weights of 0.18–0.20 g, while the three resulting crosses had standard deviations varying between 0.13 and 0.19 g even though the crossbreds were heavier than the purebreds. Increasing heterozygosity causes populations to be more uniform for heterotic traits.

TABLE 14.2
Variances and Covariances between Crossbreds

	V_A	V_{AA}	V_D		Covariance[a]	
					V_A	V_D
Two-way	F	F^2	F^2	A-:A-	$F/2$	—
Three-way	$3F/4$	$9F^2/16$	$F^2/2$	A.B-:A.B-	$5F/8$	$F^2/4$
Four-way	$F/2$	$F^2/4$	$F^2/4$	A-.B-:A-.B-	$F/4$	$F^2/16$
				AB.-:AB.-	$F/4$	—
				A-.-:A-.-	$F/8$	—

[a] Coefficients of genetic variance.

TABLE 14.3
Variance of Purebreds and Crossbreds[a]

	n	L	LY	YL	Y
		Variance between litter mates			
Eight-week weight	1308	32.24	22.69	20.10	30.99
Twenty-week weight	868	78.45	52.77	57.50	83.29
		Variance between gilts			
Eight-week litter size	619	7.13	5.26	5.16	8.0

[a] Source: Skarman (1965). L, Y: Landrace, Yorkshire.

However, rather few investigations deal with this aspect of crossbreeding, and numbers would have to be large to make differences between the variances significant.

Kidwell and Kempthorne (1966) have made an extensive experimental investigation of the behavior of genetic variances in inbred populations. In a population consisting of lines inbred to level F, the covariances between half-sibs (HS) and full-sibs (FS) have expectations as follows:

$$\text{cov HS} = \frac{1+F}{4} V_A + \left(\frac{1+F}{4}\right)^2 V_{AA} + \cdots$$

$$\text{cov FS} = \frac{1+F}{4} V_A + \left(\frac{1+F}{2}\right)^2 V_D + \left(\frac{1+F}{2}\right)^2 V_{AA}$$

$$+ \left(\frac{1+F}{2}\right)^2 V_{AA} + \frac{1+F}{2}\right) V_{AD} + \cdots$$

Drosophila populations with seven levels of F were established: 0, 0.25, 0.375, 0.5, 0.73, 0.86, 1.0. The population with $F = 1.0$ was created by the marked-inversion outcross technique specific to *Drosophila,* and the remaining lines by repeated full-sibbing. Sib covariances and residual variances were estimated for body weight and chaeta number in the populations at each level of inbreeding but no directed change was discernible, i.e., the covariances showed no increase and the residual variance no decrease. Possible explanations for the unexpected results include selective elimination of homozygotes, a fairly highly inbred and/ or selected base population, maternal age effects, and linkage. The authors concluded that linkage and selection against some genotypes is the most likely explanation.

14.3. Causes of Inbreeding Depression and Heterosis

The two phenomena can be considered as the opposing sides of the same coin—they are caused by less or by more heterozygosity. Most investigations of the effects of inbreeding have shown that it impairs survival and reproduction. Conversely, crossbreeding improves traits sensitive to inbreeding, which are defined as heterotic traits. Obviously, for such traits dominance of favorable genes is positively directed and the fundamental question posed is what causes this widespread phenomenon. R. A. Fisher conjectured that directed dominance developed during evolution. At first, following mutation, the harmful gene is unique and will remain for long periods of time nearly exclusively in heterozygotes. Selection will favor heterozygotes where the gene does no harm, i.e., if the gene is completely recessive. Therefore, selection will concentrate on modifying genes that make harmful genes recessive. S. Wright pointed out that if genes make enzymes, which was at the time of the proposal still a conjecture, the half-rate of enzyme synthesis proposed for the heterozygote should suffice for normal function. This assumption has been proved correct many times. For example, humans heterozygous at the glucose transferase locus (lack of the normal allele causes galactosemia) can be diagnosed electrophoretically even though carriers are clinically absolutely normal. Similar conditions have been found for dozens and possibly hundreds of other genes. However, one may assume that Fisher's hypothesis will be relevant to certain classes of mutants, for example, to morphogenetic mutants.

Inbreeding depression and heterosis that vary linearly with F support the one-locus hypothesis explaining the two phenomena by dominance or overdominance. A curvilinear relation between inbreeding and crossbreeding effects implies epistasis and in particular the interaction between dominance effects of different loci.

The two models, one-locus versus epistatic contributions, were tested with *Drosophila*. Tantawy and Reeve (1956) found little depression in performance at low degrees of inbreeding, but a disproportionately strong reduction at higher inbreeding levels. They explained the observation by conjecturing that homozygosis at few loci has little or no untoward consequences as long as many other loci remain heterozygous. Mukai (1979) permitted mutations to accumulate on chromosome II, with methods specific to *Drosophila*. The effect of the mutations was tested periodically in homozygotes. Even though mutations accumulated linearly with time, viability of homozygotes showed a progressive decline, pointing to reinforcing types of epistasis. Latter and Robertson (1962) found linearity of inbreeding depression and explained the difference in the outcome between Tantawy and Reeve's experiments and theirs by the different breeding history of the populations. While their *Drosophila* population had a large effective size N_e, the population used by Tantawy and Reeve had experienced a "bottleneck" (small N_e) and may therefore have rid itself of detrimental genes. Therefore, weak inbreeding caused little harm.

McGloughlin (1980) investigated heterosis in mouse crosses with various degrees of heterozygosity (F_1, F_2, backcrosses) and found a nearly perfect linear relationship between degree of heterozygosity and litter size. In contrast, inbreeding depression showed no such linearity, but this is explicable by natural selection against the more homozygous individuals. Stephenson *et al.* (1953) report similar observations from poultry. The rate of performance decline became progressively less with an increasing level of inbreeding. Closer analysis revealed that lines with a high degree of sensitivity to inbreeding became extinct through natural selection in the early stages of the experiment. Dickerson (1963) reports evidence of epistatic contributions to heterosis. Under the single-locus hypothesis of heterosis, the performance of a three-way cross can be predicted from the performance of the two nonparental single crosses:

$$E[A \times (B \times C)] = \tfrac{1}{2}[E(A \times B) + E(A \times C)]$$

As shown in Table 14.4, the three-way cross is below the average of the single crosses. Their superiority is due to previous selection among single crosses and their high performance may be caused by favorable epistatic effects in addition to dominance and genic effects. Similar observations were made in experiments with corn, but Stuber and Moll (1969) conclude that, as an explanation of heterosis, epistasis is of secondary importance.

Flock (1980) reports a nearly perfect linear relationship between laying performance and heterozygosity status of crosses (Table 14.5). In contrast, Sheridan (1980), from a summary of several crosses between Australorps and White Leghorns, found that heterosis of egg production in the F_2 was much less than half the heterosis in the F_1 (3 vs. 16%). Therefore he concluded that epistasis was important for egg production. The effect of heterozygosity on fertility in a panmictic cattle population was investigated by Hierl (1976). The degree of heterozygosity was estimated by means of marker loci (blood groups), and calving interval showed a nearly linear relation to this measure (Fig. 14.3). In contrast, Seebeck (1973) reported from Australian experiments a disproportionately large decrease in calving percentages of F_2 and F_3 Brahman crosses (61%) relative to the F_1 (81%) and to Hereford–Shorthorn crosses (70%).

TABLE 14.4
Heterosis and Epistasis in Laying Hens[a]

	Heterosis[b]		Epistasis[c]	
	$\emptyset$P		$\emptyset$2W	T$\emptyset$2W-T$\emptyset$3W
Hatchability, %	82.3	− 2.6	—	—
Viability, %				
Pullets	66	+ 6.1 L + 7.8 B	93.2	+ 1
Layers	67	+ 3 L + 2 B	78.8	+ 2.3
Percent egg production	41	+ 4 L + 5 B	65.1	+ 1.7
Egg weight, g	60	− 0.7 L − 0.7 B	53.4	+ 0.1
Eight-week weight, g	600	+ 25 L + 41 B	—	—

[a] L, Line cross. B, Breed cross. $\emptyset$P, Average of purebreeds. $\emptyset$2W, Average of two single crosses. $\emptyset$3W, average of 3-way crosses.
[b] Nordskog and Ghostley (1954).
[c] Dickerson (1963).

Advanced Afrikaner crosses had the same calving percentage as the F_1 (76%). The depression in performance can be explained by recombination loss, which should not be surprising in crosses between *Bos taurus* and *Bos indicus* populations, which have been isolated from each other for some 5000 years. However, in crosses between Sanga cattle and *Bos taurus*, which are genetically closer to each other (Manwell and Baker, 1980), such a recombination loss appears to be absent.

The largest part of the evidence indicates that inbreeding depression

TABLE 14.5
Heterosis in Laying Hens[a]

	P	F_1	F_2	BC	Percent heterosis increment
Age at sexual maturity, days	164	158	163	163	5.3
Eggs/hen housed	122	143	132	132	17.6
Body weight, kg	1.58	1.65	1.61	1.58	5.7
Egg weight, g	57.5	59.2	57.8	57.5	2.8

[a] Source: Flock (1980). P, Parent line average. BC, Backcross.

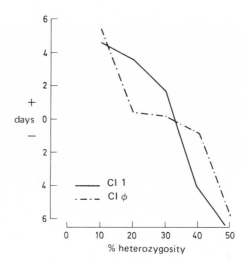

FIGURE 14.3. Connection between marker heterozygosity and calving interval. (—) First and (–·–) average calving intervals are shown. (Source: Hierl, 1976.)

and heterosis can be satisfactorily explained by the one-locus dominance hypothesis. This does not exclude epistatic effects in crosses between widely separated lines when they are highly inbred, are isolated for long periods, or are highly selected.

Another issue concerns the role of dominance versus overdominance as causes of inbreeding depression and heterosis. The existence and importance of overdominance has been much discussed. Some experimental evidence relevant to this problem will be discussed later, but the model suggested by Wright and illustrated in Fig. 2.4 deserves attention in this connection. Pleiotropic effects of alleles which are partially or completely dominant in a favorable direction will cause operational overdominance if single alleles possess jointly positive effects on one trait and negative effects on another. As outlined in Section 2.3, the locus controlling the porcine stress syndrome provides a good example.

Balanced lethals are an extreme case of overdominance. Less extreme cases, which ought to be more numerous, are rarely demonstrated. Sperlich (1973) found an improvement in viability of X-irradiated, largely homozygous *Drosophila* lines, which he attributed to overdominance. But the phenomenon is still debated. It is possible that overdominance occurs at few loci only, which, however, will continue to segregate and thus it may cause a large fraction of the genetic variance of a population. Crow (1952) pointed out that given the usual assumptions ($\mu = 2.5 \times 10^{-6}$, $s = 10^{-4}$–10^{-2}), heterotic loci will cause 100–1000 times the variance caused by other loci. Negative linkage disequilibrium, which may cause associative overdominance, and crossover suppression, which is common in

Drosophila but has also been discovered in mice (Evans and Phillips, 1975), may be favored by natural selection.

Random mating reduces the linkage equilibrium and should thereby also reduce the importance of such pseudo-overdominance. Comstock and Robinson (1952) developed a number of experimental designs to estimate the degree of dominance, which is d/a in the symbols introduced in Chapter 4. Their schemes attempt to compare the importance of dominance variance relative to the genic variance. Under the assumption of two alleles, no epistasis, and no linkage disequilibrium the two variances can be formulated as $V_D = 4 \Sigma p^2 q^2 d^2$ and $V_A = 2 \Sigma pq\alpha^2$. Assuming further that $p = 1/2$, as would be true for an F_2 derived from completely inbred lines, will yield the two quantities $\frac{1}{4}d^2$ and $\frac{1}{2}a^2$. In a diallel mating system the sire variance component σ_S^2 has the expectation $\frac{1}{8}\alpha^2$, the dam times the sire variance component σ_{SD}^2 has the expectation $\frac{1}{16}d^2$, and

$$\left(\frac{2\sigma_{SD}^2}{\sigma_S^2}\right)^{1/2} = \left(\frac{\overline{d^2}}{\overline{a^2}}\right)^{1/2} = \frac{\overline{d}}{\overline{a}}$$

The ratio will be 1 when dominance is complete and larger than 1 under overdominance at several loci. Random mating for several generations should reduce the ratio if the cause is associative overdominance. Moll *et al.* (1964) have compared the degree of dominance of an F_2 from inbred corn lines with the $d{:}a$ ratio for advanced crossbred generations. The average degree of dominance decreased over the generations of panmixis, which indicates that the high ratio of the F_2 was due to repulsion linkage of favorable and unfavorable alleles. In advanced generations $\overline{d}/\overline{a}$ indicated complete or partial dominance for several yield components, but the ratio for total yield revealed the highest degree of dominance. Overdominance may exist at a few loci while dominance at the majority of loci would be partial or complete.

Genetic correlations between purebred and crossbred performance also yields information on the possible existence of overdominance. Assume that the gene frequencies in parental strains 1 and 2 are p_1, q_1 and $p_2 q_2$, respectively. The gene substitution effect in purebreds is $\alpha_{P_1} = a - d(p_1 - q_1)$ and that in crossbreds is $\alpha_{C_1} = a - d(p_2 - q_2)$. The covariance between purebred and crossbred half-sibs is $\frac{1}{2}p_1 q_1 \alpha_{C_1} \alpha_{P_1}$ and the genetic correlation has the expectation

$$\frac{\alpha_{C_1}\alpha_{P_1}}{(\alpha_{C_1}^2\alpha_{P_1}^2)^{1/2}}$$

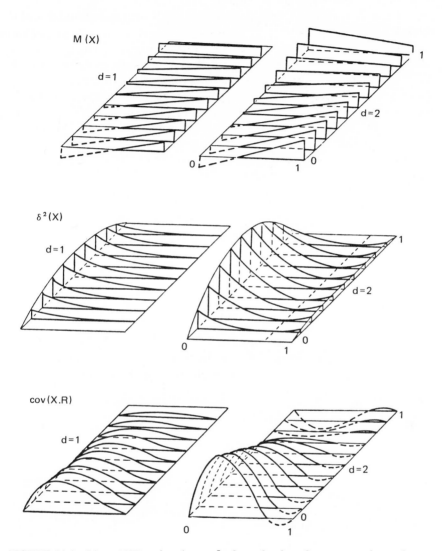

FIGURE 14.4. Mean *M(X)* and variance σ_X^2 of crossbred performance, and covariance *cov(XP)* between crossbred and purebred performance under complete dominance (*d* = 1) and under overdominance (*d* = 2). Abscissa is gene frequency.

For an individual locus this ratio will be either $+1$ or -1, depending on the signs of α_{C_1} and α_{P_1}. The signs of α_{C_1} and α_{P_1} will be opposite when gene frequencies of overdominant genes in parental strains are complementary (Pirchner and Mergl, 1977). If a trait is influenced by n' overdominant loci with complementary gene frequencies and by $m - n'$ other loci, the genetic correlation between purebred and crossbred performances is expected to be $(m - 2n')/m$. Since all other types of gene action—additive, partial or complete dominance, and overdominance at loci with noncomplementary gene frequencies—cause a genetic correlation of 1, a correlation of less than unity, $r_G < 1$, should indicate overdominance. This assumes that other causes of incomplete correlation such as $G \times E$ interaction or epistasis, can be excluded (Fig. 14.4). Continuous selection for specific combining ability which is based on performance of crossbred progeny or crossbred half-sibs (recurrent reciprocal election, RRS) should also cause a decrease in the genetic correlation. The gene frequency change in one generation of selection for crossbred performance is $\Delta p_1 = i(2C)p_1q_1\alpha_{C_1}/\sigma_C$, where σ_C is the standard deviation of the selection criterion and C is the kinship coefficient between crossbred relatives and purebred breeding animals. The analysis of 10 years of RRS in poultry revealed both a correlation deficit ($r_G < 1$) and a decline in the genetic correlation. However, the results permit no differentiation between true (one-locus) overdominance and associative overdominance. If selection had been changed from RRS to selection based on purebred performance, genetic correlation would have been expected to increase to 1.

Experiments of Abplanalp (1974) with highly inbred lines of poultry and of Falconer (1971) with highly inbred lines of mice lend support to the dominance hypothesis. In both investigations, some inbred lines that survived heavy selection between lines retained a high level of performance, for egg number in the case of poultry, for litter size in the case of mice, which would not be expected if overdominance were important.

One may conclude that epistasis has some importance for heterosis and inbreeding depression, but that its influence will be relatively minor, at least as long as large differences between mating partners are excluded. As for the two possibilities of the one-locus hypothesis, dominance has more general importance, at least insofar as it is considered to be a pure one-locus phenomenon. In practical breeding, relatively large chromosome segments should remain intact in the course of the fairly limited number of generations which an individual breeder controls. In this situation overdominance, though associative in nature, will not be without importance.

15

Consequences and Application of Inbreeding

Increased homozygosity by inbreeding affects all homozygotes equally. Consequently, the disproportionately large increase of the rare homozygotes, such as, for example, lethals, is most conspicuous. A trait with a frequency of 1% ($=q^2$) in a panmictic population will be increased to 3.25% ($q^2 + pqF$) in a line with $F = 0.25$, i.e., its frequency is tripled. The more frequent homozygote will be increased, too, by 2.25%, from, say 81% (p^2) to 83.25%. However, this increase will be hardly noticeable, which explains the popular view connecting inbreeding only with increase of rare and deleterious homozygotes. The increase of rare homozygotes is the more conspicuous the rarer they are. As shown in Table 15.1, when the gene frequency is 3.16×10^{-3} for a lethal gene, the frequency of homozygotes will be increased 40-fold in progeny from a half-sib mating but by 126-fold if the gene frequency is 10^{-3}. The consequence of this is that most lethals and detrimental genotypes in a population will result from the few inbred matings that occur usually by chance.

The number of recessive lethals from the fraction of a matings between relatives is $aN[q^2(1 - F) + qF]$, while the total number in the population is $N[q^2(1 - \overline{F}) + q\overline{F}]$. The average inbreeding coefficient of the population will be low, of the order of a few percent (1–5%). Therefore the frequency in the population will be approximately $q^2 + q\overline{F}$ and the ratio of lethals from matings between relatives to all lethals is

$$\frac{a[q^2(1 - F) + qF]}{q^2 + q\overline{F}} = \frac{a(q + pF)}{a + \overline{F}}$$

TABLE 15.1
Genetic Defects in Inbred Lines

Frequency of defect		F	Gene frequency	
			3.16×10^{-3}	10^{-3}
Panmixis q^2			10^{-5}	10^{-6}
	1/8		40×10^{-5}	126×10^{-6}
Inbreeding $q^2(1 - F) + Fq$				
	1/4		81×10^{-5}	352×10^{-6}

If a for half-sib matings is 5% and $\overline{F} = 3\%$, as Lederer and Averdunk (1973) found for Bavarian Simmentals, about 19% of all lethals caused by genes with a frequency of 3.16×10^{-3} will occur among the offspring from the comparatively rare half-sib matings. If matings between relatives are rarer yet, say 1%, still about 3% of all such homozygotes will be from the few half-sib matings.

Morton and co-workers (1956) utilized data from marriages between relatives for estimating the number of recessive lethals a human carries. They compared the mortality of offspring from cousin marriages ($F = 1/16$) with that of offspring from marriages of unrelated individuals ($F = 0$). The mortality rate during childhood and adolescence was 0.23 among offspring of cousins and 0.11 among offspring of unrelated individuals.

The increased mortality rate of offspring from parents who are cousins amounts to more than 100% over that of offspring of unrelated individuals. This increment is caused by homozygosity of an additional 1/16 of the loci. In a completely homozygous individual, 16 times as many lethal genes should become homozygous. Because at a heterozygous locus any one of the genes can become homozygous, the estimate of the lethal load must be doubled. The authors concluded that the average human carries three to five genes that cause death in the homozygous state. In domestic animals the application of this method is more problematic, since artificial selection rarely permits the estimation of mortality rates in an unbiased fashion.

Inbreeding can be used to discover recessive genes and therefore to cleanse, so to speak, a population of undesirable genes. As discussed before, most recessive genes are in heterozygotes and can only be removed by selection if inbreeding has made them homozygous.

Heterosis of crosses can sometimes be explained by the removal of deleterious recessive genes in the course of previous inbreeding. Falconer's (1971) interesting experiment to pass a selection plateau provides an example (Section 12.3).

Traits differ with regard to their susceptibility to inbreeding. Theoretically, this depends on the degree of dominance at the loci influencing the trait, which of course is unknown. However, from empirical investigations it is known that traits like butterfat-% of milk, back-fat thickness of pigs, and, to some extent, egg size of poultry show little if any depression upon inbreeding, and sometimes even a slight positive change. Growth rate is more susceptible to inbreeding depression and traits that are directly related to the selective value, such as litter size of pigs, egg number, and in particular hatchability in poultry, are very sensitive to inbreeding. To some extent milk yield of cattle can be added to this last group. The first group of traits can be considered to be relatively neutral to the fitness of the organism. Traits of this group tend to show higher heritability values and, of course, as a corollary to the small or missing inbreeding depression, no or little heterosis. Traits that are very susceptible to inbreeding depression have in general low heritability and much heterosis. However, generalizations like this have limitations, and exceptions are not infrequent. In particular, the selection history of the population must be taken into account. For example, Falconer (1971) found much less depression of litter size in lines that were inbred from a previous cross between inbred lines.

If the more homozygous individuals have reduced performance and vitality, either they are excluded from reproduction or their contribution to the next generation will be less than that of their more heterozygous sibs. The inbreeding coefficient is an expected, or an average, value. An offspring of a full-sib mating can come from gametes carrying genes differing more by descent than the inbreeding coefficient indicates, and *vice versa*. Consequently, the real homozygosity can be less than estimated from the computed F. If natural or artificial selection operates, the more heterozygous individuals are preferred and the real inbreeding is less than that computed. Briles and Allen (1961) and Briles and co-workers (1957) found that of 73 closed populations of chickens, half of which had inbreeding coefficients of 50% and more, only two were homozygous at the B locus. In additional investigations they demonstrated that the polymorphism present in most of the populations was caused by artificial and natural selection. For example, homozygotes of one kind (B_1B_1) were superior during the early period, while homozygotes of the other kind (B_2B_2) survived better in the laying house. This caused both kinds of genes to be maintained in the population.

Hierl (1976) estimated heterozygosity at ten blood group loci of the

German Hinterwälder breed and compared it to the heterozygosity expected from the parental genotypes. Progeny were more heterozygous than expected, thus indicating natural selection against the more homozygous zygotes.

Mild inbreeding offers possibilities for selection to act against fixation of undesirable genes. Thus, inbreeding depression might be avoided entirely. Dickerson *et al.* (1954) estimated that selection in swine can balance inbreeding depression if the rate of inbreeding per generation does not exceed 2.3%. In those experiments, by pursuing higher rates of inbreeding, performance decreased regardless of selection. Stephenson and co-workers (1953) estimate that mass selection for egg-laying performance can balance the depressing effect of 2% inbreeding per generation. Selection based on an optimum index of individual performance and family average cannot offset more than 5% inbreeding per generation. They concluded that strong inbreeding can decrease performance much more effectively than selection can increase it. Therefore it is barely economical to produce inbred lines for commercial use. However, using inbred lines to produce crosses is fairly widespread in commercial poultry breeding. Selection can counteract the depression caused by the inevitable inbreeding in closed populations under random mating. The population investigated by Tebb (1957) responded to selection for egg production regardless of some unavoidable inbreeding.

In addition, Abplanalp's (1974) poultry experiments, discussed in Chapter 14, would indicate that very intense selection between lines leaves surviving lines with satisfactory egg laying performance.

Inbreeding may affect direct and maternal components of a trait to different degrees and also at various times. As the dam is not inbred, at first inbreeding affects the direct component only. Examples of inbreeding effects on direct and maternal traits of pigs are given in Table 15.2. In cattle Conneally *et al.* (1963) found that early embryonic death was increased by inbreeding both of the dam and of the fetus (regression coefficients of percent early embryonic death on percent inbreeding of dam or of zygote are both about 0.4), while, surprisingly, fetal death (loss after sixth week of gestation) responded to inbreeding of the dam only ($b = 0.4$), i.e., it behaved as if it were a maternal trait.

In general, inbreeding has undesirable consequences which sometimes, as in small, closed populations, may be unavoidable. However, inbreeding also leads to differentiation of the base population by random drift between lines where entirely different genes may become abundant. As mentioned before, prediction of how lines will evolve is not possible, but speed and extent of differentiation can be estimated. Examples show-

Table 15.2
Inbreeding Depression in Swine[a]

	Depression of performance per 10% inbreeding							
	Effect on litter size after given number of weeks				Effect on body weight after given number of weeks			
Inbreeding unit	0	3	8	22	0	3	8	22
Litter	−0.20[b]	−0.35[c]	−0.38[c]	−0.44[c]	0.01	0.04	0.01	−1.56[c]
Sow	−0.17	−0.31[c]	−0.25[c]	−0.28[c]	−0.03	−0.05	0.03	−0.06

[a] Source: Dickerson et al. (1954).
[b] $p < 0.05$.
[c] $p < 0.01$.

ing different reactions of inbred lines of pigs and of cattle are given in Table 15.3.

Selection within inbred lines is impaired on two counts: The poorer reproduction of inbred animals reduces the freedom of culling and the accuracy of selection is lower. The heritability of performance differences within lines is

$$h_{IS}^2 = \frac{(1 - F)h^2}{1 - h^2 F}$$

For example, after two generations of full-sib mating $F = {}^3/_8$. In such a population a trait with $h^2 = {}^1/_4$ will have decreased the heritability to 0.172 after two generations; but no account has been taken of any im-

TABLE 15.3
Differences between Inbreeding Depression in Various Populations

	$F, \%$	Trait	Range of inbreeding depression (per 1% F)
Twelve inbred lines of swine[a]	23–51	Five-month weight	0 to −0.069 kg
Six holstein-Friesian lines[b]	25	Milk yield	−0.4 to −66 kg
		Fat content	0.014 to −0.006 kg
		Age at first calving	0.11 to −0.04 month
		Body weight	0.2 to −5.8 kg

[a] Dickerson et al. (1954).
[b] Mi et al. (1965).

pairment of physiological homeostasis. If this occurs, the environmental variance is increased and the heritability will be decreased further. If a population is subdivided into several strains and selection takes place without further cognizance of the subdivisions, heritability will be $[(1 + F)/(1 + Fh^2)]h^2$, which, with the assumptions made above, results in 0.314. This would imply both that the population consists of several lines with an inbreeding level of $^3/_8$ and that heritability is estimated from likeness of relatives without regard to the subdivisions of the population. A possible decrease of homeostasis is not considered here.

The redistribution of the genetic variance in the course of line formation makes differences between lines more distinct. These differences will be more repeatable than in the general population and the accuracy of estimating breeding value can be improved. The path diagram of Fig. 15.1 illustrates a sire–daughter mating. In a population composed of such progeny groups the genic variance is redistributed between and within inbred groups (f is the inbreeding coefficient of potential offspring of S_{11}, S_{12}):

$$\text{between lines} \quad 2fV_A \qquad\qquad = \quad (5/16)V_A$$

$$\text{within lines} \quad (1 + F - 2f)V_A = \underline{(15/16)V_A}$$

$$(20/16)V_A$$

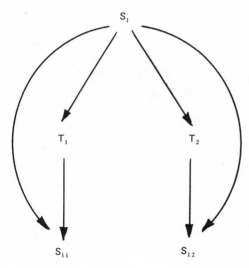

FIGURE 15.1. Sire (S)–daughter (D) mating.

As outlined previously, inbreeding affects all traits independently of heritability. Therefore, the redistribution concerns all traits. In panmictic populations the sire variance component embraces one-fourth of the genic variance. Sire lines shown in Fig. 15.1 have 5/16 of the original genic variance among them, which is more than in panmictic populations. The differences between sires within lines are correspondingly reduced. If these lines descend from highly selected sires, the advantages of such a scheme are reduced, but only with respect to the trait for which the sires were selected. Assume that 5% of sires were selected with an accuracy $r^2 = 0.5$. The distribution of the variance between and within inbred lines will be

between lines

$$2f[1 - r^2 i(i - x)]V_A = (5/16)(1 - 0.5 \times 0.8549) = 0.1789V_A$$

within lines

$$(1 + F - 2f)V_A \qquad = 15/16 \qquad\qquad = 0.9375V_A$$

$$\overline{\qquad\qquad\qquad 1.1164V_A}$$

Without inbreeding the variance distribution between and within progeny groups would be

between progeny groups $\{1 - r^2[i(i - x)]\}V_A/4 = 0.143V_A$

within progeny groups $3V_A/4$ $= 0.750V_A$

$$\overline{\qquad\qquad\qquad 0.893V_A}$$

The relatively large reduction of the "between-line" variance from originally one-fourth to barely one-sixth is due first to the intense selection between sires and second to the reduction of the within-line genic variance $(1 - F)$. If sires are intensely and accurately selected, a formation of inbred lines, as outlined above, may cause an actual reduction of the genic variance, but only for the selected trait. Other traits which may be components of the aggregate breeding value are much less affected by selection and the formation of inbred lines then may increase genic variance and may facilitate selection.

Apart from intentional inbreeding, a mild kind of inbreeding exists when the population is deliberately held at low numbers to avoid diluting genes from valuable ancestors. This form of breeding is called line breeding and combines mild inbreeding with selection. Inbreeding depression is avoided or at least minimized. Only related animals are used, and their genotypes are relatively well known. This avoids the risk of losing a high level of performance through the introduction of unknown breeding stock. In addition, any favorable epistatic combinations, if present, tend to be preserved. If such a breeding system is continued for a long time, inbreeding depression may be unavoidable and some introduction of outside genes may become necessary. To minimize risks in this case, the first crosses should be to related lines, and then only if these are successful should crosses to other lines be made.

It was mentioned in Chapter 3 that in closed populations some inbreeding is inevitable. The rate of this inbreeding is determined by the effective population size and selection causes it to increase, since reproduction is then restricted to a few animals that are similar in phenotype with a tendency to have the same genes. Particularly in traits with higher heritability, for which estimation of the breeding values is fairly accurate, reproduction will be confined to a few families. These descend from only a few superior individuals. Their genes will be spread widely. Thus, selection acts to decrease effective population numbers over and beyond the decrease caused by random inequality of progeny numbers.

Robertson (1961) suggested the following relationship between actual (N) and effective (N_e) population size in the presence of selection (remember that rate of inbreeding can be written as $F = \frac{1}{2}N_e$):

$$N/N_e = 1 + 4i^2t$$

where i is the intensity of selection (z/b) and t is the intraclass correlation of family members, $h^2/4$ for half-sibs, $h^2/2$ for full-sibs. Effective size N_e decreases relative to N with increasing i or t. If the fraction selected is 20%, $i = 1.4$, $h^2 = 0.40$, and the above relation, assuming the population to be subdivided into full-sib groups, becomes

$$N/N_e = 1 + 4 \times 1.96 \times 0.2 = 2.57$$

The rate of inbreeding is more than doubled relative to a population with random mating. When selection is based on an index involving both individual performance and family averages the ratio N/N_e is even more affected:

$$N/N_e = 1 + 4i^2(1 - t)$$

If the intraclass correlation t is low, that is, if the family average is more heavily weighted than individual performance in selection, the ratio N/N_e can become quite large and the effective population size correspondingly small. However, if t is mainly due to environmental correlations between family members, selection should be based on deviations from family averages, with the consequence that breeding animals will originate from many families, causing the effective population size to remain high and the rate of inbreeding low.

16

Crossbreeding

Crossbreeding involves mating between individuals from different populations (lines, strains, breeds). If parents arose from different gene pools, crossbreds have increased heterozygosity and therefore a heterosis increment is expected. The one-locus model of heterosis leads one to expect more heterosis the more divergent genetically the parents are. As mentioned before, epistasis may contribute to heterosis. If partner populations are very distant genetically, crossbreds may have lost the coadaptation of gene complexes and F_1 performance may be even lower than the midparent level. An observation of this nature is reported by Enfield (1980) from a long-term (100 generation) selection experiment with *Tribolium*, where the performance of the F_1 between the replicate selection lines fell below the pure lines. However, impairment of this kind is to be expected more frequently in advanced cross generations. The haploid parental gene arrangements are still intact in the F_1 but will disintegrate progressively in the successive generations. If epistasis is important, this disintegration may cause a decrease of performance. Dickerson (1973) coined the term recombination loss for this impaired productivity. He defined it as the fraction of pairs of independently segregating loci that are nonparental combinations.

Heterosis is important for the same traits as are of concern in inbreeding, with reversed sign, of course. Reproductive traits and health benefit most from crossbreeding, while traits like egg quality, back-fat thickness, etc., show no heterosis. Milk yield and growth rate have heterosis increments somewhere between these two groups. Of course, as indicated above, the extent of the heterosis increment of a given trait depends on the genetic distance between the parent populations.

The differentiation between direct and maternal effects is very useful when discussing the merits of crossbreeding. It is indicated when the general productivity to be improved is determined largely by heterotic traits. Examples include egg production, an important component of reproduction, and meat production, where heterosis is important for the reproduction of the meat animals but where some heterosis also occurs for growth. The total productivity of a cross $A \times B$ can be described by a model (Smith, 1964; Moav, 1966) such as

$$T = P\left(\frac{A}{2} + \frac{B}{2} + H\right) + R(B)$$

The total productivity T equals the sum of the direct production P and the reproductive performance of the dam line, a maternal trait (H is the individual heterosis).

The expected heterosis increment of some cross combinations for direct and maternal traits is given in Table 16.1 (Dickerson, 1973). The table also gives the fraction of nonparental haploid gene combinations that increase with increasing numbers of loci. When three loci are important, the gametes produced by an F_1 have only one-fourth of the parental gene combinations and if coadaptation between different loci is important, impairment of F_2 performance will be unavoidable.

With the exception of very wide crosses, heterosis increases with decreasing relationship between parental populations. An example of

TABLE 16.1
Gene Effects in Crossbreds[a]

♀ ♂	h_I	h_M	r_I	r_M
$F_1\ A \times B$	1	0	0	0
BC $A \times (A \times B)$	1/2	1	1/2	0
$F_2\ (A \times B) \times (A \times B)$	1/2	1	1/2	0
Three-way $C \times (A \times B)$	1	1	1/2	0
Four-way $(C \times D) \times (A \times B)$	1	1	1/2	0
Rotation $A \times B$	2/3	2/3	2/9	2/9
$C \times A \times B$	6/7	6/7	6/21	6/21
Synthetic				
Two strains	1/2	1/2	1/2	1/2
Three strains	2/3	2/3	2/3	2/3
Four strains	3/4	3/4	3/4	3/4

[a] h_I, h_M, Individual, maternal heterosis. r_I, r_M, Fraction of recombinations of two independent loci.

this is provided by Nordskog and Ghostley (1954), who reported less heterosis for line crosses than for breed crosses (Table 14.4). The importance of heterosis for maternal and for direct components is illustrated by figures from a review of Sellier (1970). These results also demonstrate the multiplicative nature of the function that relates components and total productivity (Table 16.2).

In single crosses reproductive performance reflects mainly the purebred maternal genotype, while heterosis of the newborn is relatively unimportant. However, growth of the progeny benefits from heterosis. The full utilization of heterosis demands three- or four-way crosses. The final product of such crosses benefits from heterosis of growth traits, intra- and extrauterine viability, and maternal traits.

Table 16.3 gives purebred performance of Duroc (D), Yorkshire (Y), and Hampshire (H) pigs as 300, 297, and 308 g, respectively, for daily meat gain, and as 5.3, 7.4, and 5.2 piglets per litter, respectively. Heterosis for growth is assumed to be 30 g and fetal heterosis for litter size is 3/4 piglet. The same increment is assumed for maternal heterosis. The meat gain of a three-way cross H × (D × Y) should be

$$\frac{308}{2} + \frac{297 + 300}{4} + 30 = 334$$

TABLE 16.2
Heterosis in Breed Crosses[a]

	Single cross	Three-way cross
Dam	Purebred	Cross
Litter	Cross	Cross
Maternal performance		
Number born alive	102	108
Percent weaned	106	106
Number weaned	108	116
Litter weight, birth	104	110
Weaning weight, litter	115	125
Individual performance		
Weaning weight, animal	106	107
Daily gain	106	106
Weight, 5 months	110	110
Feed conversion	103	103
Percent prime cuts	100	100

[a] Source: Sellier (1970).

TABLE 16.3
3 × 3 Diallel (Swine Crosses)[a]

			δ		
			D	H	Y
♀	D	Daily meat gain, g	300	330	322
		No. piglets weaned	5.3	6	7.2
	H	Daily meat gain, g	335	308	306
		No. piglets weaned	6	5.2	6.6
	Y	Daily meat gain, g	336	332	298
		No. piglets weaned	8	6.5	7.4

Average of crosses: 327 g; 6.7 piglets
$GC^1(D) = \frac{1}{4}(3 - 5 + 8 + 9) = 3.75$
$GC^2(H) = -1.25$
$SC(D \times H) = 5.5 - 3.75 - (-1.25) = 3.0$

[a] D, Duroc. H, Hampshire. Y, Yorkshire. GC denotes general combining ability. SC is specific combining ability.

Crossbred sows D × Y should manifest full heterosis and the same is expected of the three-way crossbred piglets. The litter size of the three-way cross is expected to be

$$\frac{5.3 + 7.4}{2} + 0.75 + 0.75 = 7.85$$

The benefits of crossbreeding to various species (performance traits) depends on the extent of heterosis for productive, reproductive, and possibly paternal traits and furthermore on the scale of breed differences for the characteristics of performance. Complementarity between breeds (populations) for productive and reproductive traits are also of great importance. Finally, the rate of reproduction of the species in question is relevant. Issues on three levels are to be decided:

The first decision concerns the choice between continuous and discontinuous crossing. Terminal crosses are discontinuous, in which the final product is to be marketed and not to be used for breeding. In continuous crossbreeding, such as criss-cross or three- or four-way rotations, the females are used for reproduction of the next generation.

The next step concerns the choice of the partners of a cross combination. Partners that promise to nick well, i.e., where progeny have much heterosis, will be desirable. Also, complementarity will be important and fertile lines, etc., will be chosen for the maternal side of a cross and fast-growing lines as the terminal sire line.

The third problem concerns the improvement of a cross once it is chosen.

Neither inbreeding followed by crossing nor selection for combining ability, as just discussed, can be used in species with low rates of reproduction, such as cattle, sheep, and horses. In these species the pure strains need to encompass a large part of the population, so relatively few females would be available for crossing. The heterosis in this group could not be expected to balance the loss of performance incurred by the inbred strains or, generally, the cost of the scheme. Therefore, the production of commercial crosses some time ago turned to rotational crossbreeding (Winters *et al.,* 1935). In this system of breeding, the females themselves are crosses, and these are mated to males from inbred lines or pure breeds, depending on whether the rotation is among breeds or inbred lines. Figure 16.1 illustrates a three-way cross and the proportions of genes from the different breeds expected in the various generations.

Carmon and co-workers (1956) attempted to predict the performance of progeny of rotational crosses from the performance of progeny of single crosses among constituent lines. For predicting the performance from a two-way rotational cross they suggested use of the expression

$$R(2) = C(2) - \frac{C(2) - P(2)}{3}$$

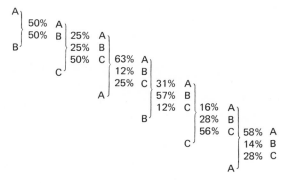

FIGURE 16.1.   Rotational cross between three lines, and the ensuing proportions of genes in the cross (*A, B,* and *C*).

where $C(2)$ and $P(2)$ denote the performance of the single cross and the average performance of the parental strains, respectively. According to this expression, the performance from a two-way rotational cross is lower than from a single cross performance by one-third of the difference between the single cross and the parental average. This reduced performance is due to the reduced heterozygosity of progeny of rotational crosses relative to single crosses. The equation for estimating a three-way rotational cross is

$$R(3) = C(3) - \frac{C(3) - P(3)}{7}$$

where $C(3)$ and $P(3)$ denote the averages of single crosses and parental strains. The reduction relative to single crosses is less than with two-way rotational crosses, as heterozygosity remains higher in a three-way rotational cross. However, an average of three single crosses is likely to be lower than the value for one single cross. The same authors tested the various crossbreeding schemes with mice. Single crosses were superior to rotational crosses, and these in turn were superior to pure strains. From theory, the performance of rotational crosses would be expected to approach the average of single cross performance as the number of strains participating in the rotation increases. However, such a breeding system becomes complicated, and, furthermore, as the number of crosses increases, their average performance will necessarily approach that of the general population.

The advantage of such crossbreeding systems consists of the built-in production of breeding females, which may be considerably more economical than the continuous purchase necessary in terminal crosses. Possibly even more important, continuous crossbreeding reduces hygienic problems which arise from the introduction of foreign stock. The disadvantages are twofold: It is impossible to utilize complementarity of strains and the composition of the cross may change drastically between generations. After attaining the equilibrium at about the fourth generation, the three-way cross illustrated in Fig. 16.1 has 31% genes of strain A, which is reduced in the next generation to 16%, to rise to 58% one generation later. This may cause difficulties in the market. Consequently, rotational crosses should ideally involve similar lines, such as Yorkshire and Landrace, for example.

Discontinuous systems are practically restricted to species, such as pigs and poultry, with highly productive breeding cycles. If sex determination of sperm becomes practical, or if twin births can be induced routinely, discontinuous crosses would become more practical in cattle.

Choice of the crossbreeding system is influenced also by the size of heterosis effects and recombination losses. If the latter are serious, a four-way cross, which is otherwise optimal because it fully exploits maternal, paternal, and product heterosis, will hardly be competitive. The discontinuous or terminal crossbreeding systems permit exploitation of complementarity by deploying strains according to their advantages. A fertile strain or breed can be used as a dam line and a fast-growing meaty strain as a sire line. Moav (1966) termed the benefit from complementarity "profit heterosis." An additional advantage of terminal crossbreeding systems is the uniformity of the final product. The disadvantage is the need for continuous acquisition of purebred animals. Terminal systems are economical for large-scale production units where all in–all out replacement procedures are desirable for hygienic reasons.

The advantage of complementary strains—the importance of position effects—can be illustrated with data given in Table 16.3 and by Fig. 16.2 (Young *et al.*, 1976). Daily meat gain and number of weaned piglets are given for three swine breeds and their reciprocal crosses. For estimation of the overall profit it is assumed that 10 g daily meat gain is worth 6 monetary units and one additional piglet 8 monetary units. The increments of profit from increased performance are not constant over the whole range of performance; in particular, fecundity affects profit in a curvilinear manner (Moav, 1966; Jakubec and Fewson, 1971). However, to simplify the presentation, linearity is assumed, which in view of the small range of performance values should cause small errors only. The profit accruing from pure Durocs is expected to be 300 × 0.6 units + 5.3 × 8 units = 222.4 monetary units. It is further assumed that the average of both reciprocal crosses is the best estimate of crossbred performance, i.e., for meat gain by the combination D × Y: (322 g + 336 g)/2 = 329 g. The profit expected from the combination D♂ × Y♀

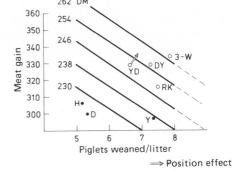

FIGURE 16.2 Position effects in pig crosses. The numbers labeling the straight lines correspond to relative monetary units. RK backcross; DM monetary units.

is expected to be 329 × 0.6 units + 7.25 × 8 units = 255.4 monetary units, while from the reciprocal cross Y♂ × D♀ it should be 329 × 0.6 units + 6.6 × 8 units = 250.2 monetary units, lower by 5.2 monetary units. The cause for this decrement is, of course, the use of the less fertile breed in the position of the dam line (litter size of purebreds is taken as the average of the two crosses where the respective breed is used as dam line). This positional effect is missing, or is much smaller if the differences between breeds are smaller, as, for example, in case of the reciprocal D × H crosses (252.3:249.4). The benefit of complementarity does not depend on heterosis; it can accrue even if traits behave in a strictly additive manner.

The problems involved in estimating crossbred performance and in choosing profitable crosses will be discussed by reference to a diallel cross. The total k^2 combinations between k strains embrace k purebred strains and $k(k - 1)$ crosses (Table 16.3). A "modified diallel" consists only of the crosses. Diallel crosses permit the evaluation of the crosses and the estimation of the genetic variances. Diallel crosses are much used in plant breeding, but also, interestingly, in behavior genetics (Ehrman and Parsons, 1976). Sprague and Tatum (1942) introduced the expressions general and specific combining ability, which proved to be very useful in the analysis of diallel cross experiments. The performance of all crosses that derive from one strain is designated the general combining ability (GCA) of this strain. The specific combining ability (SCA) is a joint attribute of two strains and signifies the deviation of a cross from the sum of the GCAs of its parent strains:

$$C(A \times B) = GCA(A) + GCA(B) + SCA(A \times B)$$

For example, the genotypic value of the cross D × H can be decomposed as follows:

$$C(D \times H) = 3.75 - 1.25 + 3 = 5.5$$

The general combining ability results from the average effects of genes of strain A if these are combined with the genes of the other strains of the diallel, in analogy to average gene effects or to transmitting ability within populations. Specific combining ability is the deviation of the cross performance from its expected value, somewhat analogous but not identical to dominance deviations in within-population analyses. In decomposing the variances caused by GCA and SCA the assumption is made that equal inbreeding of all strains and random mating between animals of different strains (no selection within strains) occurred. If these assumptions are justified the variances estimated from a modified diallel are

$$V(\text{GCA}) = FV_A + F^2V_{AA} + F^3V_{AAA}$$

$$V(\text{SCA}) = F^2V_D + F^3V_{DA} + F^4V_{DD}$$

where V_{AA} and V_{AD} denote variances caused, respectively, by epistatic interactions between average gene effects at two loci and by interactions between average gene effects at one locus and dominance deviations at the other. A comparison of the two quantities indicates that SCA is more important if strains are more inbred and consequently more differentiated. Increased inbreeding makes complex gene combinations more stable and thus such gene combinations are repeated to a greater extent in crossbred individuals. Therefore, between generations of such crossbred individuals an increasing part of the genotype, proportional to functions of F of the parental strains, will recur. This contrasts with panmictic populations, where successive generations share only the transmitting value, or with full-sibs only this value plus one-fourth of dominance deviations.

As pointed out before, not all cross combinations are expected to be superior to the base population or even to the parent strains. Therefore, in crossbreeding selection between crosses (Y_{ij}) also becomes important. Progress by such a selection is, analogous to mass selection,

$$\Delta G = (Y_{ij} - \overline{Y})\frac{\text{cov }YG}{V(Y_{ij})}$$

i.e., the product between the selection differential $Y_{ij} - \overline{Y}$ and the regression coefficient of future cross performance on observed performance or, generally, on a selection criterion. The covariances between the selection criterion and the genotypic value of the cross, i.e., its future performance, are given in Table 14.2. In the case of two-way crosses the covariance is $FV_A + F^2V_D$, in the case of four-way crosses $\frac{1}{2}FV_A + \frac{1}{4}F^2V_D$, etc. The variance of observed cross performance will be $V_G + V_R/n$, where V_R is the total variance within cross combinations and V_G the variance of the genotypic values of crosses. In truncation selection the selection differential will be $i\sigma_{Y_{ij}}$, where i is the superiority in standard deviations of the k best entries of an ordered sequence (e.g., Table 6.2). The genetic progress by truncation selection will be

$$\Delta G = i\frac{\text{cov }YG}{\sigma_{Y_{ij}}}$$

The estimation of the performance of simple crosses from parental performance, or of the performance of complicated crosses (e.g., four-way) from that of simpler crosses (e.g., two-way) is a traditional problem of crossbreeding theory (Jenkins, cited by Cockerham, 1967). The number of potential crosses between even a moderate number of strains prohibits their exhaustive testing and makes prediction from related crosses, etc., a necessity. Between k strains $k(k - 1)/2$ single crosses are possible if reciprocals are neglected. The numbers of three- and four-way crosses, again neglecting reciprocals, are $k(k - 1)(k - 2)/6$ and $k(k - 1)(k - 2)(k - 3)/24$, respectively [the number of combinations taken three (resp. four) at a time]. Where $k = 10$ these amount to 45, 120, and 210 different combinations. The testing of even the 45 single crosses will be a formidable task in terms of numbers, space, and expense and the complete testing of all crosses of higher orders is practically impossible.

Some covariances are given in Table 14.2. It is evident that purebred performance predicts the genic component of crossbred performance only. The dominance component of crossbred performance can only be estimated from another cross where the parent strains appear on separate sides in the other cross. For example, the covariance between dominance deviations of a three-way cross $(A.B \times C)$ and a two-way cross $(A \times B)$ is F^2V_D but it is lacking between $(A.B \times C)$ and the single cross $(B \times C)$.

The accuracy of selection of crosses can be increased by a selection index that combines all the information contained in a diallel, for example (Henderson, 1963; Cockerham, 1967). The index coefficients are derived from a set of linear equations such as

$$V_{Y_{ij}}b_i = V_{GCA}$$

where $V_{Y_{ij}}$ denotes the variance–covariance matrix of observed cross means, V_{GCA} is the vector of variances of the GCA of the strains involved, and b_i is the vector of regression coefficients.

Jenkins suggested three estimators for double crosses. Estimator A is the mean of all possible single crosses between the four parental strains (six), estimator B is the mean of the four nonparental single crosses (for $A \times B$ and $C \times D$ these are $A \times C$, $A \times D$, $B \times C$, and $B \times D$), and estimator C involves all crosses where the parent strains are involved. Cockerham (1967) showed that estimator C is optimal if GCA is important but estimator B is if dominance mainly determines the quality of the cross, that single crosses are accurately tested, and that the single-locus hypothesis is valid. The difference in the qualities of the three predictors and the differences between these and a selection index were found to be relatively small. If the one-locus model is valid, the efficiency of estimating a four-way cross yield from single crosses appears to be surprisingly good.

Empirical observations often agree fairly well with theoretical expectations.

The correlation between performance of parental inbred lines and the offspring of their crosses is generally high for traits with high general combining ability and large additive gene effects. For such traits, prediction of the cross performance can be made from performance of the parental lines with reasonable accuracy. For example, Bradford and co-workers (1958) found that growth of the offspring of crossed lines of swine during the fattening period can be predicted rather well from the performance of the parental lines. Blyth and Sang (1960) also found a rather close connection between parental lines and offspring performance for a number of production traits in poultry. Surprisingly, laying performance, which generally has low heritability, was also predictable from parental performance. However, of the six lines crossed, only one had been previously selected for egg production. Therefore, the other lines may have had considerable additive-genetic variance for egg production, making it possible for general combining ability to be more important than usual.

Where specific breeding value is important, only test crosses can reveal the very best combination of lines. For example, Merritt and Gowe (1960) found general breeding value important for traits, such as egg weight and body weight, that are generally found to have high heritability. Conversely, specific combining ability caused the largest part of the genotypic variance in traits like egg production and age at sexual maturity. In an experiment with six inbred lines of swine, Hetzer and co-workers (1961) found the variance caused by general combining ability to be more important than the variance due to specific combining ability. The tested lines, however, had little inbreeding. In contrast, Henderson (quoted by Hetzer et al., 1961) found, in crosses among more highly inbred lines, general combining ability to account for about 5% and specific combining ability for 5–12% of the total variance of eight litter traits.

The estimation of the quality of cross combinations could be improved if genes and their frequencies in pure strains were known. Of course this knowledge is far off in the future, although gene markers can be used to estimate the genetic distance of strains. If those distances are representative of the whole genotype, in particular of the genotype relevant to performance, heterosis in crosses is expected to be directly proportional to the square of the genetic distance between parent strains.

Hierl (1976) reports a nearly linear relation between marker heterozygosity and fertility of cattle in one breed. However, only 1–2% of the variance was associated with differences in the marker heterozygosity, not enough to make mass selection worthwhile, but possibly useful when partner strains for crossbreeding are to be chosen. Hohenbrink (1970) related the heterosis increment of pig crosses to the marker distance and observed some connection. However, in this and in Hierl's investigation rather few markers were used, which may be unrepresentative. Use of more and better markers could provide a useful tool for selecting optimal crossbred combinations. Sarkissian and McDaniel (1967) used the metabolic activity of mixtures of parental mitochondria to predict heterosis of corn crosses and Dzapo *et al.* (1973) applied this method with some success to rabbits.

In crosses between inbred lines and particularly if the lines are highly inbred, the single cross combinations cannot be improved further. Since most of the loci are homozygous in a highly inbred line, selection will be ineffective in changing gene frequencies. The only way for improvement is to replace one cross combination with another, that is, selection between crosses. Therefore, a number of selection schemes have been suggested that aim at improving the specific combining ability of a particular combination. The lines used for such breeding schemes are not at all, or are only mildly, inbred and should contain considerable genetic variation. The two breeding schemes to be discussed have been described as recurrent selection and reciprocal recurrent selection.

Recurrent selection was proposed by Hull (1945) for the improvement of specific combining ability in corn. In this method, male plants of one line are crossed to an inbred tester line. In animals, females are usually crossed to inbred males, minimizing the deleterious effects of inbreeding on fertility. The tested dams are selected on performance of their testcross progeny. Such females are mated to males of their own line to reproduce the segregating population under selection for combining ability. This cycle is repeated in the next and the following generations. At heterotic loci, the frequencies of those alleles that are complementary to the alleles of the tester line are increased. For example, if the tester line is AA, recurrent selection will increase the frequency of a in the segregating population, which eventually will become aa. Ultimately, the segregating population will become homozygous in a way complementary to the inbred tester. If that state is reached, performance cannot be increased further by selection.

The second breeding method that aims at improving the specific combining ability of two strains is recurrent reciprocal selection (RRS). Bonnier (1939) suggested improving the performance of a breed cross by selecting the two constituent breeds for improved performance. Comstock and co-workers (1949) proposed crossing two segregating populations and basing selection within these on the performance of their progeny. If both additive and overdominant (true and apparent) gene action are important, this method should permit the greatest long-term progress. Recurrent reciprocal selection, as proposed by these authors, is illustrated in Fig. 16.3. Parent individuals of both populations are crossed for production of test progeny. Based on the test progeny performance, parents are selected and mated within their own line to produce the next generation of pure lines. The cycle is repeated in the next generation. This is the same as progeny testing breeding animals by their testcross progeny. Therefore, the system has the disadvantage of progeny testing—the long generation interval, which can, however, be shortened if selection is based on the performance of testcross half-sibs. (Selection between purebred breeding animals is based on the family average of their crossbred half-sibs.) The breeding value of the purebred individuals is estimated from crossbred animals using a regression coefficient based on the heritability of crossbred performance. The heritability is based on the genetic variance as estimated from the variation between half- and full-sib groups of crossbred individuals, similarly to the estimation procedure used for purebred individuals. However, the sire component so estimated does not depend exclusively on additive effects, but

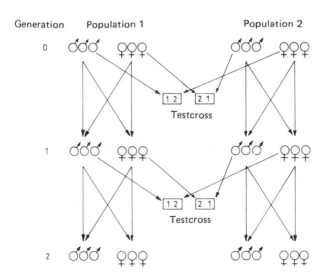

FIGURE 16.3. Recurrent reciprocal selection.

is also influenced by gene effects, showing dominance in the crosses. Therefore, the "cross heritability" is usually higher than heritability of the same trait in pure lines.

Nevertheless, the comparison with mass selection will only be favorable to RRS if traits have low heritabilities and/or if the genetic correlation between purebred and crossbred performance is small. This is expected if gene frequencies of overdominant loci are complementary (Section 14.3). In populations with a long history of selection genetic variability for purebred performance may be smaller than for crossbred performance.

The rates of genetic progress from RRS (Δ_C) and from pureline selection (Δ_P) are, respectively,

$$\Delta_C = i_C \, \mathrm{cov}_C/(\sigma_{\bar{C}} t_C)$$

$$\Delta_P = i_P \, \mathrm{cov}_P/(\sigma_{\bar{P}} t_P)$$

The subscripts C and P refer to crossbreds and to purebreds, respectively, $\sigma_{\bar{C}}$ is the standard deviation of the crossbred information, and t_C is the generation interval when using RRS. The covariances have the following composition:

$$\mathrm{cov}_C = 2C_C V_{A_C}, \qquad \mathrm{cov}_P = 2C_P \sigma_{A_C} r_G \sigma_{A_P}$$

where σ_{A_C} is the genic standard deviation of crossbred performance and C_C the coefficient of kinship between crossbred test animals and breeding animals. The ratio of genetic gains of RRS to pure line selection, assuming equal culling percent, variability, kinship coefficients, and generation interval, will be

$$\frac{\Delta_C}{\Delta_P} = \frac{\sigma_{A_C} \sigma_{\bar{P}}}{\sigma_{A_P} r_G \sigma_{\bar{C}}} = \frac{h_C}{h_P r_G}$$

In pure line selection, individual performance can be considered, and where mass selection is possible, selection intensity can be much higher. For traits of low heritability, mass selection has little application and progeny testing or sib testing become necessary. If so, i_P and i_C, C_P and C_C, and the generation intervals are similar between pureline selection and RRS.

Cole and Hutt (1973) successfully improved poultry crosses by pure line selection. Bell (1972) concluded, from an extensive review of the literature, that two-thirds of the experiments on RRS proved pure line selection to be superior. However, in the comparison, traits to be improved were not differentiated as to their heritability and one may assume that had the comparison been restricted to heterotic traits, RRS would have appeared more advantageous. White Leghorn strains under long-term RRS yielded the following parameters: $h_P^2 = 0.10$, $h_C^2 = 0.14$, $r_G = 0.7$. These resulted in a ratio of (Pirchner and von Krosigk, 1973)

$$\frac{\Delta_C}{\Delta_P} = \frac{0.37}{0.32 \times 0.7} = 1.76$$

indicating a sizable superiority of selection for combining ability.

Synthetic populations harbor more heterozygosity than the parental strains (Table 14.1) and they should show some heterosis gain. The heterozygosity may, and often will, be reduced by inbreeding subsequent to a reduction in the synthetic population size and selection. Also, if recombination losses are important, these will be evident in later generations of the synthetic population.

17

Genetic Polymorphism and Single Genes

17.1. Paternity Control

Studies of blood groups in domestic animals have been common for about three decades. Such investigations provide an important component of the infrastructure necessary for efficient animal improvement. Since the introduction of starch gel electrophoresis by Smithies (1955) a large number of polymorphic systems have been discovered, and this has profoundly influenced the development of theoretical population genetics. In Table 1.1 the number of blood-group genotypes known in 1958 for cattle is given. Additional blood-group factors and the many protein loci discovered since then have increased the number of combinations by several magnitudes. Therefore, it is possible in principle to identify every individual unambiguously, with the exception of monozygotic twins.

The usefulness of various blood-group and serum protein systems for pedigree control depends on the mode of inheritance and the gene frequencies, and, naturally, on the type of problem. In domestic animals, most problems requiring pedigree checking are either of disputed paternity or of suspicion that animals have been interchanged. In the latter type of problem, both the officially recorded sire and dam may be wrong; in the former, only the sire is in question. In certain situations, paternity can be excluded regardless of the blood type (or serum protein type) of the dam. Such an unconditional exclusion is impossible with blood-group systems that show complete dominance (unless the genotype of the sire is known from progeny test). The genotypes of the dam and of the alleged sire are usually known and the paternity of a certain sire can be excluded when the genotype of the dam is considered (conditional exclusions). Assume that there is a locus, such as the FV blood-group locus in cattle, at which only two alleles are present in most populations and that the heterozygotes can be recognized. If a cow with genotype FF is mated (supposedly) to a sire with the same genotype, the probability of exclusion of alleged paternity is equal to the probability that FV offspring result from such a mating. In a population with gene frequencies p_F and q_V, this equals q_V. If the putative sire has the genotype FV, it is impossible to prove that he is not the sire by using the FV system. In Table 17.1 the possible mating combinations and the probabilities of excluding paternity are given. These sum to $pq(1 - pq)$, the maximum of which occurs at $p = q = 0.5$, where it is equal to 0.1875. Jamieson (1965) has given a general formula for computing the proportion of wrong pedigrees detectable by the use of a locus with n allelic genes. If the frequencies of the genes are denoted by p_i, the probability for detecting a false pedigree is given by

$$\sum_i p_i(1 - p_i^2) - \sum_{i > j} (p_i p_j)^2 [4 - 3(p_i + p_j)]$$

Genes with intermediate or, in the terminology of immunogenetics, codominant inheritance permit greater exclusion probabilities.

If dominance is complete, the only possible exclusion is that a mating

TABLE 17.1
Probability of Discovering False Paternity

Genotype		Mating frequency	Exclusion probability of false "sires"	
Of dam	Of putative sire		In family	In population
FF	FF	p^4	q	$p^4 q$
	FV	$2p^3 q$	0	0
	VV	$p^2 q^2$	p	$p^3 q^2$
FV	FF	$2p^3 q$	$q/2$	$p^3 q^2$
	FV	$4p^2 q^2$	0	0
	VV	$2pq^3$	$p/2$	$p^2 q^3$
VV	FF	$p^2 q^2$	q	$p^2 q^3$
	FV	$2pq^3$	0	0
	VV	q^4	p	pq^4
Sum		1		$pq(1 - pq)$

of two recessive genotypes could produce an offspring with the dominant phenotype. The probability of recognizing an error in the imputation of paternity in such a mating is p; the total probability of recognizing such an error with the help of a locus that has a dominant allele is q^4p. The maximum is reached at $q = 0.8$, and it then equals 0.082. Segregation at several independent loci can be combined to increase the probability of recognizing wrong pedigrees. If the first locus gives the probability P_1 and the second the probability P_2, the total probability P of recognizing the wrong sire is given by the following expression:

$$P = P_1 + P_2 - P_1P_2 = 1 - (1 - P_1)(1 - P_2)$$

For cattle the exclusion probabilities range between 80 and 90% if test sera for 10–12 blood-group systems are used. Inclusion of additional systems increases the exclusion probability. These problems were first treated in human genetics (A.S. Wiener and co-workers, 1930), and Rendel (1958) and Gahne (1961) were the first to apply them to cattle breeding.

Early investigators (for example, Neimann-Sørensen, 1958) determined that 4–5% of heifers at progeny testing stations had wrong pedigrees and even larger discrepancies were found in field data. Assume that n' of n progeny of a sire have the correct pedigree and for $n - n'$ the putative sire identification is wrong. This reduces the correlation between breeding value and progeny average to

$$\frac{n'h/2}{n + n'(n' - 1)t^{1/2}}$$

Ten percent pedigree errors in progeny groups of 50 animals (five wrong pedigrees) reduce the correlation from 0.88 to 0.85, and 20% errors to 0.82. Therefore, it has been argued (Van Vleck, 1970) that mistaken pedigrees will not affect genetic progress as long as the fraction of false pedigrees is not large. This of course is a consequence of large progeny groups, which even with low heritabilities provide accurate measures of genetic merit. However, one can ask how many progeny are necessary to achieve the accuracy of n true progeny if the fraction of false pedigrees is $(n - n')/n$? If the error rate is 20%, 80 progeny are necessary to achieve the same correlation as 50 progeny with correct pedigree—a 60% increase in progeny numbers. Rational and economic breed improvement implies economy of means. Therefore, a reduction in wrong pedigrees by paternity control would seem to be an obvious way to utilize testing resources efficiently.

17.2. Connection with Performance

Progress in immunogenetics and biochemical genetics has been stimulated by the hope of finding direct connections between biochemical and performance traits. If such connections, which could arise for several reasons, could be found, the costly recording of performance traits could be reduced, restrictions due to sex-limited traits could be circumvented, and so on. It is probable that blood-group genes and genes influencing other biochemical traits can be linked to performance effects, or may show pleiotropic effects on performance traits. If heterosis is important, blood-group genes may aid in the recognition of the degree of heterozygosity.

Neimann-Sørensen and Robertson (1961) elaborated the use of genetic markers for improving performance. Let V_B and V_G denote, respectively, the variance connected with marker genes and the genic variance, and V_P the phenotypic variance. B is contained in G and G in P, and therefore

$$\text{cov } BG = \text{cov } BP = V_B$$

$$\text{cov } GP = V_G$$

The correlation between marker information and genetic merit is given by

$$r_{BG} = \frac{\text{cov } BG}{(V_B V_G)^{1/2}} = \left(\frac{V_B}{V_G}\right)^{1/2}$$

Smith (1967) investigated further the possibility of applying marker genes to animal improvement. Genetic progress arising from the use of markers relative to that from mass selection will be

$$\frac{k_B}{k_M} \frac{r_{BG}}{h}$$

where $k_B = i_B/t_B$, the standardized selection differential divided by generation interval for marker selection, and k_M is the analogous ratio for mass selection. Stepwise selection and marker selection followed by mass selection results in

$$1 + \frac{i_B}{i_M} \frac{r_{BG}}{h}$$

The ratio of genetic progress from marker selection to that from sib selection is

$$\frac{k_B}{k_C} \frac{r_{BG}}{h} \left[\frac{1 + (n - 1)t}{2nC} \right]^{\frac{1}{2}}$$

where t, n, and C are, respectively, the correlation between sibs, their number, and the kinship coefficient, and k_C refers to sib selection. The expressions indicate that marker selection has more advantages at low heritabilities and that it may be more efficient if the relationship between test animals and the candidate is weak. Marker selection can be accomplished in the young animal and it should be particularly useful if traits are sex-limited. Smith (1967) showed that if sib selection is necessary for a moderately heritable trait ($h^2 = 0.4$), marker information with $r_{BG}^2 \leq$ 1/5 may increase the accuracy and thus the genetic progress by some 25%. The same increase in efficiency can be achieved by stepwise selection with r_{BG}^2 values as low as 0.03. Sex-limited traits, such as litter size, with low heritability, would benefit 25% from using marker information if r_{BG}^2 is as low as 0.07. The variance of r_{BG}^2 if information from one locus is used is given by

$$V(r_{BG}^2) \approx \frac{2}{Nh^2} \left(2r_{BG}^2 + \frac{m}{Nh^2} \right)$$

N denotes the number of tested animals. If a few loci are used, the variance depends on r_{BG} and h^2. An accurate estimate of r_{BG} for poorly heritable traits will require large numbers of tested animals.

Neimann-Sørensen and Robertson (1961) investigated dairy progeny groups. The correlation between markers and genetic merit was 0.23 for milk and 0.18 for fat content. Jamieson and Robertson (1967) found that the transferrin locus accounted for 1.1% and 0.4%, respectively, of the genic variance of milk yield and fat-%. Many other investigations are based on blood-group information. Since blood groups are properties of the cell surface, it seems improbable that close connections exist between them and a performance trait and probable that correlations discovered so far are mostly due to linkage if not fortuitous.

Close associations between disease resistance in poultry and blood-group markers have been reported. Lymphoid leucosis resistance has long been considered as a trait with low heritability, yet Waters and Burmester (1961) showed that the resistance of inbred lines and their crosses was influenced by major genes. Intensive work on this problem, availability of test sera, and, last but not least, models from bacterial genetics culminated in the identification of a series of loci which control resistance to specific virus groups. Crittenden *et al.* (1970) were able to show that resistance of the *B* group of leucosis virus is influenced by genes of the blood-group locus *R*. Cole (1968) was able to rapidly change resistance to Marek's disease by selection. In field investigations, after test infection, heritabilities of resistance against Marek's disease turned out to be around 0.20 (Table 13.2). Use of inbred lines again led to the identification of major genes. Allele B^{21} causes a dramatic improvement in resistance to Marek's disease. Collins *et al.* (1977) reported mortality rates after tumor induction in B^2B^1, B^2B^5, and B^5B^5 genotypes of 5, 26, and 93%, respectively. In retrospect, the results seem less surprising, as cell wall constituents are intimately connected with recognition of foreign material and thus with defense against infection.

Relationship caused by linkage differs mainly from that due to pleiotropy by the fact that the direction of the liaison between marker and performance may change from population to population. For example, allele $B^{BO_1Y_1D'}$ is connected in Scandinavian breeds with a decrease of milk fat-% (Neimann-Sørensen and Robertson, 1961; Rendel, 1961), but in Holstein-Friesians with an increase. If linkage breaks occur in a population, blood group–sire interactions should ensue. As long as linkage remains intact, marker–performance relationships due to pleiotropy and those due to linkage cannot be differentiated.

Rasmusen and Christian (1976) discovered a connection between halothane sensitivity and blood-group locus *H*, which Andresen (1971) has reported as linked with loci for 6-phosphogluconate dehydrogenase (6-PGD) and phosphohexose isomerase (PHI). Andresen and Jensen (1977) suggest the following sequence and recombination fraction:

$$HAL-PHI\overset{2.6}{-}H\overset{3.4}{-}6\text{-}PGD$$

In Denmark only the combination PHI^B-HAL^n was observed, so that for some time the hypothesis that *HAL* and *PHI* loci were identical could not be rejected. However, another combination was discovered (PHI^A-HAL^n) and the linkage between the two loci fully established (Guerin *et al.*, 1978). The close linkage between the halothane locus and markers opens the possibility of using marker selection at least for screening.

If marker loci influence performance, either directly via pleiotropy or by linkage association, selected lines should differ in the frequency of such loci. Of course, drift will cause differentiation, too. However, Brown and Nordskog (1962) reported that after four generations lines selected either for high egg weight or for high body weight had similar gene frequencies, which differed from those of lines selected in opposite directions by more than could have been explained by drift. Garnett and Falconer (1975) investigated frequencies at 22 marker loci of mouse selection lines and found only one where differences were larger than expected from drift.

17.3. Heterozygosity

If genetic markers are representative and sufficiently numerous, they can be used to characterize the total genome, for example, with respect to the fraction of loci that are heterozygous.

Most investigations which involve markers are limited to polymorphic loci, for obvious reasons. Such loci, however, do not represent a random sample of all loci. Therefore, Harris (1966) and Lewontin and Hubby (1967) chose loci not according to "genetic" considerations—i.e., whether or not a locus shows segregation—but solely from analytical considerations—whether products can be identified by electrophoresis. This approach led to the discovery of an unexpectedly large fraction of polymorphic loci. In populations of humans and of mice, between 25 and 40% of loci are polymorphic. Of these, about 30–40% are heterozygous in each individual. Therefore, about 1/12 of the loci of an individual genome represented by marker loci seems to be heterozygous.

Most of the customarily used marker loci are polymorphic. If 10–12 loci are studied, 30–40% are heterozygous per individual. Gruhn and Dinklage (1971) found the degree of heterozygosity in five pig breeds to be between 29 and 42%. When breeds were ranked according to the fraction of heterozygous loci, there seemed to be a correlation with fertility. The possibility of using marker information to select partner strains for crossbreeding has been pointed out in Chapter 16.3.

17.4. Incomplete Penetrance

Timoféeff-Ressovsky (1927) defined penetrance as the fraction of individuals that manifest the genotype. For example, incomplete penetrance (w) or the presence of recessives causes a genotypic distribution of normal phenotypes as follows:

$$
\begin{array}{ccc}
AA & Aa & aa \\[4pt]
\dfrac{p^2}{1 - wq^2} & \dfrac{2pq}{1 - wq^2} & \dfrac{(1 - w)q^2}{1 - wq^2}
\end{array}
$$

The recessive trait is expressed in wq^2 individuals and the probability that an animal of normal phenotype is homozygous recessive is $(1 - w)q^2/(1 - wq^2)$. Lauvergne and Lefort (1973) demonstrated that under random mating, w and q can be estimated from the following relationships:

$$q^2 w = (1 + qw)Q$$

$$1 - q(1 - e^{-nwq/2}) = (1 - S)^{1/2}$$

where Q is the frequency of the recessive trait and S is the fraction of carrier sires that have been diagnosed by abnormal progeny.

Smith and Bampton (1977) applied the maximum likelihood method for estimating frequency and penetrance of the halothane gene. They concluded that the porcine stress syndrome and malignant hyperthermia upon administration of halothane were caused by a recessive homozygote with high penetrance, provided the diagnostic technique was appropriate. Rather large differences between breeds in the frequency of the recessive gene became apparent. Even though their material was limited, they arrived at precise estimates.

The causes of incomplete penetrance may be in the environment or in the residual genotype. French investigators have found, for example, that the syndrome of arthrogryposis and palatoschisis (SAP) is caused by a recessive gene with a penetrance of less than 1/4, i.e., less than 25% of recessive homozygotes display abnormality. Goonewardene and Berg (1976) observed that in progeny from a mating between a Charolais bull with crossbred daughters of another Charolais bull and Hereford cows, a ratio of eight SAP to eight normal occurred. This can be explained by assuming that both Charolais were homozygous for the SAP gene without showing the syndrome, and that penetrance was fully restored in the crossbreds—1/4 Hereford, 3/4 Charolais—probably by disruption of the modifier complexes that are common in the pure Charolais. Environmental influences on penetrance are evident in "quacky" mice and pastel minks. The gene causes an impairment of zinc absorption. Therefore homozygotes display the syndrome if Zn supply is suboptimal, but oversupply of Zn will lead to "impenetrance" of the defective genotype. Even before the biochemistry of "quacky" was clarified, Lyon (1954) demonstrated that penetrance of the recessive genotype increased with increasing birth litter size—explicable now by poor neonatal Zn supplies in offspring from large litters.

The one-locus model with incomplete penetrance is in many situations an alternative to the polygenic threshold model (Chapter 13).

Discrimination between the two models is difficult. Reich *et al.* (1972) demonstrated that a differentiation between them requires the existence of a least two thresholds and consequently of three distinct classes. However, use of the one-locus model with incomplete penetrance to problems where that model is correct should facilitate such objectives in breeding work as the eradication of a lethal, and it represents a real advance in the application of genetics.

17.5. Testing for Recessive Genes

The detection of heterozygous carriers of undesirable recessive genes may become an urgent problem. Practically all animals are heterozygous for some undesirable genes (Chapter 2), but in general these differ from animal to animal. An undesirable gene becomes a problem only when concentration of breeding upon one animal has caused a wide diffusion of its genes and consequently has increased the probability of homozygotes. This may culminate in an epidemic of the particular lethal and thus create a need for the identification of heterozygous carriers. Advances in biochemistry and cytology may serve this need in some cases. However, even in humans, where much more work along these lines has been initiated, most lethal genes cannot yet be recognized in heterozygotes, and comparable studies in domestic animals have barely begun. Therefore, the diagnosis of heterozygosity requires progeny testing. The only exception would be offspring from the mating of alternative homozygotes *(CC × cc)*. Berge (1931) was the first who, following Bernstein, calculated probabilities of correct diagnosis in domestic animals.

For every breeding animal there exists *a priori* some notion about the probability of the genotype at the locus of interest. The animal may have a heterozygous parent, in which case the *a priori* probability w_1 of it being homozygous normal equals 1/2. An animal may be an offspring of two heterozygotes with an *a priori* probability of homozygosity for the normal gene of 1/3 ($\frac{1}{4}/\frac{3}{4}$). If no information about its parents (except that they were not *cc*) is available, it can be considered as a random member of the population and the probability of it harboring no recessive genes equals $p/(1 + q) \approx p$.

The testing of the animal is accomplished by mating it to partners, some or all of which are carriers, possibly even homozygous, but mostly heterozygous, for the undesirable gene. If a homozygous recessive is born, the test animal is diagnosed as a carrier. However, the probability w_2 remains, even if normal progeny only are born, that nevertheless the test animal is a heterozygote. The probability that a test animal with only (phenotypic) normal offspring is itself homozygous normal can be computed as *a posteriori* probability by the Bayes formula:

$$P(A_1/B) = \frac{P(A_1)P(B/A_1)}{\sum_i P(B/A_i)P(A_i)}$$

The probability of event A_1 (homozygous normal) happening given event B (no abnormal progeny) is given by the product of the *a priori* probability (A_1) of having no recessive genes times the conditional probability of having only normal offspring (= 1 if normal genes are dominant) divided by the sum of the probabilities of mutually exclusive events. These are $P(B/A_1)P(A_1)$ and $P(B/A_2)P(A_2)$, where $P(A_2)$ is the *a priori* probability of heterozygosity and $P(B/A_2)$ the probability of normal offspring from a heterozygote test animal:

$$P(CC/B) = \frac{P(B/A_1)P(A_1)}{P(A_1)P(B/A_1) + P(A_2)P(B/A_2)} = \frac{1}{1 + P(B/A_2)w_2/w_1} \quad (17.1)$$

The probability of all normal offspring from a heterozygous test animal $P(B/A_2)$ depends on the genotype and on the number of mates of the test animal. If the undesirable recessive is viable and fertile, such animals can be used for test mating: $C/? \times cc$. If the test animal is heterozygous, the offspring from such a mating are expected in the ratio of one-half normal *(Cc)* to one-half homozygous recessive *(cc)*, but only an offspring of this latter sort yields a diagnosis. The probability of two phenotypically normal offspring resulting from such a mating equals $1/2 \times 1/2$, and the probability of n only normal progeny is $(1/2)^n = P(B/A_2)$.

In most situations, for example, with lethals, homozygous recessive animals cannot reproduce. The next best mating partners, from the point of view of efficiency of testing, are known heterozygotes (known because they have produced at least one recessive offspring). The probability of normal offspring from such animals when mated with another heterozygote equals 3/4, which is reduced to $(3/4)^n$ when n normal offspring are born. Heterozygosity for a detrimental recessive gene is economically undesirable. Therefore, it is frequently difficult to assemble enough known heterozygotes for test matings and other mating partners must be chosen.

When the heterozygote frequency in the test population is known, the number of matings necessary to test for the presence of an undesirable recessive can be derived from the following expression (Johansson *et al.*, 1966):

$$P(B/A_2) = [D + H(\tfrac{3}{4})^n]^k = w^*$$

where D is the proportion of homozygous dominant individuals, H the proportion of heterozygotes in the population, n the litter size, and k the number of mating partners. When $D = 0$ and $n = 1$ the formula reduces to the simple one already given for the mating carrier × carrier in monoparous species, that is, to $(\tfrac{3}{4})^k$.

Wriedt and Mohr (1928) suggested testing the candidate by mating him to his own daughters or to daughters of known heterozygotes. Approximately half of the mates would then be heterozygotes and half homozygotes (with sire–daughter matings this would hold, of course, only if the sire were a carrier). Therefore, applying the formula just given to monoparous births gives

$$P(B/A_2) = (\tfrac{1}{2} + \tfrac{1}{2} \times \tfrac{3}{4})^k = (\tfrac{7}{8})^k$$

and applying it to multiparous births gives

$$P(B/A_2) = [\tfrac{1}{2} + \tfrac{1}{2} \times (\tfrac{3}{4})^n]^k$$

These formulas hold only if $q = 0$ in the original population, as has been pointed out by Rasch (1967). If q is not zero—as is likely—some of the daughters will have received the recessive gene from their dams, thus increasing the proportion of heterozygotes to more than one-half. Considering this, the probability of detecting a carrier among monoparous animals becomes

$$P(B/A_2) = \left(\frac{7 + 3q}{8 + 4q}\right)^k$$

The final decision to keep or reject a sire takes a long time if it depends on the outcome of sire–daughter matings. Furthermore, the offspring of these matings are 25% inbred, which causes decreased performance in all traits susceptible to inbreeding depression. On the other hand, this type of mating tests not only for one but for all undesirable recessive genes that the sire might carry. This, however, is not as much of an advantage as it might at first appear. We must assume that all animals carry some undesirable genes. As long as such genes are rare in the population, this is not too important, since the probability that mating partners are heterozygous for the same gene is very small. Therefore, it appears to be of little value to know that the sire is a carrier of certain rare genes. The conclusion is different when an undesirable gene has become sufficiently frequent as to cause serious economic consequences. In such cases it should be possible to find daughters of known carriers that can be used as testers. This makes it possible to test a candidate without lengthening the generation interval and causing inbreeding depression.

Another possibility is to test a candidate by mating it to animals of the general population. Animals that are homozygous for recessive lethal genes are, of course, excluded from reproducing. Heterozygotes and dominant homozygotes will be present in the population in the ratio $2q/(1 + q):(1 - q)/(1 + q)$. The probability among monoparous animals that normal progeny will be produced from k matings of heterozygous candidates with partners chosen at random from the population is

$$P(B/A_2) = \left[\frac{1 - q}{1 + q} + \frac{2q}{1 + q}\frac{3}{4}\right]^k = \left(\frac{2 + q}{2 + 2q}\right)^k$$

Incomplete penetrance increases w^*. For example, a penetrance of 1/5 will increase the probability of normal progeny in a backcross $Cc \times cc$ from 1/2 to 0.9. Incomplete recording of abnormal progeny is analogous.

Assume that an outstanding sire is belatedly recognized as a heterozygote and that he has produced with a carrier cow a young male for which $w_1 = 1/3$. This young sire has then produced, with ten daughters of a known carrier, only normal offspring. The probability of his being homozygous normal is given by

$$\frac{1/3}{1/3 \times (2/3)(7/8)^{10}} = \frac{1}{1 + 2 \times 0.263} = 0.66$$

The number of test matings necessary to reach a given probability of normal homozygosis results from rearranging formula (17.1):

$$n = \frac{\log(1 - P) - \log(P) - \log(w_2/w_1)}{\log w^*}$$

$$= \frac{\log[(1 - P)/(w_2/w_1)P]}{\log w^*}$$

The number of progeny required to be 95% certain that the test animal is homozygous normal is given in Table 17.2. Several test situations are

TABLE 17.2
Number of Test Matings of Monoparous Animals Required to Achieve a Probability of 95%

Mating partner	w^*	w_1 1/2	1/3
Homozygous recessive	1/2	4.2	5.2
Heterozygous	3/4	10.2	12.6
Progeny of heterozygotes	7/8	22.1 (4.9[a])	27.2 (6.1[a])
Sample of population $q = 0.2$	11/12	33.8	41.8
$q = 0.1$	21/22	63.3	78.2

[a] Number of matings required of multiparous animals, $k = 8$.

considered: *a priori* probabilities of 1/2 (test animal out of $Cc \times CC$) and 1/3 (from $Cc \times Cc$); and different genotypes of test mating partners: cc, Cc, progeny of Cc animals, and random animals from the population. If the *a priori* probability is 1/2, relatively few test matings are necessary to achieve a fairly certain diagnosis. However, the time and expense involved may be considerable.

We must assume that every animal is heterozygous for some undesirable genes. The outstanding sires will spread the undesirable genes they possess, and their sons, which have an above average probability of being used, will further disseminate these genes. Two to three generations after use of the outstanding sires homozygotes start to appear and this may aggravate management for some time. Such epidemics of lethals will recur in animal populations, although even at their apex they will rarely involve more than a few percent of the births. Well-known examples of these epidemics include the rampant eruption of dwarfs in U.S. Herefords in the 1950s and the fairly frequent occurrence of dwarfs of a different origin in segments of European Brown cattle in the 1960s.

The mating of candidates to daughters of known carriers appears to strike the best compromise between numbers required and probability of a recessive birth. On the one hand, the location of suitable daughters should not be too difficult, and on the other hand, all births could be monitored, as only a moderate number of test matings is necessary. Test mating with larger numbers of random mates may well be impractical, since it would necessitate a considerable effort in supervision.

18

Breeds and Breed Formation

18.1. Breed Identification

A description of breeds should ideally consist of the characterization and enumeration of genotypes in terms of the molecular genetics of a library of all DNA sequences. Of course, at present this prospect is a considerable distance away. Breed description, traditionally and also at present, consists mostly of the specification of color and type traits, which in general are highly heritable and independent of environmental influences. Such traits usually permit an unequivocal differentiation of breeds and leave little room for overlapping and ambiguity. However, it would be a serious mistake to assume that breeds differ only in these obvious traits when breed differences in economically very important traits are common. Unfortunately, many of these traits are strongly influenced by environment and have a large variance. Therefore, in spite of their economic importance, they are poorly suited for classification.

Discrimination between breeds, or, generally, between groups of more or less related animals, is greatly facilitated by the simultaneous consideration of several traits, employing methods appropriate to multivariate analysis. Fisher (1948) introduced discriminant functions to tackle such problems. Traits 1 to n are used to differentiate two groups which differ in the n traits by D_1 to D_n. The coefficients of the discriminant function are estimated from a group of n linear equations:

$$b_1 V_{X_1} + b_2 \text{ cov } X_1 X_2 + \cdots + b_n \text{ cov } X_1 X_n = D_1 C$$

$$b_2 \text{ cov } X_1 X_2 + b_2 V_{X_2} + \cdots + b_n \text{ cov } X_2 X_n = D_2 C$$

where

$$C = \frac{n_1 n_2}{n_1 + n_2}$$

In matrix notation the equations are

$$S_{ij} b = D$$

The numbers of animals in the two groups are n_1 and n_2; b and D are, respectively, vectors of regression coefficients to be estimated and observed breed differences; and S_{ij} is the phenotypic variance–covariance matrix. The multiple correlation coefficient is $r^2 = b_1 D_1 + b_2 D_2 + \cdots + b_n D_n$.

Witt and Döring (1955), for example, used discriminant functions of body measurements to differentiate between strains of German Friesians. These strains were from different areas, so differences between them were due to both genetic and environmental factors. Discriminant functions used from two to seven body measurements and needed as few as four or five animals to identify a strain. In contrast, when the best single trait ("best" with respect to differentiation) was used, twice as many individuals were needed. Since the differences between the groups were slight compared to those between breeds, this illustrates the power of the discriminant function for problems involving the differentiation of groups using quantitative traits.

Such methods can also be used to characterize and identify various types or strains within breeds. Ollivier (1968) employed several slaughter traits to construct a discriminant function which distinguished among the carcasses of Pietrains, Yorkshires, and their crossbreds. Discriminant analyses permit the nearly unambiguous identification of double-muscled cattle and of their crosses as distinct from normal animals. Festing (1974) suggested that discriminant functions could be used to identify inbred mouse lines used in laboratory animal breeding.

Traits used in problems of identification should be largely independent of the environment, and therefore in their solution marker genes and marker genotypes are a natural choice. Sokal and Sneath (1973) present the various methods of quantitative taxonomy, which rest largely on computing correlations or their complement, the differences between taxonomic units. Correlations or differences are then used to construct genealogic trees. Nozawa *et al.* (1976) applied this approach to investigate the relationship between Japanese horse breeds, which enabled them to revise previously held notions.

Marker genes are appropriate tools for the study of the relationship between breeds and for breed development. These genes can be considered neutral, at least over the relatively short time periods spanned by breed history, and changes in gene frequency largely reflect drift. A problem arises because correlations or variances between population gene frequencies depend on the frequency of binomially or multinomially distributed traits. Cavalli-Sforza and Edwards (1967), following Fisher, suggested transformation of the frequencies to arcsines. Thus, if populations A and B have gene frequencies p_a and p_b at a certain locus, a suitable transformation will be

$$\text{cov } \phi = \sum_i (p_a p_b)^{1/2}$$

This transformation stabilizes the variances between frequencies 0.05 and 0.95. The genetic distance d between the two populations is given by the length of the chord between the two points on the surface that characterizes the population gene frequencies, i.e., $d = (1 - \cos \phi)^{1/2}$. If several loci are used, the individual distances must be combined, either additively, $D = \sum d$, or by constructing a sort of a geometric mean, $D = (\sum d^2)^{1/2}$. Distances can be standardized and Cavalli-Sforza and Edwards (1967) suggest expressing these as fractions of a complete gene substitution between breeds. Such a substitution would span one-fourth of the whole circle, and therefore the distance is to be divided by $\pi/2$:

$$d' = d\frac{2\sqrt{2}}{\pi} = \frac{2[2(1 - \cos \phi)]^{1/2}}{\pi}$$

Table 18.1 gives frequencies of the β-lactoglobulin gene and of casein haplotypes in several cattle breeds; the computation of the breed distances from the gene frequencies is illustrated in Table 18.2. The distance M–P at the Lg locus is computed as

$$2\{2[1 - (0.3 \times 0.34)^{1/2} - (0.7 \times 0.66)^{1/2}]\}^{1/2}/\pi = 0.0273$$

TABLE 18.1
Gene Frequencies (%) of European Cattle Breeds

Breed[a]	β-Lactoglobulin		Casein[b]				
	Lg A	Lg B	BA	CA	BB	BC	CB
M	30	70	71	19	7	4	0
P	34	66	69	20	5	6	0
G	56	44	61	25	10	4	0
B	56	44	69	12	15	1	3
F	50	50	66	10	17	4	3
R	34	66	74	12	12	0	2
S	40	60	89	7	4	0	0
J	41	59	61	5	31	0	3

[a] M, Murbodner; P, Pinzgau; G, Tyrolian grey; B, Brown; F, Simmental; R, Red-and-White lowland; S, Friesian; J, Jersey. The frequencies of M, P, and G were provided by Dr. Grosclaude, CNRZ, Jouy-en-Josas, France; the frequencies of other breeds are reported by Aschaffenburg.
[b] The first letter of casein haplotypes denotes α_{S1}-casein, the second β-casein.

The standardized variance of the gene frequency (Wahlund variance, Chapter 2) can easily be computed from these distances. Gene frequencies can be transformed to sines:

$$\sin \phi = (p_a p_b)^{1/2} = \sin^2 \phi$$

and $\phi = \arcsin p^{1/2}$. The term ϕ is a function of the gene frequency, the variance of which can be deduced as follows:

$$V_\phi = V_{f(q)} = V_p \left(\frac{d\phi}{dp}\right)^2 = V_p \left(\frac{d \arcsin p^{1/2}}{dp^{1/2}} \frac{dp^{1/2}}{dp}\right)^2$$

$$= V_p \left(\frac{1}{(1-p)^{1/2}} \frac{1}{2p^{1/2}}\right) = \frac{V_p}{4pq}$$

The Wahlund variance V_p/pq is four times as large as the variance of the transformed variable:

$$\frac{V_\phi}{pq} = 4(1 - \cos \phi) = F = 4d^2$$

TABLE 18.2
Genetic Distances and F Values between Breeds[a]

	M	P	G	B	F	R	S	J	
M	—	27	169	169	131	27	67	74	d(Lg)
		66	232	328	295	201	252	354	D
P			142	142	104	0	40	46	
	23		222	331	289	207	267	367	
G				0	38	142	102	96	
	32	26		169	211	333	336	384	
B					38	146	102	96	
	53	55	34		**107**	220	297	247	
F						104	64	58	
	44	45	31	6		244	297	227	
R							40	46	
	31	42	56	38	30		**189**	210	
S								6	
	38	44	64	48	58	29		280	
J	83	103	91	32	31	29	74		

(*F*, lower-left triangle values)

	G	B'	R'	J
M'	227	310	232	362
G		**190**	334	384
B'			230	237
R'				245

	B''	R'	J
M'	268	**232**	362
B''		282	310
R'			245

	B''	J
M''	275	304
B''	310	

$d(M'G) = \tfrac{1}{2}[d(MG) + d(PG)] = \tfrac{1}{2}(232 + 222) = 227$

$d(M'B') = \tfrac{1}{4}[d(MB) + d(MF) + d(PB) + d(PF)] = 310$

[a] All distances given are sums of chord length × 1000. d(Lg), Distance estimated from Lg locus. D, Distance estimated from both loci. F, Inbreeding coefficient relative to common base population.

In the case of k multiple alleles, $F = 4(1 - \cos \phi)/(k - 1)$, and if several loci are combined,

$$F = \frac{4 \sum (1 - \cos \phi)}{\sum (k - 1)}$$

The F coefficient is relative to the base population from which the various subpopulations developed. The F coefficient of Friesians relative to the population from which both lowland breeds—Friesians and Red-and-White cattle—originated is 0.029, but it is 0.058 relative to the European population from which Friesians and Simmentals are derived. The population differences in gene frequencies are caused by drift, by selection, which is different in direction and/or intensity, differences in rate and source of immigration, and by mutations. If the last two factors can be excluded, drift and selection remain as the causes. Selection should affect the single loci in a way which differs from locus to locus. So the genetic difference between populations should change significantly if computed from different loci. In contrast, if population differences are mainly caused by drift, differences estimated from different loci should vary only randomly. Robertson (1973) has suggested a modification of a test for differences between different loci which was proposed by Lewontin and Krakauer (1973).

Selection will have been different in each breed. For example, Pinzgaus and Friesians, though rather similar with respect to genetic markers, surely have in the past been selected for different goals and by somewhat different methods. The differences caused by selection will have affected their genotypes relevant to performance without any noticeable difference in their marker genotypes. This of course implies that the markers used are practically neutral, a characteristic of markers rendering them eligible for the analysis of the evolution of breeds.

Several techniques are known for the construction of genealogical trees from breed distances. A simple one presented by Sokal and Sneath (1973) begins at the periphery and combines groups that have the least distance between them. The higher order groups arising from the original ones are again combined according to the same principle until a tree has been constructed. The first steps in the construction of a genealogical tree are illustrated in Table 18.2. The smallest value in each row or column is in bold print. If two columns (respectively, rows) share the smallest figure, the breeds heading the column are combined in a derived higher order group. This occurs to B and F; M and P; S and R in Table 18.2. The procedure is repeated in the next round; in our example, G and B' are combined into the group of western Alpine breeds. The tree constructed from the data of Table 18.1 is shown in Fig. 18.1. With the exception of the group MP, it is generally similar to trees which have been constructed from more loci.

The divisive cluster method suggested by Cavalli-Sforza and Edwards (1967) begins at the root of a tree and divides the data in such a way that variances between groups are maximized and those within groups minimized. This procedure is continued until a tree is complete. Of course, many possibilities of dividing sets of data exist, but this approach

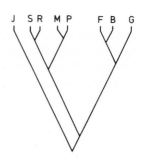

FIGURE 18.1. Relationship of European cattle breeds. J, Jersey; S, Friesian; R, Red-and-White lowland; M, Murbodner; P, Pinzgau; F, Simmental; B, Swiss Brown; G, Tyrolean Grey.

can be used only if moderate numbers of groups are to be classified and ordered. Kidd and Sgaramella-Zonta (1972) have developed approximations for the process in which the various possible clusters of groups are assessed. In the method of minimum paths, clusters are compared by the total length of those paths between them. The closeness of fit of constructed trees, or rather, of the distances in such trees, can be compared to real distances between populations.

At present no possibilities exist for testing the quality of constructed trees, i.e., to decide which is the best. Extensive simulation work has not provided a clear tool for this evaluation either. Therefore, it appears that procedures should be selected which are most suitable for a particular problem and which can be handled with available programs and computer space.

Examples of the application of methods for retracing the genealogy of domestic animal breeds include papers by Kidd and Pirchner (1971) on Central European cattle, Kidd *et al.* (1980) on Iberian cattle, and Nozawa *et al.* (1976) on Japanese horses.

The estimation of F from distances between population gene frequencies permit estimation of the time since separation of the groups (Chapter 3):

$$F = 1 - e^{-t/2N_e}$$

If t is small, one may use the approximation $F = t/2N$, and when t is large, $\log(1 - F) = -t/2N$. Under the influence of random drift the F coefficient is proportional to time since isolation began, or, if time is considerable, $-\log(1 - F)$ is proportional to time. The F coefficients relevant to pairs of breeds are given in Table 18.2. If N_e were known, the time could be estimated. Kidd and Cavalli-Sforza (1974) were able to collect information about the numbers of cattle in Iceland since colonization. Their application of the above expression yielded t values which agreed fairly closely with the time passing since Icelandic cattle were separated from Scandinavian stock. However, they excluded Celtic breeds and therefore the reliability of the close agreement between estimated and historical time scales may be somewhat less than if these breeds had been considered (marker studies in humans have indicated a significant immigration of Celts to Iceland).

Marker genes make possible the investigation of breed mixtures and quantification of components of a given hybrid population. Markers may also reveal breed admixture which had been forgotten through lack of known records. For example, a study of Kidd *et al.* (1974) made it probable that German Gelbvieh received genes from South Devon cattle, which has been confirmed by historical records. Farrel *et al.* (1971) discovered the same rare casein allele in both Holstein-Friesians and in Red Danes, which points to the Holstein component of the Holstein-Friesians and its origin from areas bordering Jutland on the one hand and on the other to the migration of cattle from Jutland to Holland and Friesland after the great Rinderpest epidemic of the 18th century.

These kinds of study are largely of heuristic value. However, there is a linear relationship between the genetic distance of populations and heterosis in a cross between them (Chapter 16). Therefore it appears that the information gained could be useful in selecting mating partners for crossbreeding.

18.2. Creation of New Breeds

Old breeds such as Friesians or Swiss Brown were formed during long periods of pure-breeding. Lörtscher (1947) relates the breeding history of Simmentals in the 19th century. Line breeding and a fairly narrow base became evident, so it must be assumed that even Simmentals, which were kept traditionally in many small herds, had a small effective number in their formative decades. These periods of breed consolidation are generally preceded by periods with varying degrees of hybridization, well documented for some recently formed breeds, but for old breeds meagre evidence is extant.

Examples for which evidence exists include the Haflinger horse, originating from a cross between the native Alpine pony and Arabian stallions, and the German Gelbvieh, into which, more or less continuously, until well into the 19th century, animals of various breeds—Shorthorns, Devons, Simmentals, Red Friesians—were introduced. The Pietrain breed of pigs originated from a population of pigs in SE Belgium into which pigs of various breeds were introduced during World War I.

Scientific interest in the creation of new breeds was fostered by imports into North America of European dual-purpose cattle beginning in the 1960s. Preceding these imports, interest in the utilization of germ plasma resources became evident with regard to both plants and animals (Hodgson, 1961). Dickerson (1969, 1973) discussed problems and their solution entailed in the exploitation of these resources, which Fewson (1973) has further elaborated. Utilization of the variability conserved, so to speak, in the breeds may involve initial use of terminal and rotational crosses but also of synthetics of various kinds and degrees. This use may vary from occasional introduction of genes from other breeds to synthetics where contributing breeds are either equally represented or even may contribute nearly 100% of the total genes in the upgrading of native stock by an imported breed. If a grading up policy is pursued, native hybrid sires are frequently used after 3–4 generations of crossing to import sires. These native sires then harbor 1/16–1/8 of the original genes. However, Essl (1976) has pointed out that the fraction of base population genes varies considerably due to Mendelian segregation, and thus selection for the traditional type can maintain it at a level higher than the pedigree suggests. Most of the European Friesian, Simmental, and Brown Alpine strains, which differ somewhat between countries, have a history of this kind of upgrading, which is equally true for most Yorkshire and Landrace strains of pigs.

When making a synthetic, breeders attempt to create something new which combines the advantages of the parental breeds. One may ask when does a synthetic deserve to be considered a new breed?

Lush (1948) suggested that a population of this nature should not be called a breed until the transient heterosis increments of the first crossbred generation(s) have disappeared. As discussed in Chapter 14, this is the F_2 in the case of single crosses and the F_3 in four-way crosses and backcrosses, after which one-half and three-fourths of the heterosis increments remain, respectively.

The extent of the introduction of foreign genes can be estimated if genic breed difference g, heterosis increment h, and importance of epistatic effects r are known (Fewson, 1973). Unfortunately these parameters frequently can be estimated only when the cross is already well advanced. This is particularly true of r, while for g and h information is often available in the literature.

The number of epistatic effects is beyond our grasp. Therefore, only the recombination loss between two or at most three loci is considered. Recombination loss (or, more generally, recombination effect) is the fraction of nonparental combinations of independent genes in gametes from both parents. There is thus no recombination loss in the F_1, both gametes of the F_1 carrying parental combinations. The backcross individuals harbor one-fourth and the F_2 individuals one-half nonparental combinations and the recombination loss is expected to correspond to these. Even though important with respect to the dissolution of parental coadapted gene complexes, recombination loss certainly represents only a minute fraction of the possible epistatic combinations of the genome.

Coefficients of g, h, and r for various kinds of cross are given in Table 18.3. The coefficients of recombination loss increase in later generations. Cross performance in later generations then falls back relative to the F_1, not only due to the disappearance of the transient heterosis increment, but also because of dissolution of coadapted gene complexes. The advantage ΔC of a certain cross when k traits are considered can be estimated from

$$C_i = \sum_k v_k \sum_j c_{ij} p_j = \sum_k v_k (c_{i1} g + c_{i2} h + c_{i3} r)$$

where v_k is the economic weight of trait k, p_j represents the genetic parameter g, h, or r, and c_{ij} is the coefficient of the genetic parameter.

An example is given in Table 18.3. It was assumed that the breed difference g between a dairy and a dual-purpose breed amounts to 600 kg milk, that the heterosis increment is 120 kg, and that no recombination

TABLE 18.3
Parameters Relevant for Crossbreeding[a]

Fraction of genes imported	g	h		r_1	r_2	Milk[b]	Beef[c]	Economic value[d]
100	1	0		0	0	600	0	180
75	$3/4$	$1/2$[e]	$3/8$[f]	$1/4$	$3/8$	510[e]	$2 1/2$	190.5[e]
						495[f]		186[f]
50	$1/2$	1[e]	$1/2$[f]	0[e] $1/2$[f]	$1/2$	420[e]	5	201[e]
						360[f]		183[f]
25	$1/4$	$1/2$[e]	$3/8$[f]	$1/4$	$3/8$	210[e]	$7 1/2$	175.5[e]
						195[f]		171[f]
0	0	0	0	0	0	0	10	150

[a] g, Genic breed difference. h, Heterosis increment. r_1, Fraction of nonparental two-locus combinations in first crossbred generation. r_2, Fraction of nonparental two-locus combination in advanced crossbred generation.
[b] g = 600 kg, h = 120 kg, r = 0.
[c] g = 10 kg, h = 3 kg, r = 0.
[d] Economic weights: 1 kg milk, 0.3 unit; 1 kg (boneless) beef, 15 units.
[e] In F_1 or in first backcross generation.
[f] In F_2 or in advanced generation.

loss occurs. On the meat side, the dual-purpose breed is superior by 10 kg meat per carcass ($=g$) and both h and r are zero. Milk and meat have economic weights of 0.3 and 15 monetary units, respectively. A synthetic with 50% genes from each breed is expected to show an economic advantage of

$$0.3(\tfrac{1}{2} \times 600 + \tfrac{1}{2} \times 120) - 15(\tfrac{1}{2} \times 10) = 33 \text{ units}$$

An F_1 with the same gene composition but more heterosis would be more profitable (51 units), while grading up to the dairy strain would be less advantageous than production of the synthetic (30 units).

If both parental breeds and the F_1 are available, the heterosis increment in the first generation and recombination loss in the second can be estimated. However, if males or sperm only are used on the one side, data from the second generation are required to estimate the heterosis increment and in some instances one may even require data from the third generation. Estimation from second generation data presupposes F_2 animals and/or backcrosses to both the import breed (B) and local breed (A).

Frequently, more types of crossbreds are available than are required for parameter estimates. Use of multiple regression methods permits efficient estimation of these parameters. For example, breed difference, heterosis increment, and recombination effects can be estimated from crossbred data by supposing the model

$$y = a + c_1 g + c_2 h + c_3 r + e$$

In matrix notation

$$y = X\beta + e$$

The performance of the local breed is described by a, and c_1, c_2, and c_3, respectively, are the coefficients of the breed difference, heterosis increment, and recombination loss, and e denotes the errors.

Assume that, in addition to the local breed, an F_1, both backcrosses and an F_2 are available. Performance of five animals, one from each category, can be described by the following five equations:

$$
\begin{bmatrix} y_1 \\ y_2 \\ y_3 \\ y_4 \\ y_5 \end{bmatrix}
=
\begin{bmatrix}
1 & 0 & 0 & 0 \\
1 & 1/2 & 1 & 0 \\
1 & 3/4 & 1/2 & 1/4 \\
1 & 1/4 & 1/2 & 1/4 \\
1 & 1/2 & 1/2 & 1/2
\end{bmatrix}
\begin{bmatrix} a \\ g \\ h \\ r \end{bmatrix}
+
\begin{bmatrix} e_1 \\ e_2 \\ e_3 \\ e_4 \\ e_5 \end{bmatrix}
$$

The estimates are found from

$$\hat{\beta} = (X'X)^{-1} (X'y)$$

and their variances from

$$V\hat{\beta} = \sigma_e^2(X'X)^{-1}$$

where σ_e^2 represents the residual variance.

The errors of these estimates depend on the numbers of animals as well as on the relationship structure of the groups to be compared. The variance of a group mean of v sire progenies (unselected sires) is

$$V_{\bar{x}} = \frac{\sigma_v^2}{v} + \frac{\sigma_x^2 - \sigma_v^2}{nv} = \frac{h^2\sigma_x^2}{4nv}\,(n + k), \qquad k = \frac{4 - h^2}{h^2}$$

Assuming that $nv = N = $ const, i.e., the testing capacity is given, the smallest variance results from $n = 1$, which seems to be impractical. The reduction in variance from increasing the number of sires can be estimated by differentiating the variance with respect to the number of sires (Hill, 1976):

$$v^{-1}\frac{dV_{\bar{x}}}{dv} = -\left[v\left(1 + \frac{k}{v}\right)\right]^{-1}$$

Increasing v from 10 to 11 reduces the variance of the mean by 7% ($k = 15$, $N = 300$). In reality, a variation will occur in numbers of sires selected and in numbers of their progeny produced. Nevertheless, in order to reach a reasonably good estimate of differences between offspring from imported sires and those from local stock, several sires, say about half a dozen, should be used.

Animals are usually few in the first stages of crossbreeding and information is frequently lacking to set up the equations given above, or, alternatively, groups may have been subjected to a variety of management practices, etc. Therefore, estimation of reliable values for g, h, and r can present a formidable task (Gravert, 1975; Schulte-Coerne, 1977; Clement, 1978; Teehan,1980). It must be admitted that decisions in practice are frequently made on the basis of patchy information.

In the first panmictic generations after a cross, loss of performance relative to the first or second (in the case of a four-way cross) crossbred generation is expected because heterosis declines and coadapted gene complexes disintegrate. The heterosis loss equals one-half if the synthetic is developed from an F_2, but only one-fourth if the base generation is a four-way cross or a backcross (this has only one-half the heterosis of an F_1). Recombination losses accumulate during the first generation of a synthetic and selection is necessary to neutralize this untoward development and to isolate favorable gene combinations. In contrast, the initial heterosis loss cannot be avoided and a fairly large population size is necessary to avoid accretion of inbreeding depression in the course of further generations. Persistent heterosis in a synthetic of reasonable numbers will be one-half of the F_1 heterosis and three-eighths of the heterosis in a backcross. A possible example of recombination loss and its successful neutralization may be represented by the Australian Milking Zebu (AMZ), a synthetic between Jerseys and Sahiwals, where milk yields of F_1, F_2, F_3, and later generations were respectively 1944, 758, 895, and 1985 kg milk. Hayman (1974) reports that improvement was largely achieved by culling the cows with poor milking temperament. In the F_1 this trait obviously behaved as a recessive but reappeared in the succeeding generations.

Synthetics are expected to harbor more genetic and genic variability than the parental pure breeds. Therefore, selection should be more successful. Neglecting heterosis, synthetics should be inferior to parental breeds in some traits. If parental breeds are inbred to the degree F and if both were selected with equal intensity, it is expected that the better breed of the two will be superior by $0.56(2F)^{1/2}\sigma_A$ units (the better of two populations is above the mean by 0.56σ). In later generations the ratio of selection progress of a synthetic to that in the parental breed is $(1 - F)^{-1/2}$ if selection intensities, etc., are equal. After t generations of selection, the superiority relative to the mean of the parental breeds in generation zero is expected to be $tih\sigma_A(1 - F)^{1/2}$ for the synthetic and

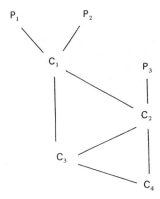

FIGURE 18.2. Creation of composite population.

$[0.56(2F)^{1/2} + tih]\sigma_A$ for the superior parental breed. Assuming $i = 1$, $h = 1/2$, and $F = 1/10$, it will require 9.8 generations of selection before the synthetic passes the better parent. However, component breeds of a cross are certainly not randomly chosen and each of them will possess specific advantages which the synthetic is to combine, at least to a medium degree.

Synthetics can be built from different numbers of parental breeds. If these are each to contribute one-half, one-fourth, or one-eighth of the genes, the various two-way, three-way, four-way, and backcrosses can be chosen as base population. For gene contributions of one-third, Lauprecht (1961) has proposed the scheme shown in Fig. 18.2. Members of different generations are crossed, but after the second generation the synthetic becomes closed. Therefore, the initial generations must be sufficiently numerous to avoid inbreeding depression.

19

Breeding Plans

Genetic improvement can be brought about in many ways. Practical animal breeding then must strive to combine components of genetic progress in such a way that, given biological, economic, and political constraints, genetic advantage and, even more, economic profit is maximized.

19.1. Economic Aspects in Selection

In a selection program the rate of genetic advance is determined by the accuracy and intensity of selection, generation interval, and the fraction of the population that undergoes performance recording. Accuracy and selection intensity are to some extent antagonistic, but consideration of their product permits optimization of a breeding program if the size of the recorded population is given. However, this may vary. Skjervold and Langholz (1964) investigated the influence of the size of the recorded population upon genetic progress in AI cattle populations with progeny testing. It appeared that large yearly rates of genetic progress demand that a large fraction of the population be progeny-tested. Potous and Vissac (1962) and Soller *et al.* (1966) included the time element in their investigations and concluded that interest rates to be paid reduce the value of long-term gains relative to gains in the near future. This affects the relative contributions of paths on which genetic improvement flows from one generation to the next, as Brascamp (1973) has pointed out. It also affects the relative values of improvement for various traits. For example, improvements in meat performance can be harvested sooner than improvements of dairy performance.

In many investigations which deal with economic aspects of breeding, relatively high interest rates were assumed, partly to neutralize inflation, partly to take account of the risk which attaches to a chosen breeding goal. Smith (1977) argued that inflation affects costs and profits equally and that therefore the interest rate relevant to breed improvement plans should be lower. The current rate of interest r is composed of a rate free of inflation s and a component to neutralize inflation t: $1 + r = (1 + s)(1 + t)$. The interest rate without inflation should be $s = (r - t)/(1 + t)$. If $r = 0.10$ and $t = 0.05$, the interest rate corrected for inflation is $s = 0.048$. It might be subdivided further into a risk component c and into a truly risk-free rate of interest d, if such indeed is possible: $1 + s = (1 + c)(1 + d)$.

The risk differs depending on the extent of the reference population or unit. If the national herd is to be improved, the view will be long term and the risk involved will be much less than if a private flock or herd is concerned. Here the risk through vacillating customer preferences is much higher than the risk for the national herd, where improvement of an economic trait almost always will benefit the whole economy in the long run.

Breed improvement has an economic motive and its aim is not necessarily to improve any one trait to a maximum. Economic considerations demand that cost–benefit ratios of breeding efforts be reckoned with, which may lead to different decisions than if only large genetic gains are sought.

Investigations of breeding plans can and should deal with all their components: fraction of recorded population for various traits, fraction under AI, culling rate, selection intensity, accuracy of selection, generation interval, etc. The number of possible combinations is very great and interactions abound. An investigation where all components vary simultaneously is well nigh impossible and the results would be difficult to comprehend. Usually, therefore, some components are held constant so that the effect of varying other factors can be studied more easily. In the short term, some components of genetic improvement can scarcely be changed. For example, the extent of milk recording remains fairly stable over reasonably extended time periods and questions of interest may concern effects which sperm conservation has on genetic progress relative to that of retaining test bulls until progeny results have become available. Another question may concern the choice of testing bulls for meat performance or of concentrating entirely on milk.

Such problems have been treated mainly in the context of AI breeding of cattle. Here, it is of interest to compare the effects of various culling rates, the accuracy of testing, and the varying contributions of individual sires on genetic progess. McClintock and Cunningham (1974) have proposed the method of "discounted gene flow," which appears very appropriate to problems of this kind. It estimates the consequences which an insemination of a cow has in the course of some years, in the example to be considered here, 12. The expected performances in milk and beef traits are discounted to the date of insemination or to another date. One may just as effectively compare the consequences of a single calving, as Bar-Anan et al. (1966) have done.

Following McClintock and Cunningham (1974), we assume that insemination involves up to three services (e.g., up to three services per lactation period are billed as one; the fourth, if necessary, is billed separately). On average, a cow will have 1.43 services (nonreturn rate to service equals 0.7). The probability of a productive offspring from these services (either producing milk, or beef and milk) is 0.8, which implies 1.8 services per viable and productive offspring. This animal has the probability p_F of being fattened (Fig. 19.1). If the herd size is constant, p_M corresponds to the reciprocal of the number of offspring k: $p_M = 1/k$, and if $k = 4$, $p_M = 1/4$. The probability of a beef animal is $p_F = 0.8 - 0.25 = 0.55$. The subsequent fate of the insemination is illustrated in Fig. 19.1. The meat value of the offspring can be reaped about 1 year after birth or 2 years after insemination. The dairy value is repeated with every lactation. Its mean time of realization is about $4\frac{1}{2}$ years after birth or $5\frac{1}{2}$ years after insemination, but this will vary, depending on age at first calving, calving interval, and number of lactations. When finally the cow is sold for slaughter, on average about 5–6 years after her first calf, the slaughter value is realized.

The first generation receives half of its genes from the sire, the grand-progeny one-fourth. An F_1 daughter is assumed to have four calving periods. Therefore the probability of a dairy animal in the F_2 is $4p^2_M$, and that of a beef animal is $4p_F = 4p_M(0.8 - p_M) = 4(1/4)(0.8 - 0.25) = 4 \times 0.1375 = 0.55$. However, the male of the base generation contributes only half as many genes to an F_2 as to the F_1. Consequently its total contribution to the F_2 is halved. On the beef side the development is analogous, except that beef animals do not reproduce but originate from the dairy animals.

The probability of a test bull p_T results from the ratio of all test bulls introduced to the population to the total number of inseminations by sperm of tested sires $N(1 - pt)$, where N, t, and p are, respectively, population size, fraction of test bulls, and fraction of population recorded. In the example of Table 19.1, p_T is assumed to be 1/1000. Such

Years after selection

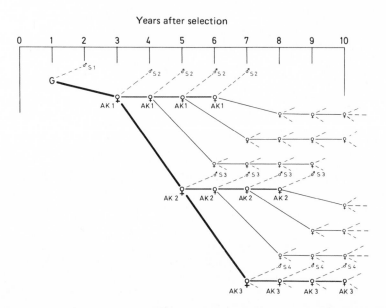

FIGURE 19.1. Consequences of an insemination. G, Birth of progeny; S_2, male slaughter animal in F_2; AK_1, calving of first dairy progeny in F_3 female descendants only; AK_2, calving of dairy progeny of 2nd generation; AK_3, calving of progeny of 3rd generation.

a test bull is expected to beget 100 dairy cows, which produce simultaneously the F_2 animals out of F_1 daughters. The probability of a granddaughter via a son equals 0.1 and she has one-fourth of the genes of the original male. From test bulls progeny-tested sires are selected whose daughters commence production in year 10. It is assumed that 0.20 of the bulls are selected from progeny testing. If progeny-tested bulls leave on average 1500 dairy cows, the probability of a dairy animal will be 0.3, and each will have on average one-fourth of the grandsire's genes.

The contributions of the various classes and generations of offspring are illustrated in Fig. 19.2. The lag between the use of a selected animal and the realization of its genetic merit in the improved performance of its progeny is apparent. Also, a considerable discontinuity exists which in reality will be unpronounced and will be evident only in the early stages of a breeding program to be followed by a more continuous sequence of progeny contributions (Hill, 1974). The genetic contribution of a breeding animal approaches an equilibrium which in Fig. 19.2 is achieved at about generation ten. The equilibrium is caused by two opposing tendencies: As generations advance, the fraction of genes of the base animal decreases in a geometric progression, but at the same

TABLE 19.1
Consequence of an Insemination[a]

♀ Generation	Proportion of genes	n	p_M	l	t_M	d_M	w_M	p_F	t_F	p_F	w_F	p_S	t_S	d_S	w_S
1	1/2	1	0.25	4	$5\frac{1}{2}$	0.76	0.380	0.55	2	0.91	0.2503	—	$7\frac{1}{2}$	0.69	0.087
2	1/4	4	0.25^2	4	9	0.64	0.160	0.1375	5.5	0.76	0.1045	—	11	0.58	0.036
3	1/8	16	0.25^3	4	10.9	0.59	0.074	0.0344	9	0.64	0.0440	—	11.75	0.57	0.018
4	1/16	32	—	—	—	—	—	0.0086	10.9	0.59	0.0103	—	—	—	—
Sum							0.614				0.434				0.142
♂ P 2	1/4	100	1/1000	4	$7\frac{1}{2}$	0.69	0.07	0.22	4	0.82	0.0450	—	$9\frac{1}{2}$	0.63	0.016
A 2	1/4	1500	1/(5 × 1000)	3	11	0.58	0.10	0.66	9	0.64	0.1056	—	12	0.56	0.042
Sum							0.17				0.151				0.058

[a] p_M, p_F, p_S, Probability of "dairy progeny," "meat progeny," cull cows.
t_M, t_F, t_S, Time to realization of dairy, beef, cull values.
d_M, d_F, d_S, Discount factors of dairy, beef, cull values.
w_M, w_F, w_S, Discounted values of dairy performance, beef performance, cull value.
P, A, Progeny of test bulls and of selected bulls.

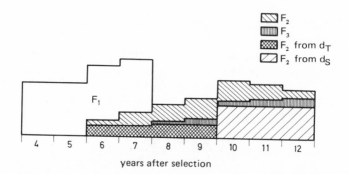

FIGURE 19.2. Realization of dairy merit of progeny-tested sires. F_1, F_2, F_3, Female descendence. d_T, test sons; d_S, selected sons.

time the genes are distributed to an equally increasing number of progeny. This results in a constant fraction of the total gene pool being derived from the individual animal, as has been pointed out by Hinks (1971) and Hill (1974).

Various methods have been used in investigations of the economic consequences of breeding plans. The payoff period defines the time period within which the investment must be returned. Let Y_t and C_t denote yields and costs, respectively. The payoff period is completed when accumulated yields and costs are equal:

$$\Sigma \frac{Y_t}{(1 + i)^t} = \Sigma \frac{C_t}{(1 + i)^t}$$

The internal interest rate is the discount rate at which the current value of returns is equal to the current value of costs. In consequence of diminishing returns on investments both the internal interest rate and payoff period will favor low-cost plans.

Obviously, low-cost breeding plans are justified if limited resources are to be exploited in an optimal way and if other factors such as competition or long-term genetic improvement can be neglected.

The assessment of the value of the capital accumulated would seem to be more appropriate for comparing various plans for breed improvement. The value of the assets (VA) is the difference between cumulative yields and costs, discounted to the same time:

$$VA = \Sigma \frac{Y_t}{(1 + i)^t} - \Sigma \frac{C_t}{(1 + i)^t}$$

The time need not be limited, but long-term yields and expenses are of little consequence.

The time factor can change the relative importance of the components of an aggregate economic value, such as dairy and beef merits. In Table 19.1 the computations extend to year 12 and include three generations of dairy animals and also three and part of the fourth generation of beef animals. The probability that the "dairy" merit of the base animals is expressed in the third generation results from the product of the number of progeny n times the probability that one of these is a dairy animal p_M^3 times the fraction of genes of the base animal (one-eighth) times four lactations:

$$np_M^3(\tfrac{1}{2})^3 \times 4 = 16 \times (\tfrac{1}{4})^3 \times \tfrac{1}{8} \times 4 = {}^1\!/_8$$

The corresponding probability of a beef animal is $p_M^2 p_B = 0.0344$. Beef and dairy merits are realized at different times, the lactation value on average at year 10.9, the slaughter value on average at year 9. At an interest rate of 5% the corresponding discount factors are 0.59 and 0.64, respectively. Over all generations, up to year 12, one insemination (on average 1.43 services) results in 0.61 lactations, 0.43 beef animals, and 0.14 culled cows.

The 4000 semen doses of a bull suffice to inseminate 2778 cows and result in 1694 lactations, 1136 beef animals, and 295 cull cows, all in present-day values. Improvements in dairy and beef merits of bulls should be weighted according to these figures. If in addition progeny of F_1 sons are included, the numbers of discounted expressions of merit increase by the equivalent of 472 lactations, 419 beef animals, and 161 bull cows.

The computation of the value of accumulated assets of different versions of a breeding program can be illustrated with the figures given in Table 19.2: population size 200,000 animals; 80,000 of these are milk-recorded, 80,000 are in AI, and 40,000 are both in AI and under milk recording. Two versions of sire testing are considered: laying off bulls until results become available or storing frozen semen and slaughtering bulls. The costs per bull up to the time when progeny test results become available are assumed to be equal and amount to 16,000 monetary units. For the version that involves bull slaughter, storage of 4000 portions of semen is assumed. The number of bulls required for semen production is derived from $N(1 - p) \times 1.8/d$, where d denotes number of doses per tested sire (1.8 inseminations per conception). In the version involving semen storage and bull slaughter, d is assumed to be 4000, but 10,000 when bulls are laid off.

Assume that six inseminations are required to produce a daughter

TABLE 19.2
Accummulated Assets in AI Breeding
Program[a]

		$P = 0.3$	$P = 0.5$
		4000 doses/proven sire; 1694 discounted lactations	
	T	2000	3333
	S	31	27
$n = 20$	b	1/3	1/6
	Mi	406	574
	Δ	74	76
	Σ	2.29	2.05
$n = 50$	b	$1/2^b$	1/2.5
	Mi	356	427
	Δ	75	88
	Σ	1.47^b	2.38
	S	12	11
		10000 doses/proven sire; 4233 discounted lactations	
$n = 20$	b	1/8	1/16
	Mi	1549	1834
	Δ	337	294
	Σ	4.04	3.23
$n = 50$	b	1/3.3	1/6
	Mi	1285	1664
	Δ	280	403
	Σ	3.36	4.43

[a] p, Fraction of test inseminations. T, Test offspring. S, Proven sires. n, Number of tested progeny. Mi, Genetic gain of tons of milk/proven sire. Δ, Difference between income over feed cost and expenses. Σ, Sum of net profit. S Δ, in millions of monetary units. b (fraction selected) = 1/2, i = 0.80; b = 1/2.5, i = 0.95; b = 1/3, i = 1.06; b = 1/3.3, i = 1.15; b = 1/6, i = 1.49; b = 1/8, i = 1.60; b = 1/16, i = 1.90. $r(n=20)$ = 0.76, $r(n=50)$ = 0.88.
[b] Semen production covers 65% of semen demand.

with records. Thirty percent ($=p$) of the 40,000 cows that are both milk-recorded and under AI will produce 2000 test daughters (one-sixth of 12,000). For insemination of the remaining 68,000 cows, 31 tested sires are required if semen is stored and 12 if the selected sires remain alive. If sires were each tested with 20 daughters, 100 bulls could be tested and the best one-third selected to obtain the 31 sires required for general use. The genetic improvement resulting from the selection of one out of three tested sires, accumulating over the course of 20 years, amounts to 406 tons milk, discounted to the time of selection ($1694 \times 1.06 \times 0.76 \times 0.3$ = number of lactations $\times$ $ir\sigma_A$). If feed costs are assumed to equal 40% of the producer price of milk, the income over feed cost per kg milk will be about 0.3 monetary units. Three test bulls are required per selected sire, each costing about 16,000 units, for a total of 48,000 units. The difference between additional income (406 tons $\times$ 300 units) and costs (48,000 units) is 74,000 monetary units and represents the value accumulated by the genetic improvement. Neither costs of milk recording nor of calculating breeding values were debited to the profit of the genetic improvement.

Table 19.2 illustrates that accumulated assets need not parallel the genetic gain. For example, the version with laidoff bulls and 50 daughters per bull is more profitable than that using only 20 daughters, although the latter improves the genetic merit equivalent to some 264, or 170 tons over and above the improvement when using 50 daughters. This results from the many fewer bulls to be laid off per bull eventually selected. Comparison of the figures also indicates that laying off bulls is more profitable than the semen storage–bull slaughter policy, but this depends somewhat on the assumptions made here. Nevertheless, Hinks (1971) found it true also for a variety of situations. The table makes it evident that increased use of proven sires is a very effective way to improve profitability of breeding programs, simply because additional genetic improvement can be achieved without additional cost. However, long-term consequences, for instance, the reduction in genetic variability, are neglected in this approach. The effect of increasing the culling rate among progeny-tested sires is illustrated in Table 19.3. Forty tested bulls are available for selection and the top 30%, 20%, or 10%, respectively, are selected. Altogether, 122,000 doses of semen are required, which implies either 10,000, 15,000, or 31,000 doses per proven sire. The figures in the table demonstrate that the income rises with increasing selection intensity.

Hinks (1971) suggested that one examine the feasibility of changing from one breeding plan to another by comparing the additional income with the additional cost. The ratio of the two increments increases 2.7 times when 20% instead of 30% are selected, while changing from 20% to 10% improves income 1.4 times as much as it increases costs. However, fixed costs were not considered and even though realistic quantities were assumed for variable costs, they were not thoroughly examined. Nevertheless, a reduction in the fraction of tested bulls retained for service from 30% to 20% looks very efficient.

TABLE 19.3
Profitability of Various Selection Intensities[a]

S/P	12/40	8/40	4/40	8/80
n	50	50	50	25
i	1.15	1.37	1.70	1.75
Tons of milk/proven sire	1.285	1.531	1.900	1.756
Additional income[b]	386	459	570	527
Additional expenses[b]	52.8	80	160	160
Additional profit[b]	333	370	410	367
ζ	2.69	—	1.39	0.85
Semen doses/Y/S	10,000	15,000	31,000	15,000

[a] S, Proven sires. P, Test bulls. n, Number of daughters with which each bull is tested. i, Standardized selection differential. ζ, Ratio of increments in income and expenses relative to the 8/40 version. Y/S, Number of semen doses per year and per proven selected sire.
[b] In thousands of monetary units.

The consequences of changing the intensity of selection can be examined also. Assume that eight proven sires are desired. If the number of test daughters is 2000, these eight sires can be selected from 80 bulls, each tested with 25 daughters, or from 40 bulls, each tested with 50. The 8/80 variant of the testing scheme implies a standardized selection differential of 1.75 and a correlation with genetic merit of 0.79. This results in a discounted improvement in milk yield of 1.76 tons per proven sire, which is more than the 1.53 tons resulting from the 8/40 scheme. However, in the 8/80 variant, ten test bulls must be retained per chosen bull until selection, each costing 16,000 monetary units. The difference between additional income (527,000 units) and additional cost (160,000 units) is 367,000 monetary units per bull and the ratio of the increments in income to cost when changing from the 8/40 to the 8/80 scheme is only 0.85, so that the latter scheme is less profitable.

So far, only progeny of daughters of proven sires have been considered. As is evident from Table 19.1, 0.17 discounted lactations are expected by the sire–son path in the time span of 12 years. Assuming that the genetic superiority of bull sires equals 1.5–1.75 times the genetic superiority of all selected sires, these 0.17 lactations should be equal in value to 0.25–0.30 lactations from the sire–daughter path, or half as many as the contribution via the female descendence (sire–daughter). This additional improvement comes about with no additional expense, and therefore it is of great importance. In many cases it is this additional increment in genetic improvement which justifies an expensive breeding program. If the potential genetic improvement of the population derived from the production of superior sons is neglected, it appears inadvisable to contemplate breeding programs at all.

It was pointed out in Chapter 12 that the path sire–son contributed more to genetic improvement than the path sire–daughter. However, if the genetic improvement is discounted, the ratio is reversed and sires of females appear more important, in accord with the conclusions of Brascamp (1973). The balance is tilted very much in favor of cow sires in the examples of this chapter, as the contribution of sons remains limited during the time span of only 12 years considered here. The contribution of discounted lactations of bull dams remains more important than that of cow dams, but the ratio is much narrower than if economic considerations are entirely omitted.

Economic deliberations become particularly important if various traits are to be improved. Changes in them can only be contemplated in conjunction with a knowledge of their monetary values. Bar-Anan *et al.* (1966) have investigated the consequences of various culling rates of bulls for meat and milk merit in order to determine optima. Niebel (1974) and Haring (1972) also examined the optimal selection intensity of dual-purpose bulls for beef traits from considerations of the economic consequences. Adelhelm *et al.* (1972) have estimated the economic importance of various performance characteristics of cattle.

Assume that the income over feed cost of 1 kg milk is 0.3 monetary unit and that the meat/bone (M/B) ratio of the carcass is the beef trait. Its improvement by 0.1 of a unit raises the income from a beef animal of 500–600 kg by 20 monetary units. Further, it is assumed that the M/B ratio can be estimated on the live animal by scoring for muscularity, or with ultrasonic devices, etc. Selection for dairy traits only permits intense selection of bull dams (3% of best cows). Selection of the best 40% young bulls ($i = 1$) in terms of beef performance necessitates 2.5 times as many bull dams, so that they are to be chosen among the 7.5% best cows, or among the top 15% if the bulls selected are among the 20% best for M/B. The less intense selection among cows reduces the genetic superiority of bull dams. It is further assumed that the genetic correlation between milk yield and M/B ratio is -0.3 (Mason, 1964), and that the other parameters given in Table 19.4 result in a genetic regression coefficient of milk yield on M/B of -4.5 kg/0.1 unit or in the reciprocal coefficient of -0.02 (unit of 0.1 M/B per 1 kg milk). Half the genetic selection differential among dams—direct for milk, indirect for M/B—is repeated in their sons. Individual selection among the latter adds the full genetic selection differential—direct for M/B, indirect for yield—to the inherited differences. Reference to Table 19.4 reveals that an inseminated cow has 0.78 discounted lactations and nearly as many

TABLE 19.4
Selection for Milk and Meat Performance

b		Milk breeding value			Meat/bone breeding value			Economic value[a]				
Milk	M/B	Direct: one-half dam's deviation	Indirect: young bull selection	Sum	Indirect: one-half dam's deviation	Direct: young bull selection	Sum	I	II	P_I	P_{II}	E
0.03	0	170	0	170	−0.034	0	−0.034	34.5	27.2	95.8	75.6	0
0.075	0.4	142	−54	88	−0.028	0.12	0.092	34.9	36.0	97.0	100	5
0.15	0.2	116	−76	40	−0.023	0.17	0.147	32.3	37.0	89.7	103.9	10

	h^2	σ_A	r_G
Milk	0.25	300 ⎫	−0.3
M/B	0.40	0.2 ⎭	

[a] Price variant I, $(0.3 \times \text{milk kg} + 200 \times \text{M/B}) \times 0.78$. II, $(0.2 \times \text{milk kg} + 200 \times \text{M/B}) \times 0.78$. P_I, P_{II}, Expected profit of alternatives I, II in thousands of monetary units. E, Expenses, in thousands of monetary units, per selected young bull.

expressions of the beef merit. The improvement in the two traits is weighted by both the expected numbers of discounted progeny expressions and the economic value, and the resulting sums are given in Table 19.4. To these sums the contributions from selection among progeny-tested bulls have to be added, but this is not done in Table 19.4. If the price variant I is assumed, weak selection for beef performance increases the income slightly, but additional costs are incurred, so that profitability would be slightly decreased. If selection for beef performance were intense, the reduced milk improvement is not balanced by the increased beef improvement and the income is decreased compared with selection for milk only. However, if the price relations of milk to beef are altered, the profitability of genetic improvement in beef traits is quite evident. But even here, intense selection ($b = 1/5$) among test bulls appears not to be more profitable than less intense selection ($b = 2/5$).

Increased costs incurred by beef testing must be deducted from the increased income. If the costs per young bull amount to 2000 monetary units, a culling intensity of 2/5 implies a cost of 5000 monetary units per selected test bull, and if $b = 1/5$, of 10,000 monetary units. The costs per proven sire will be multiples of this. If one-third of test bulls are selected, the costs of beef testing per proven sire would sum to 15,000 and 30,000 monetary units, respectively, for the two selection intensities. The consequence is that, if the economics of meat and milk are consistent with version 1, beef testing of dual-purpose bulls is not a profitable proposition, while version 2 would justify a weak selection ($b = 2/5$) for beef performance but not the intense selection. The relation of increments in income to those of costs when moving from no selection for beef to choosing 2/5 (40%) is 1.63 (24,500/15,000), but it is only 0.27 when selection of the best 20% is compared to selection of the best 40%. The low efficiency of beef performance testing in the context of dairy progeny testing has been pointed out by Hinks (1971). It can be explained by the increase of costs with increasing selection intensity on the one hand and on the other by the loss of the short generation intervals characteristic of performance testing when a progeny testing scheme is involved.

19.2. Hierarchy in Populations of Domestic Animals

With respect to breeding, domestic animals show a hierarchic organization. This has been described for cattle by Robertson and Asker (1951) and subsequently by many others, for example, by Wiener (1961) for sheep and by Jonsson (1965) for pigs. The classic scheme involves three tiers: the elite stratum or the nucleus, the multiplier stratum, and the producer stratum, although animals of a lower stratum may be exchanged for animals of the stratum above and there may be more or less than three tiers. The genetic superiority of higher strata is transmitted by males or sperm to the lower strata. Various combinations of strata are possible; for example, semen of nucleus animals may be deployed in the producer stratum and the producer females may be replaced by females of the multiplier stratum. In modern poultry production, the physical replacement of animals of lower strata by those of higher strata is common and in many cases multipliers are franchised by the breeding organization. Similar efforts are made in the swine industry, but here, as in other domestic mammals, transfer of genetic merit by sale of males or by semen is still the rule. In cattle breeding the advent of AI has revolutionized the structure, which has changed from a more or less rigid three-tier scheme to a two-tier, or even a $1\frac{1}{2}$-tier, scheme, as semen from elite bulls is frequently available also to producers. If embryo transfer should become routinely possible, complete physical replacement of lower tiers can be envisaged.

The genetic progress of hierarchic populations is illustrated in Table 19.5. In general, the genetic merit of tiers corresponds to their position in the hierarchy and equal differences occur between successive tiers. If one assumes full replacement and equal genetic progress (ΔG) in each generation, then the genetic superiority of the nucleus stratum over the multiplier herds will be ΔG and over the producer herds $2\Delta G$. Frequently, the best animals of the nucleus population are used for their own replacement and the remainder in the recruitment of multiplier animals, implying a negative selection differential for the latter. However, this is not acumulative. After t generations the genetic difference between the elite and the multiplier populations will be $\Delta n - \Delta v$. Sometimes animals introduced into the multiplier herds are selected again, thus creating a positive selection differential, which, however, also does not accumulate.

For animal improvement the transfer of genetic merit via males or via semen has more significance than that by females. The lower strata replace only half of their genes, while the other half derives from their own females. The genetic difference is thus twice the difference under complete replacement, so that the superiority of the nucleus over the multiplier population is $2\Delta G$ and over the producer herds $4\Delta G$. A negative selection differential in the nucleus group when producing multiplier animals ($-\Delta V$) will not accumulate, but remains constant at $-\Delta V$. A similar situation pertains to a positive selection differential resulting from selection in the multiplier stratum.

TABLE 19.5
Genetic Progress in Hierarchic Breeding System[a]

Generation	Nucleus population	Multiplier herd	Producer herd
a. Complete replacement			
0	0		
1	Δn	$-\Delta v$	$+\Delta p$
2	$2\Delta n$	$\Delta n - \Delta v$	$-\Delta v + \Delta p$
3	$3\Delta n$	$2\Delta n - \Delta v$	$\Delta n - \Delta v + \Delta p$
$\cdots$		$3\Delta n - \Delta v$	$2\Delta n - \Delta v + \Delta p$
t	$t\,\Delta n$	$(t-1)\,\Delta n - \Delta v$	$(t-2)\,\Delta n - \Delta v + \Delta p$
b. Sale of male breeding animals			
0	0	0	0
1	Δn	$-\tfrac{1}{2}\Delta v$	$+\tfrac{1}{2}\Delta p$
2	$2\Delta n$	$\tfrac{1}{2}(\Delta n - \Delta v) - \tfrac{1}{4}\Delta n - \tfrac{1}{4}\Delta v = \tfrac{1}{2}\Delta n - \tfrac{3}{4}\Delta v$	$-\tfrac{1}{4}\Delta v + \tfrac{3}{4}\Delta p$
3	$3\Delta n$	$\tfrac{1}{2}(2\Delta n - \Delta v) + \tfrac{1}{4}\Delta n - \tfrac{3}{8}\Delta v = 5/4\,\Delta n - \tfrac{7}{8}\Delta v$	$\tfrac{1}{4}\Delta n - \tfrac{3}{8}\Delta v - \tfrac{5}{8}\Delta p$
4	$4\Delta n$	$(17/8)\Delta n - \tfrac{15}{8}\Delta v$	$\Delta n - \tfrac{11}{16}\Delta v + \tfrac{15}{16}\Delta p$
$\cdots$			
t	$t\,\Delta n$	$[(t-2) + (\tfrac{1}{2})^{t-1}]\,\Delta n$ $- [1 - (\tfrac{1}{2})^{t}]\,\Delta v$	$[(t-4) + (t+2)(\tfrac{1}{2})^{t-1}]\,\Delta n$ $- \{[2^{t} - (t+1)]\,(\tfrac{1}{2})^{t}\}\,\Delta v$ $+ [1 - (\tfrac{1}{2})^{t}]\,\tfrac{1}{2}\Delta p$
		$\approx (t-2)\,\Delta n - \Delta v$	$\approx (t-4)\,\Delta n - \Delta v + \Delta p$

[a] n, Genetic gain in nucleus population. $-\Delta v$, Genetic selection differential in nucleus population of parents of multiplier animals. Δp, Genetic selection differential in multiplier herds of parents of producer animals.

Bichard (1971) has investigated other ramifications of hierarchic systems. One possibility involves the use of nucleus sires in the multiplier stratum. This reduces the difference between the two strata if $\Delta n_m / \Delta t_m > \Delta G$, i.e., if the ratio of genetic advance in nucleus males to the generation interval is greater than the yearly genetic gain. The genetic difference between strata can be considerable. In the traditional three-tier system of cattle breeding the genetic merit of nucleus and producer strata differs by the equivalent of some 15 years of genetic progress. Modern AI breeding can reduce this gap to the equivalent of 2 years of genetic progress. Nevertheless, only gains made in the nucleus stratum are cumulative.

The economic consequences of selection in hierarchic breeding populations can be investigated by the methods discussed in the previous section. For example, it turns out that maternal traits, which are expressed earlier than the productive traits, deserve more attention when genetic progress is not the sole concern. Also, selection in the multiplier stratum should receive relatively more weight than in a purely genetic context, as its improvement is realized sooner than that of the nucleus. Danell (1977) has investigated several versions of possible hierarchic structures under a variety of price relationships in sheep breeding.

20

Final Remarks

Population genetics has provided to animal breeding a theoretical framework which furthers both the understanding of processes and phenomena and permits the planning of breed improvement in a rational way. However, success is limited largely to the realm of genic variation. The understanding of dominance and epistasis, as they affect both differences between animals and genetic progress, is much less developed. Concepts such as genetic or additive-genetic, dominant, and epistatic effects, as used in quantitative genetics, are largely of a predictive nature and they do not imply any causal explanation. Incongruities appear when the theory is tested empirically, some of which have not been satisfactorily explained (Kidwell and Kempthorne, 1966; Kempthorne, 1976). The impact of concepts developed through animal breeding theory on scientific endeavors in other fields can be considered as a gauge of its scientific maturity. In this context it appears to be of some significance that Wright's genetic drift concept has received much attention in the theory of evolution. The concept originated from Wright's efforts to analyze the model breeding experiments with guinea pigs which the USDA had undertaken before World War I. The same significance pertains to the inbreeding coefficient. The concept of heritability, in a way central to animal breeding theory, has found wide application in theoretical plant breeding and has attracted much attention in human genetics, even though discussions have not always been impartial and objective (Jensen, 1972; Kamin, 1974; Feldman and Lewontin, 1975).

When Mendelism became accepted, efforts were first directed toward explaining performance differences by one or few loci. Soon the polygenic hypothesis (Fisher, 1930; Wright, 1921) superseded such views.

However, at present renewed efforts are being made to rationalize major gene effects, stimulated both by increased knowledge concerning the physiology and biochemistry which underlie a trait, and by improved insight into the genetic architecture of traits. Some surprising discoveries have encouraged such endeavors. For example, resistance to fowl leucosis was considered to be a classic case of a trait with a very low heritability and the slow progress made through selection and by the efforts of breeders seemed to imply many genes with individually small effects. Meanwhile, it turned out that rather few loci affect the first line of defense, i.e., the resistance of cells to infection, and that a major locus is of great importance in the extent of the immune response of infected birds. A similar situation has been revealed in the case of resistance to Marek's disease, and in addition the porcine stress syndrome, a fitness trait of a kind, also depends on a major locus. It appears, therefore, that major loci will receive more attention henceforth.

In successful breeding, existing genotypes are exchanged for better ones. Modern methods have accelerated this progress. These entail an increased rate of disappearance of less productive strains and breeds and increased uniformity among the surviving populations. This loss of genetic heritage is acutely realized in plant breeding, where successful genotypes are frequently homozygous. In animal breeding, where more or less panmictic populations prevail, the danger of gene loss is less imminent. Nevertheless it exists and techniques such as AI, and even more so embryo transfer, if it becomes routinely possible, will accelerate it.

It is impractical to retain unproductive strains indefinitely, in particular of large domestic mammals, although medium-term gene banks are possible if deep-frozen semen and/or deep-frozen embryos are used. The problem is acute, in particular where production is capital- and labor-intensive, but it is less pressing for low-cost forms of production. If these entail utilization of natural resources, adaptation to variable ecological niches may favor genetic diversity.

Changes in consumer habit, of disease vectors, etc., can make hitherto neglected genotypes interesting again. Examples of changing preferences are the fairly sudden demand for large-framed continental cattle breeds in the U.S. caused by an increased consumer preference for lean meat and the availability of cheap concentrates. However, if concentrate feeding becomes too expensive, the trend may be reversed. Nevertheless it appears that the testing of endangered breeds and species and of their crosses under alternative production systems, for example, in developing countries, would be desirable before their genotypes have disappeared.

Technical innovations in animal production may affect not only production as such but also breeding itself. An example from the remote past is the artificial hatching of poultry—common in Fayumis of ancient Egypt—and a recent example is artificial insemination, which has changed the whole breeding structure of dairy cattle and forced a change of breeding plans. At present much work is being devoted to the development of routine procedures for embryo transfer. It already has been used commercially for the propagation of very valuable genotypes, for instance, where health quarantine prevents more than a few animals of a new breed from being imported. If embryo transfer becomes routinely possible, it would make cattle genetically equal to a multiparous species and permit intensive selection among females.

As is evident from Table 20.1, only routine embryo transfer on a large scale will have a reasonably great impact on genetic progress in dairy cattle breeding. The reason for this is the notable efficiency of present-day AI breeding and progeny testing. Surprisingly, embryo transfer could affect beef cattle breeding more, because intense selection is possible on the female side without prolonging the generation interval.

Breeding will be profoundly influenced if direct indicators of the genetic merit of individuals for sex-limited traits become available. If, for example, the dairy merit of a bull could be estimated by direct measures on him, individual selection could reduce the need for expensive progeny testing. However, the great deal of effort in this direction

TABLE 20.1
Effects of Embryo Transfer on Genetic Progress[a]

| | Fraction selected[b] | | | | | | | | | | |
| | I | | | | II | | | | III | | |
	CC	CS	ΔG	%Δ	CC	CS	ΔG	%Δ	C	ΔG	%Δ
Conventional	80	3	0.185	—	89	17	0.39	—	100	0.35	—
A_1	80	0.75	0.196	6	89	8.6	0.31	3			
A_2	80	0.04	0.216	17	89	2	0.34	13			
B_1	20	0.75	0.221	19	44	8.6	0.35	17	25	0.50	33
B_2	1	0.04	0.271	46	10	2	0.42	40	1.25	0.66	76

[a] CC, Dam to daughter. CS, Dam to Son. ΔG, Genetic gain per year, in genic standard deviations. %Δ, Increment of genetic gain relative to natural reproduction (including AI). $A_1 B_1$, 4 embryos transferred for bulldams only, in total population, respectively. $A_2 B_2$, 80 embryos transferred per year for bulldams and in population, respectively. Selection differential for sire to son (SS), sire to daughter (SC): 2.06, 0.93. Correlations with breeding value for SS, SC are: milk 0.88; meat 0.70. Generation interval SS, SC, CS, CC: 7, 5.3, 5, 4 years.
[b] Selection: I, For milk, $h^2 = 0.25$. II, For milk and meat, $h^2_M = 0.25$ and $h^2_F = 0.36$; $r_A = 5/6$ genic correlation between male and female expression of meat trait F). Equal economic weights per genic standard deviation. III, Selection for meat. i for sires (mass selection) 1/2%, $r = 0.6$, $t = 2\frac{1}{2}$ years (bulls and cows).

has not yet resulted in sufficiently accurate yardsticks (Schönmuth *et al.*, 1966; Joakimsen 1974; Tilakaratne *et al.* 1980).

Sex determination of sperm has understandably attracted much attention (Schilling and Thormählen, 1976). The successful sex control of offspring would favor specialized breeds, such as Jerseys, where all surplus calves could be produced as males from beef sires while pure-breeding was reserved for the production of replacement females (Cunningham, 1977).

The recent advances of cell biology and molecular genetics have been rapid, and techniques and methods of genetic improvement may become available which until now have been considered to be in the realm of science fiction.

Endeavors of this kind are subsumed under genetic engineering, although Dickerson (1973) used the same term for more or less conventional techniques of animal breeding improvement. Genetic engineering as it is generally understood involves techniques such as transplantation of nuclei into enucleated ova, but more typically the application of restriction endonucleases to isolate specific stretches of DNA, which are then transferred into other cells through the agency of plasmids. Ova transplantation and their subsequent development, or the development of diploid somatic cells into intact individuals, would permit vegetative reproduction and thus circumvent entirely the Mendelian lottery. Restriction endonuclease techniques and transfer via plasmids permit transfer of isolated genes free from other possibly undesirable genes, an unavoidable corollary of introduction by crossing. Nonetheless application of the recombinant DNA technique presupposes knowledge of desirable genes, which is a partly premature assumption in animal breeding.

The theoretical models of breed improvement developed in the science of animal breeding favor large populations and large-scale breeding organizations, in particular if progeny and sib testing are required for accurate selection. Breeding animals, mostly sires, having been selected at high cost, need to be utilized to the fullest possible extent, which in turn also entails large populations. However, large organizations are beset by some degree of inflexibility, simply due to scale. In view of possibly fast-changing scenarios—consumer habits, disease, costs of feed—flexibility would be desirable, which could be provided by a larger number of smaller units, each having slightly different goals and employing somewhat different techniques.

The efficiency of resource utilization by the animal industry in the production of edible energy and protein sources for the human population is much less than the efficiency in plant production. In view of the precarious situation of the global supply of human food, animal production is therefore often criticized as wasteful and undesirable, but animals can use resources which are not directly suitable for human consumption. Animal products provide essential elements which are difficult to provide in an entirely vegetarian diet and these products are of special value to the majority of people. Animals remain an important source of agricultural energy in many parts of the world, and provide much pleasure in sports and as pets. Breeding strives to improve the efficiency of animal production by increasing and exploiting the genetic potential of our domestic animals and thus it contributes to a competitive animal production.

Animal genetics as discussed in this text has broadened our understanding of the variability extant in the populations of our domestic animals, and has greatly helped breed improvement. The numerous and extensive explorations into gene action and molecular genetics using the tools of cell biology, physiology, and biochemistry make it probable that future breed improvement work will be more directly involved in multiplying and combining desirable genotypes and genes than has been possible heretofore.

Literature

General Sources

American Society of Animal Science. 1969: *Techniques for and Procedures in Animal Science Research*. American Society of Animal Science, Champaign, Illinois.

Bulmer, M. G. 1980: *The Mathematical Theory of Quantitative Genetics*. Clarendon Press, Oxford.

Cavalli-Sforza, L. L., and Bodmer, W. F. 1971: *The Genetics of Human Populations*. W. H. Freeman, San Francisco.

Crow, J. F., and Kimura, M. 1970: *An Introduction to Population Genetics Theory*. Harper and Row, New York.

Falconer, D. S. 1981: *Introduction to Quantitative Genetics*. Longman, London.

Fisher, R. A. 1930: *The Genetical Theory of Natural Selection*. Oxford University Press, Oxford.

Jacquard, A. 1977: *Genetique des Populations Humaines*. Presses Universitaires de France, Paris.

Johansson, I. 1961: *Genetic Aspects of Dairy Cattle Breeding*. University of Illinois Press, Urbana.

Kempthorne, O. 1957: *An Introduction to Genetic Statistics*. Wiley, New York.

Lerner, I. M. 1950: *Population Genetics and Animal Improvement*. Cambridge University Press, Cambridge.

Lerner, I. M. 1958: *The Genetic Basis of Selection*. Wiley, New York.

Lerner, I. M., and Donald, H. P. 1966: *Modern Developments in Animal Breeding*. Academic Press, London.

Li, C. C. 1976: *Population Genetics*. Boxwood Press, Pacific Grove, California.

Lush, J. L. 1945: *Animal Breeding Plans*. Iowa State College Press, Ames, Iowa.

Lush, J. L. 1948: *The Genetics of Populations*. Iowa State College Press, Ames, Iowa.

Malécot, G. 1948: *Les Mathématiques de l'Heredité*. Masson, Paris.

Ollivier, L. 1981: *Eléments de Génétique Quantitative*. Masson, Paris.

Stahl, W., Rasych, D., Siler, R., and Vachal, J. 1969: *Populationsgenetik für Tierzüchter*. VEB Deutscher Landwirtschaftsverlag, Berlin.

Turner, H. N., and Young, S. S. Y. 1969: *Quantitative Genetics in Sheep Breeding*. Cornell University Press, Ithaca, New York.

Wright, S. 1968/1969/1977/1978: *Evolution and the Genetics of Populations.* Vol. I. *Genetic and Biometric Foundations* (1968). Vol. II. *The Theory of Gene Frequencies* (1969). Vol. III. *Experimental Results and Evolutionary Deductions* (1977). Vol. IV. *Variability within and among Natural Populations* (1978). University of Chicago Press, Chicago, Illinois.

References

Abplanalp, H. 1961: Linear heritability estimates. *Genet. Res.* **2,** 439.

Abplanap, H. 1972: Selection of extremes. *Anim. Prod.* **14,** 11.

Abplanalp, H. 1974: Inbreeding as a tool for poultry improvement. In *1st World Congress on Genetics Applied to Livestock Production,* Volume 1, p. 897. Editorial Garsi, Madrid.

Abplanalp, H., Lowry, D. C., Lerner, I. M., and Dempster, E. R. 1964: Selection for egg number with X-ray induced variation. *Genetics* **50,** 1083.

Adelhelm, R., Boeckenhoff, E., Bischoff, T., Fewson, D., and Rihler, A. 1972: *Die Leistungsmerkmale beim Rind—wirtschaftliche Bedeutung und Selektionswürdigkeit, A. Hohenheimer Arbeiten Nr. 64,* University of Hohenheim, Federal Republic of Germany.

Allaire, F. R., and Henderson, C. R. 1965: Specific combining abilities among dairy sires. *J. Dairy Sci.* **48,** 1096.

Allaire, F. R., and Henderson, C. R. 1966a: Selection practiced among dairy cows. I. Single lactation traits. *J. Dairy Sci.* **49,** 1426.

Allaire, F. R., and Henderson, C. R. 1966b: Selection practiced among dairy cows. II. Total production over a sequence of lactations. *J. Dairy Sci.* **49,** 1435.

Allaire, F. R., and Henderson, C. R. 1967: Selection practiced among dairy cows. III. Type appraisal and lactation traits. *J. Dairy Sci.* **50,** 194.

Alsing, I., Krippl, J., and Pirchner, F. 1980: Maternal effects on the heritability of litter traits of pigs. *Z. Tierz. Zuechtungsbiol.* **97,** 241.

Anderson, D. E. 1960: Studies on bovine ocular squamous carcinoma ("Cancer eye"). V. Genetic aspects. *J. Hered.* **51,** 51.

Andresen, E. 1971: Linear sequence of the autosomal loci *PHI, H* and *6-PGD* in pigs. *Anim. Blood Groups Biochem. Genet.* **2,** 119.

Andresen, E., and Jensen, P. 1977: Close linkage established between the *HAL* locus for halothane sensitivity and the *PHI* (phosphohexose isomerase) locus in pigs of the Danish Landrace breed. *Nord. Veterinaermed* **29,** 502.

Andresen, E., Hojgard, N., Jylling, B., Larsen, B., Moustgaard, J., and Neiman-Sørensen, A. 1960: Blut- und Serumgruppenuntersuchungen bei Rind, Schwein und Hund in Dänemark. *Zuechtungskunde* **32,** 306.

Aschaffenburg, R. 1968: Review of the progress of dairy cattle research. *J. Dairy Sci.* **35,** 447.

Ashton, G. C., and Hewetson, R. W. 1969: Transferrins and milk production in dairy cattle. *Anim. Prod.* **11,** 533.

Bailey, D. W. 1959: Rates of subline divergence in highly inbred strains of mice. *J. Hered.* **50,** 26.

Baker, C. M., and Manwell, A. 1980: Chemical classification of cattle. 1. Breed groups. *Anim. Blood Groups Biochem. Genet.* **11,** 127.

Baker, L. M., Hazel, L. N., and Reinmiller, C. F. 1943: The relative importance of heredity and environment in the growth of pigs at different ages. *J. Anim. Sci.* **2,** 3.

Bar-Anan, R., and Bowman, J. C. 1974: Twinning in Israeli-Friesian dairy herds. *Anim. Prod.* **18,** 109.

Bar-Anan, R., and Sacks, J. M. 1974: Sire evaluation and estimation of genetic gain in Israeli dairy herds. *Anim. Prod.* **18**, 59.

Bar-Anan, R., Soller, M., and Bowman, J. C. 1976: Genetic and environmental factors affecting the incidence of difficult calving and perinatal mortality in Israeli-Friesian dairy herds. *Anim. Prod.* **22**, 299.

Barker, J. S. F., and Robertson, A. 1966: Genetic and phenotypic parameters for the first three lactations in Friesian cows. *Anim. Prod.* **8**, 221.

Baxa, H. 1973: Gene frequencies in stray cat populations from Vienna. *Genetica* **44**, 25.

Bell, A. E. 1972: More on reciprocal recurrent selection. *In Proceedings of the 21st National Poultry Breeders Roundtable*, p. 18, Poultry Breeders of America, Kansas City, Missouri.

Bereskin, B., and Lush, J. L. 1965: Genetic and environmental factors in dairy sire evaluation. III. *J. Dairy Sci.* **49**, 356.

Berge, S. 1931: Die Wahrscheinlichkeit der Homozygotie der Dominanttype bei vollständiger Dominanz in monohybriden Spaltungen. *Z. Indukt. Abstammungs-Vererbungsl.* **58**, 157.

Berger, P. J., and Harvey, W. R. 1975: Realized genetic parameters from index selection in mice. *J. Anim. Sci.* **40**, 38.

Bernstein, F. 1930: Fortgesetzte Untersuchungen aus der Theorie der Blutgruppen. *Z. Indukt. Abstammungs-Vererbungsl.* **56**, 233.

Bichard, M. 1971: Dissemination of genetic improvement through a livestock industry. *Anim. Prod.* **13**, 401.

Blyth, J. S. S., and Sang, J. H. 1960: Survey of line crosses in a Brown Leghorn flock. *Genet. Res.* **1**, 408.

Bohlin, O., and Rönningen, K. 1974: Inbreeding and relationship within the North-Swedish horse. *Acta Agric. Scand.* **25**, 121.

Bohren, B. B., Mckean, H. E., and Yamada, Y. 1961: Relative efficiencies of heritability estimates based on regression of offspring on parent. *Biometrics* **16**, 481.

Bohren, B. B., Hill, W. G., and Robertson, A. 1966: Some observations on asymmetrical correlated responses to selection. *Genet. Res.* **7**, 44.

Bonnier, G. 1934: Production av Brukshönor. *Med.* **20**, 56. Institutet Husdjursförädling, Stockholm, Sweden.

Botkin, M. P., and Stratton, P. O. 1967: Factors limiting selection effectiveness in small flocks of Columbia nad Corriedales. *J. Anim. Sci.* **26**, 971.

Bouw, J., Buys, C., and Schreuder, I. 1974: Further studies on the genetic control of the blood group system C of cattle. *Anim. Blood Groups Biochem. Genet.* **5**, 105.

Bowman, J. C., and Powell, J. C. 1962: The effect of different levels of management on the assessment of differences between varieties of chickens. I. *Anim. Prod.* **4**, 319.

Bradford, G. E., Chapman, A. B., and Grummer, R. H. 1958: Effects of inbreeding, selection, linecrossing, and topcrossing in swine. *J. Anim. Sci.* **17**, 426.

Brascamp, E. W. 1973: The economic value of genetic improvement in milk yield. *Z. Tierz. Zuechtungsbiol.* **90**, 1.

Brascamp, E. W. 1980: Selection index for desired gains. *In Proceedings 30th Ann. Meet. EAAP, Harrogate, 1979. Ann. Genet. Sel. Anim.* **12**, 133.

Briles, W. E., and Allen, C. P. 1961: The B-blood group system of chickens. II. The effects of genotype on livability and egg production in seven commercial inbred lines. *Genetics* **46**, 1273.

Briles, W. E., Allen, C. P., and Millen, Z. W. 1957: The B-blood group system of chickens. I. Heterozygosity in closed populations. *Genetics* **42**, 631.

Brown, R. V., and Nordskog, A. W. 1962: Correlated response in blood group frequencies with selection for body weight and egg weight in the fowl. *Genetics* **47**, 945.

Bucio-Alanis, L. 1966: Environmental and genotype–environment components of variability. I. Inbred lines. *Heredity* **21**, 387.

Bulmer, M. G. 1971: The effect of selection on genetic variability. *Am. Nat.* **105**, 201.

Bulmer, M. G. 1974: Linkage disequilibrium and genetic variability. *Genet. Res.* **23**, 281.

Buschmann, H. 1962: Blutgruppengenetische Untersuchungen an süddeutschen Rinderrassen. *Z. Tierz. Zuechtungsbiol.* **78**, 12.

Buschmann, H., and Schmid, D. O. 1972: *Serumgruppen bei Tieren.* Paul Parey, Hamburg.

Buschmann, H., Kraeusslich, H., Meyer, J., Radzikowski, A., and Osterkorn, K. 1976: Weitere Untersuchungen an Mäusen welche auf hohes und niedriges Phagozytosevermögen selektiert worden sind. *Zentralbl. Veterinaermed. Reihe B* **23**, 331.

Butcher, D. F., and Freeman, A. E. 1968: Heritabilities and repeatabilities of milk and milk fat production by lactation. *J. Dairy Sci.* **51**, 1387.

Butler, I. von 1981: Vergleich der Effizienz verschiedener Methoden bei Selektion auf antagonistische Eigenschaften bei der Hausmaus. Thesis, Technische Universität München.

Carmon, J. L., Stewart, H. A., Cockerham, C. C., and Comstock, R. G. 1956: Prediction equations for rotational crossbreeding. *J. Anim. Sci.* **15**, 930.

Castle, O. M., and Searle, S. R. 1957: Repeatability of dairy cow butterfat records in New Zealand. *J. Dairy Sci.* **40**, 1277.

Cavalli-Sforza, L. L., and Bodmer, W. F. 1971: *The Genetics of Human Populations.* W. H. Freeman, San Francisco.

Cavalli-Sforza, L. L., and Edwards, A. W. F. 1967: Phylogenetic analysis: Models and estimation procedures. *Evolution* **21**, 550.

Ceppelini, R., Siniscalco, M., and Smith, C. A. 1955: The estimation of gene frequencies in a random mating population. *Ann. Hum. Genet.* **20**, 97.

Clayton, G. A., and Robertson, A. 1957: An experimental check on quantitative genetical theory. II. The long term effects of selection. *J. Genet.* **55**, 152.

Clayton, G. A., and Robertson, A. 1964: The effects of X-rays on quantitative characters. *Genet. Res.* **5**, 410.

Clayton, G. A., Morris, J. A., and Robertson, A. 1957: An experimental check on quantitative genetical theory. I. Short term response to selection. *J. Genet.* **55**, 131.

Clement, A. 1978: Rassendifferenz, Heterosis und genetische Varianz bei Braunvieh—Brown Swiss Kreuzungen. Thesis, Technische Universität München.

Cochran, W. G. 1951: Improvement by means of selection. In *Proceedings Second Berkeley Symposium on Mathematical Statistics Probability,* p. 449. University of California Press, Berkeley, California.

Cockerham, C. C. 1954: An extension of the concept of partitioning hereditary variance for analysis of covariances among relatives when epistasis is present. *Genetics* **39**, 859.

Cockerham, C. C. 1967: Prediction of double crosses from single crosses. *Zuechter* **37**, 160.

Cockerham, C. C. 1973: Analyses of gene frequencies. *Genetics* **74**, 679.

Cole, R. K. 1968: Studies on genetic resistance to Marek's disease. *Avian Dis.* **12**, 9.

Cole, R. K., and Hutt, F. B. 1973: Selection and heterosis in Cornell White Leghorns: A review with special consideration of interstrain hybrids. *Anim. Breed. Abst.* **41**, 103.

Collins, W. M., Briles, W. E., Zsigray, R. M., Dunlop, W. R., Corbett, A. C., Clark, K. K., Marks, J. L., and McGrail, T. P. 1977: The *B* Locus (MHC) in the chicken: Association with the fate of RSV-induced tumors. *Immunogenetics* **5**, 333.

Comstock, R. E., and Moll, R. H. 1963: Genotype–environment interactions. In W. D. Hanson and H. F. Robinson, eds., *Symposium on Statistical Genetics and Plant Breeding,* p. 164. National Academy of Sciences–National Research Council, Washington, D.C.

Comstock, R. E., and Robinson, H. F. 1952: Estimation of average dominance of genes. In J. W. Gowen, ed., *Heterosis,* p. 494. Iowa State College Press, Ames, Iowa.

Comstock, R. E., Robinson, H. F., and Harvey, P. H. 1949: A breeding procedure designed to make maximum use of both general and specific combining ability. *Agron. J.* **41,** 360.

Conneally, P. M., and Stone, W. H. 1965: Association between a blood group and butterfat production in dairy cattle. *Nature* **206,** 115.

Conneally, P. M., Stone, W. H., Tyler, W. J., Casida, L. E., and Morton, N. E. 1963: Genetic load expressed as fetal death in cattle. *J. Dairy Sci.* **46,** 222.

Cooper, D. W., Bailey, L. F., and Mayo, O. 1967: Population data for the transferrine variants in the Australian merino. *Aust. J. Biol. Sci.* **20,** 959.

Cotterman, C. W. 1940: A calculus for statistico-genetics. Thesis, Ohio State University, Columbus, Ohio.

Cotterman, C. W. 1954: Estimation of gene frequencies in nonexperimental populations. *In* O. Kempthorne, ed., *Statistics and Mathematics in Biology*, p. 449. Iowa State College Press, Ames, Iowa.

Crawford, R. D. 1964: Fertility of walnut-combed domestic fowl. *Can. J. Anim. Sci.* **44,** 184.

Crettenand, J. 1975: Le testage de taureaux sur la base d'équations linéaires. Thesis, Federal Institute of Technology, Zürich, Switzerland.

Crittenden, L. B., Briles, W. E., and Stone, H. A. 1970: Susceptibility to avian leukosis-sarcoma virus: Close association with an erythrocyte isoantigen. *Science* **169,** 1324.

Crow, J. F. 1952: Dominance and overdominance, *In* J. W. Gowen, ed., *Heterosis*, p. 282. Iowa State College Press, Ames, Iowa.

Crow, J. F. 1954: Breeding structure of populations. II. Effective population number. *In* O. Kempthorne, ed., *Statistics and Mathematics in Biology*. p. 543. Iowa State College Press, Ames, Iowa.

Crow, J. F. 1958: Some possibilities for measuring selection intensity in man. *Hum. Biol.* **30,** 1.

Crow, J. F., and Denniston, C. 1981: The mutation component of genetic damage. *Science* **212,** 888.

Crow, J. F., and Kimura, M. 1970: *An Introduction to Population Genetics Theory.* Harper and Row, New York.

Cruden, D. 1949: The computation of inbreeding coefficients in closed populations. *J. Hered.* **40,** 248.

Cunningham, E. P. 1965: The evaluation of sires from progeny test data. *Anim. Prod.* **7,** 221.

Cunningham, E. P. 1975a: Multi-stage index selection. *Theor. Appl. Genet.* **46,** 55.

Cunningham, E. P. 1975b: Genetic studies in horse populations. *In Proceedings International Symposium on Genetics and Horse Breeding*, Royal Dublin Society, Dublin, Ireland.

Cunningham, E. P. 1977: Züchterische Konsequenzen der Anwendung neuer biotechnischer Verfahren beim Rind. *Zuechtungskunde* **49,** 435.

Czekanowski, J. 1933: Mendelistisches "Law of Ancestral Heredity." *Z. Indukt. Abstammungs-Vererbungsl.* **64,** 154.

Danell, O. 1978: Effect of population type on the definition of breeding goals. *In Proceedings 28th Ann. Meet. EAAP, Brussels, 1977.* Ann. Genet. Sci. Anim. **12,** 133.

Danell, Ö., and Rønningen, K. 1981: All-or-none traits in index selection. *Z. Tierz. Zuechtungsbiol.* **98,** 265.

De Fries, J. C., and Touchberry, R. W. 1961: The variability of response to selection. *Genetics* **46,** 1519.

Dempfle, L. 1974: Bestimmung der optimalen Selektionsintensität bei Selektion über mehrere Generationen. *Z. Tierz. Zuechtungsbiol.* **90,** 160.

Dempfle, L. 1975: A note on increasing the limit of selection through selection within families. *Genet. Res.* **24,** 127.

Dempfle, L. 1977: Relation entre BLUP et estimateurs bayesiens. *Ann. Genet. Sel. Anim.* **9**, 27.

Dempfle, L. 1978a: On the properties of two iterative methods of sire evaluation. *Z. Tierz. Zuechtungsbiol.* **94**, 307.

Dempfle, L. 1978b: Methode zur Berechnung der Indexgewichte bei der Zuchwertschätzung von Kühen. *Z. Tierz. Zuechtungsbiol.* **95**, 28.

Dempfle, L., and Hagger, C. H. 1979: Untersuchungen ueber Effizienz und Korrekturfaktoren bei der Zuchtwertschaetzung von Bullen. *Z. Tierz. Zuechtungsbiol.* **96**, 135.

Dempster, E. R., and Lerner, I. M. 1950: Heritability of threshold characteristics. *Genetics* **35**, 212.

Dickerson, G. E. 1960: Techniques for research in quantitative animal genetics. *Tech. Bull. Am. Soc. Anim. Sci.* **56**, 36.

Dickerson, G. E. 1962: Implications of genetic–environmental interaction in animal breeding. *Anim. Prod.* **4**, 47.

Dickerson, G. E. 1963: Experimental evaluation of selection theory in poultry. *In* S. J. Geerts, ed., *Genetics Today. Proceedings of the XIth International Congress of Genetics,* Volume 3, p. 747, Pergamon, Oxford.

Dickerson, G. E. 1973: Inbreeding and heterosis in animals. *In Proceedings Animal Breeding and Genetics Symposium in Honor of Dr. J. L. Lush,* p. 54. American Society of Animal Science, Champaign, Illinois.

Dickerson, G. E., and Hazel, L. N. 1944: Effectiveness of selection on progeny performance as a supplement to earlier culling of livestock. *J. Agric. Res.* **69**, 459.

Dickerson, G. E., Blunn, C. T., Chapman, A. B., Kottman, R. M., Krider, J. L., Warwick, E. J., and Whatley, J. A. 1954: Evaluation of selection in developing inbred lines of swine. *Res. Bull. MO. Agric. Exp. Sta.,* No. 551.

Doering, H., Walter, E. 1959: Ueber die Berechnung von Inzucht und Verwandtschaftskoeffizienten. *Biometrische Zeitschrift* **1**, 150.

Doolittle, D. P., Wilson, S. P., and Hulbert, L. L. 1973: A comparison of multiple trait selection methods in the mouse. *J. Hered.* **63**, 366.

Dunlop, A. A. 1962: Interactions between heredity and environment in the Australian Merino. 1. Strain × location interaction in wool traits. *Aust. J. Agric. Res.* **13**, 503.

Dunn, L. C. 1957: Evidence of evolutionary forces leading to the spread of lethal genes in wild populations of house mice. *Proc. Natl. Acad. Sci. USA* **43**, 158.

Dzapo, V., Reuter, H., and Wassmuth, R. 1973: Heterosis und mitochondriale Komplementation. *Z. Tierz. Zuechtungsbiol.* **90**, 169.

Ehrmann, L., and Parsons, P. A. 1976: *The Genetics of Behavior.* Sinauer Associates. Sunderland, Massachusetts.

Eisen, E. J. 1981: Predicting selection response for total litter weight. *Z. Tierz. Zuechtungsbiol.* **98**, 55.

Emik, L. O., and Terrill, C. E. 1949: Systematic procedures for calculating inbreeding coefficients. *J. Hered.* **40**, 51.

Enfield, F. D. 1980: Long term effects of selection; the limits to response. *In* A. Robertson, ed., *Selection Experiments in Laboratory and Domestic Animals,* Commonwealth Agriculture Bureau, Slough, U.K.

Erlacher, J. 1970: Blutgruppen und biochemischer Polymorphismus beim Tiroler Grauvieh. Thesis, Veterinary School, Vienna.

Essl, A. 1976: Zur theoretischen Verteilung des Fremdgenanteiles in verschiedenen Kreuzungsgenerationen. *Z. Tierz. Zuechtungsbiol.* **96**, 135.

Essl, A. 1981: Index selection with proportionality restriction. *Z. Tierz. Zuechtungsbiol.* **98**, 125.

Evans, E. P., and Phillips, R. J. S. 1975: Inversion heterozygosity and the origin of XO daughters of *Bpa/ +* female mice. *Nature* **256**, 40.

Falconer, D. S. 1955: Patterns of response in selection experiments with mice. *Cold Spring Harbor Symp. Quant. Biol.* **20**, 178.

Falconer, D. S. 1964: Maternal effects and selection response. *In* S. J. Geerts, ed., *Genetics Today. Proceedings of the XIth International Congress on Genetics*, Volume 3, p. 763, Pergamon, Oxford.

Falconer, D. S. 1965: The inheritance of liability to certain diseases, estimated from the incidence among relatives. *Ann. Hum. Genet.* **29**, 51.

Falconer, D. S. 1971: Improvement of litter size in a strain of mice at a selection limit. *Genet. Res.* **17**, 215.

Falconer, D. S. 1973: Replicated selection for body weight in mice. *Genet. Res.* **22**, 291.

Falconer, D. S. 1981: *Introduction to Quantitative Genetics*, 2nd ed. Longman, London.

Farrel, H. M., Thompson, M. P., and Larsen, B. 1971: Verification of the occurrence of the α_{s1}-casein *A* allele in Red Danish cattle. *J. Dairy Sci.* **54**, 423.

Fechheimer, N. S. 1977: Sources of chromosomal abnormalities in chickens. *In 19th British Poultry Breeders Roundtable*, p. 62, British Poultry Breeders' Roundtable Committee, Edinburgh, Scotland.

Fehlings, R., Grundler, C., Wauer, A., and Pirchner, F. 1983: Inzuchtzuwachs und effektive Populationsgrössen deutscher Pferderassen. *Z. Tierz. Zuechtungsbiol.* **100**, 81.

Feldman, M. W. C., and Lewontin, R. C. 1975: The heritability hangup. *Science* **190**, 1163.

Felsenstein, J. 1965: The effect of linkage on directional selection. *Genetics* **52**, 349.

Felsenstein, J. 1976: The theoretical population genetics of variable selection and migration. *Annu. Rev. Genet.* **10**, 253.

Festing, M. F. W. 1974: Genetic monitoring of laboratory mouse colonies in the Medical Research Council accreditation scheme for suppliers of laboratory animals. *Lab. Anim.* **8**, 291.

Festing, M. F. W., and Nordskog, A. W. 1967: Response to selection for body weight and egg weight in chickens. *Genetics* **55**, 219.

Fewson, D. 1973: Beitrag zur Methodik von Einkreuzungen in Reinzuchtpopulationen. *Z. Tierz. Zuechtungsbiol.* **90**, 113.

Fewson, D., and Le Roy, H. L. 1959: Körpermessungen am lebenden Schwein als Basis zur Abschätzung des Fleisch- und Fettanteiles. *Z. Tierz. Zuechtungsbiol.* **73**, 297.

Fisher, R. A. 1918: The correlation between relatives on the supposition of Mendelian inheritance. *Proc. R. Soc. Edinb.* **52**, 399.

Fisher, R. A. 1930: *The Genetical Theory of Natural Selection.* Oxford University Press, Oxford.

Fisher, R. A. 1948: *Statistical Methods for Research Workers*, 10th ed. Oliver and Boyd, Edinburgh.

Fisher, R. E., and Yates, F. 1963: *Statistical Tables.* Oliver and Boyd, Edinburgh.

Flock, D. K. 1970: Genetic parameters of German Landrace pigs estimated from different relationships. *J. Anim. Sci.* **30**, 839.

Flock, D. K. 1977: Genetic analysis of part-period egg production in a population of White Leghorns under long term RRS. *Z. Tierz. Zuechtungsbiol.* **94**, 89.

Flock, D. K. 1980: Heterosisschätzungen in einer Population von Weissen Leghorn nach langjähriger RRS. *In Proceedings VIth European Poultry Conference Hamburg*, Vol. II, pp 57–63, World's Poultry Science Association, German branch (Fed.), Hamburg.

Flock, D. K., Hausmann, H., and Fewson, D. 1971: *Tabellen für die Zuchtwertschätzung bei landwirtschaftlichen Nutztieren.* Schriftenreihe Max Planck Institut, Mariensee, No. 54.

Förster, M., Stranzinger, G., and Hellkuhl, B. 1980: X-chromosome assignment of swine and cattle. *Naturwissenschaften* **67**, 48.

Frankham, R. 1980: Origin of genetic variation in selected lines. *In* A. Robertson, ed., *Selection Experiments in Laboratory and Domestic Animals.* Commonwealth Agriculture Bureau, Slough, U.K.

Fredeen, H. T., and Jonsson, P. 1957: Genic variance and covariance in Danish Landrace swine as evaluated under a system of individual progeny test groups. *Z. Tierz. Zuechtungsbiol.* **70**, 348.

Freeman, G. H., and Perkins, J. M. 1971: Environmental and genotype–environmental components of variability. VIII. Relations between genotypes grown in different environments and measures of the environments. *Heredity* **27**, 15.

Gahne, B. 1961: Studies of transferrins in serum and milk of Swedish cattle. *Anim. Prod.* **3**, 135.

Gahne, B., Bengtsson, S., and Sandberg, K. 1970: Genetic control of cholinesterase activity in horse serum. *Anim. Blood Groups Biochem. Genet.* **1**, 207.

Garnett, I., and Falconer, D. S. 1975: Protein variation in strains of mice differing in body size. *Genet. Res.* **25**, 45.

Glodek, P., Bade, B., and Schormann, H. 1975: Die Entwicklung von Selektionskriterien für die Reitpferdezucht. III. *Zuechtungskunde* **47**, 164.

Goodwin, K., Dickerson, G. E., and Lamoreux, W. F. 1960: An experimental design for separating genetic and environmental changes in animal populations under selection. *In* O. Kempthorne, ed., *Biometrical Genetics*, Pergamon Press, New York.

Goonewardene, L. A., and Berg, R. T. 1976: Arthrogryposis in Charolais cattle, a study on gene penetrance. *Ann. Genet. Sel. Anim.* **8**, 493.

Gowe, R. S. 1962: Selection for increases in egg production based on part-year records in two strains of leghorns. *In 4th British Poultry Breeders Roundtable, British Poultry Breeders' Roundtable Committee*, Edinburgh, Scotland.

Gowe, R. S., Robertson, A., and Latter, B. D. H. 1959: Environment and poultry breeding problems. 5. The design of poultry control strains. *Poult. Sci.* **38**, 462.

Gravert, H. O. 1958: Untersuchungen über die Heritabilität der Butterfettleistung. *Z. Tierz. Zuechtungsbiol.* **71**, 155.

Gravert, H. O. 1975: Möglichkeiten der Kreuzungszucht beim Zweinutzungsrind Holstein × Schwarzbunt. *Zuechtungskunde* **47**, 404.

Greaves, J. H., and Ayres, P. B. 1977: Inheritance of Scottish-type resistance to warfarin in the Norway rat. *Genet. Res.* **28**, 231.

Griffing, B. 1956: Concept of general and specific combining ability in relation to diallel crossing systems. *Aust. J. Biol. Sci.* **9**, 463.

Griffing, B. 1960: Theoretical consequences of truncation selection based on the individual phenotype. *Aust. J. Biol. Sci.* **13**, 307.

Gruhn, R., and Dinklage, H. 1971: Blutgruppen- und Proteinpolymorphismus im Göttinger Miniaturschwein. *Z. Versuchstierkd.* **13**, 179.

Gúerin, G., Ollivier, L., and Sellier, P. 1978: Desequilibres de linkage entre les locus *Hal* (Hyperthermie maligne), *PHI* et *6-PGD* dans deux lignées Pietrain. *Ann. Genet. Sel. Anim.* **10**, 125.

Guiard, V., and Herrendörfer, G. 1977: Estimation of the genetic correlation coefficient by halfsib analysis if the characters are measured in different offspring. *Biometr. J.* **19**, 31.

Gustavsson, I. 1969: Cytogenetics, distribution and phenotypic effects of a translocation in Swedish cattle. *Hereditas* **63**, 67.

Haldane, J. B. S. 1946: The interaction of nature and nurture. *Ann. Eugen.* **13**, 197.

Haldane, J. B. S., and Jayakar, S. D. 1965: The nature of human genetic loads. *J. Genet.* **59**, 53.

Hammond, J. 1947: Animal breeding in relation to environmental conditions. *Biol. Rev.* **22**, 195.

Hammond, K., and Nicholas, F. W. 1972: The sampling variance of the correlation coefficients estimated from two-fold nested and offspring–parent regression analyses. *Theor. Appl. Genet.* **42**, 97.

Hansen, M. P., Van Zandt, J. N., and Law, G. R. L. 1967: Differences in susceptibility to Marek's disease in chickens carrying two different B locus blood group alleles. *Poult. Sci.* **46**, 1268.

Haring, F., Steinbach, J., and Scheven, B. 1966: Untersuchungen über den mütterlichen Einfluss auf die prä- und postnatale Entwicklung von Schweinen extrem unterschiedlicher Grösse. *Z. Tierz. Zuechtungsbiol.* **82**, 37.

Haring, H. J. F. 1972: Zuchtplanung in der Rinderzucht aus ökonomischer Sicht. Thesis, Universität Göttingen.

Harris, D. L. 1963: The influence of errors of parameter estimation upon index selection. *In* W. D. Hanson and H. F. Robinson, eds., *Symposium on Statistical Genetics and Plant Breeding*, p. 491. National Academy of Sciences–National Research Council, Washington, D.C.

Harris, H. 1966: Enzyme polymorphisms in man. *Proc. R. Soc. B* **164**, 298.

Harrison, B. J., and Mather, K. 1949: The manifold effects of selection. *Heredity* **3**(1), 131.

Hartmann, W. 1959: Erbliche Untersuchungen über die Milchleistungseigenschaften in der Schwarzbunt Herde Mariensee/Mecklenhorst des Max Planck Institutes. *Schriftenr. Max Planck Institut Tierzucht* **6/7**, 103.

Hartmann, W. 1972: Relationship between genes at the pea and single comb locus and economic traits in broiler chicken. *Br. Poult. Sci.* **13**, 305.

Harvey, R. W. 1972: Direct and indirect response in two-trait selection experiments in mice. *In Proceedings of the 21st National Poultry Breeders Rountable*, p. 71, Poultry Breeders of America, Kansas City, Missouri.

Hayman, R. H. 1974: The development of the Australian Milking Zebu. *World Anim. Rev.* **11**, 31.

Hazel, L. N. 1943: The genetic basis for constructing selection indexes. *Genetics* **28**, 476.

Hazel, L. N., and Lush, J. L. 1943: The efficiency of three methods of selection. *J. Hered.* **33**, 393.

Hazel, L. N., Baker, M. L., and Reinmiller, C. F. 1943: Genetic and environmental correlations between the growth rates of pigs at different ages. *J. Anim. Sci.* **2**, 118.

Heidhues, T., and Henderson, C. R. 1962: Beitrag zum Problem des Basisindex. *Z. Tierz. Zuechtungsbiol.* **77**, 297.

Heidhues, T., Van Veck, L. D., and Henderson, C. R. 1961: *Approximative weighting factors for selection indexes.* Mimeograph, Department of Animal Husbandry, Cornell University, Ithaca, New York.

Henderson, C. R. 1963: Seiection index and expected genetic advance. *In* W. D. Hanson and H. F. Robinson, eds., *Symposium on Statistical Genetics and Plant Breeding*, p. 141. National Academy of Sciences–National Research Council, Washington, D.C.

Henderson, C. R. 1973: Sire evaluation and genetic trends. *In Proceedings of the Animal Breeding and Genetics Symposium in Honor of Dr. J. L. Lush*, American Society of Animal Science, Champaign, Illinois.

Henderson, C. R. 1975: Rapid method for computing the inverse of a relationship matrix. *J. Dairy Sci.* **58**, 1727.

Hetzer, H. O., and Harvey, W. R. 1967: Selection for high and low fatness in swine. *J. Ani. Sci.* **26**, 1244.

Hetzer, H. O., and Miller, R. H. 1972: Rate of growth as influenced by selection for high and low fatness in swine. *J. Anim. Sci.* **35**, 730.

Hetzer, H. O., and Miller, L. R. 1973: Selection for high and low fatness in swine: Correlated responses of various carcass traits. *J. Anim. Sci.* **37**, 1289.

Hetzer, H. O., Comstock, R. E., Zeller, J. H., Hiner, R. L., and Harvey, W. R. 1961: Combining abilities in crosses among six inbred lines of swine. Technical Bulletin 1237, Agriculture Research Service, U.S. Department of Agriculture, Washington, D.C.

Hickman, C. G. 1971: Response to selection of breeding stock for milk solids production. *J. Dairy Sci.* **54**, 191.

Hickman, C. G., and Bowden, D. M. 1971: Correlated genetic responses of feed efficiency, growth and body size in cattle selected for milk solids yield. *J. Dairy Sci.* **54**, 1848.

Hierl, H. F. 1976: Beziehungen zwischen dem Heterozygotiegrad, geschätzt aus Markergenen, und der Fruchtbarkeit beim Rind. 1. Schätzung und Ausmass der Heterozygotie in deutschen Rinderrassen. *Theor. Appl. Genet.* **47**, 69; 2. Heterozygotiegrad und Fruchtbarkeit, *Theor. Appl. Genet.* **47**, 77.

Hill, W. G. 1972a: Estimation of genetic change. I. General theory and design of control populations. *Anim. Breed. Abst.* **40**, 1.

Hill, W. G. 1972b: Estimation of realized heritabilities from selection experiments. I. Divergent selection. *Biometrics* **28**, 747. II. Selection in one direction. *Biometrics* **28**, 767.

Hill, W. G. 1974: Prediction and evaluation of response to selection with overlapping generations. *Anim. Prod.* **18**, 117.

Hill, W. G. 1977: Aspects of experimental design and sampling for evaluation of breeds and crosses and their interaction with feeding regimes. *In Crossbreeding Experiments and Stategy of Beef Utilization to Increase Coordination of Research on Beef Production.* Scientific and Technical Information and Information Management. Commission of the European Communities Directorate General, Kirchberg, Luxemburg, p. 348.

Hill, W. G. 1980: Design of quantitative genetic selection experiments. *In* A. Robertson, ed., *Selection Experiments in Laboratory and Domestic Animals.* Commonwealth Agriculture Bureaus, Slough, U.K.

Hill, W. G. 1982: Dominance and epistasis as components of heterosis. *Z. Tierz. Zuechtungsbiol.* **99**, 161.

Hill, W. G., and Nicholas, F. W. 1975: Estimation of heritability by both regression of offspring on parent and intra-class correlation of sibs in one experiment. *Biometrics* **30**, 447.

Hindemith, A. 1978: Genetisch-statistische Auswertungen an Milchleistungsmerkmalen beim deutschen Gelbvieh. *Bayer. Landwirtsch. Jahrb.* **55**, 128.

Hinkelmann, K. 1971: Estimation of heritability from experiments with inbred and related individuals. *Biometrics* **27**, 183.

Hinkelmann, K., Simon, D. L., and Flock, D. K. 1976: Zuchtwertschätzung unter Berücksichtigung genetischer Unterschiede zwischen Vergleichswerten. II. *In Proceedings of the Annual Meeting Deut. Ges. Züchtungskunde*, Bonn.

Hinks, C. J. M. 1970a: The selection of dairy bulls for artificial insemination. *Anim. Prod.* **12**, 569.

Hinks, C. J. M. 1970b: Performance test procedures for meat production amongst dairy bulls used in AI. *Anim. Prod.* **12**, 577.

Hinks, C. J. M. 1971: The genetic and financial consequences of selection amongst dairy bulls in artificial insemination. *Anim. Prod.* **13**, 209.

Hinks, C. J. M. 1972: The effects of continuous size selection on the structure and age composition of dairy cattle populations. *Anim. Prod.* **15**, 103.

Hintz, R. L., and Van Vleck, L. D. 1978: Factors influencing racing performance of the standardbred pacer. *Anim. Sci.* **46**, 60.

Hodgson, R. E. 1961: *Germ Plasm Resources*. American Association for the Advancement of Science, Washington, D.C.

Hogsett, M. L., and Nordskog, A. W. 1958: Genetic–economic value in selecting for egg production rate, body weight and egg weight. *Poult. Sci.* **37**, 1404.

Hogsett, M. L., von Krosigk, C. M., and McClary, C. F. 1964: Genetic and phenotypic variance–covariance estimates and their application to index–reciprocal recurrent selection for egg production. *Poult. Sci.* **43**, 1329.

Hohenbrink, R. 1970: Der Heterozygotiegrad von Schweinen aus Reinzucht und Kreuzung. Thesis, Universität Göttingen.

Hull, F. H. 1945: Recurrent selection for specific combining ability in corn. *J. Am. Soc. Agron.* **37**, 134.

Jaap, G. N. 1947: Cited by A. Heisdorf, Twenty years experience with RRS. *In Proceedings of the 18th National Poultry Breeders Roundtable*, p. 113. Poultry Breeders of America, Kansas City, Missouri.

Jakubec, V., and Fewson, D. 1971: Ökonomische und genetische Grundlagen für die Planung von Gebrauchskreuzungen beim Schwein. *Z. Tierz. Zuechtungsbiol.* **87**, 2.

James, J. W. 1966: Correlations between relatives when intermediates are fittest. *Aust. J. Biol. Sci.* **19**, 301.

James, J. W., 1971: The founder effect and response to artificial selection. *Genetical Research* **16**, 241.

James, J. W. 1972: Optimum selection intensity in breeding programmes. *Anim. Prod.* **14**, 1.

Jamieson, A. 1965: The genetics of transferrins in cattle. *Heredity* **20**, 419.

Jamieson, A., and Robertson, A. 1967: Cattle transferrins and milk production. *Anim. Prod.* **9**, 491.

Jensen, A. R. 1972: *Genetics and Education*. Methuen, London.

Jerome, F. N., Henderson, C. R., and King, S. C. 1956: Heritabilities, gene interactions, and correlations associated with certain traits in the domestic fowl. *Poult. Sci.* **35**, 995.

Joakimsen, O., Steenberg, K., Lien, H., Theodorsen, L. 1971: Genetic relationship between thyroxine degradation and fat-corrected milk yield in cattle. *Acta Agric. Scand.* **21**, 121.

Johansson, I. 1965: Studies on the genetics of ranch bred mink. III. *Z. Tierz. Zuechtungsbiol.* **81**, 73.

Johansson, I., and Hansson, A. 1940: Causes of variation in milk and butterfat yield of dairy cows. *Kungl. Lantbruksak. Tidskr.* **1940** (6½).

Johansson, I., and Korkman, N. 1950: A study of the variation in production traits of bacon pigs. *Acta Agric. Scand.* **1**, 62.

Johansson, I., and Rendel, J. 1971: Studies on the variation in dairy traits of intact and split pairs of cattle twins under farm conditions. *Acta Agric. Scand.* **21**, 89.

Johansson, I., and Robertson, A. 1952: *Progeny testing in the breeding of farm animals*, Publication No. 2, European Association for Animal Production, Rome.

Johansson, I., Lindhé, B., and Pirchner, F. 1974: Causes of variation in the frequency of monozygous and diszygous twinning in various breeds of cattle. *Hereditas* **78**, 201.

Johansson, I., Rendel, J., and Gravert, H. O. 1966: *Haustiergenetik und Tierzuechtung*, Paul Parey, Hamburg.

Jonsson, P. 1959: Investigation on group feeding versus individual feeding and on the interaction between genotype and environment in pigs. *Acta Agric. Scand.* **9**, 204.

Jonsson, P. 1965: Analyse of egenskaber hos swin of Dansk Landrace med en historisk indledning, 340. Beretning Forsogslaboratorium, Kopenhagen.

Jonsson, P., and King, J. W. B. 1962: Sources of variation in Danish Landrace pigs. *Acta Agric. Scand.* **12**, 68.

Jørgensen, P. F., 1977: Studies on the association between halothane sensitivity, blood group and enzyme systems in the Danish Landrace pigs. *Arsberetning Institut Sterilitetsforsk.* **20**, 101, Copenhagen.

Kamin, L. J. 1974: *The Science and Politics of I.Q.* Wiley, New York.

Kempthorne, O. 1957: *An Introduction to Genetic Statistics.* Wiley, New York.

Kempthorne, O. 1977: Status of quantitative genetic theory. *In* E. Pollack, O. Kempthorne, and T. B. Bailey, eds., *Proceedings of the International Conference on Quantitative Genetics,* p. 362. Iowa State University Press, Ames, Iowa.

Kempthorne, O., and Nordskog, A. W. 1959: Restricted selection indices. *Biometrics* **15**, 10.

Kempthorne, O., and Pollak, E. 1970: Concepts of fitness in Mendelian populations. *Genetics* **64**, 125.

Kempthorne, O., and Tandon, O. B. 1953: The estimation of heritability by regression of offspring on parent. *Biometrics* **9**, 90.

Kidd, K. K., and Cavalli-Sforza, L. A. 1974: The rate of genetic drift in the differentiation of Icelandic and Norwegian cattle. *Evolution* **28**, 381.

Kidd, K. K., and Pirchner, F. 1971: Genetic relationships of Austrian cattle breeds. *Anim. Blood Group Biochem. Genet.* **2**, 145.

Kidd, K. K., and Sgaramella-Zonta, L. A. 1972: Genetic relationships among cattle breeds. *In Proceedings of the XII International Conference on Animal Blood Groups and Biochemical Genetics,* Akadeai Kiado, Budapest, p. 241.

Kidd, K. K., Osterhoff, D., Erhard, L., and Stone, W. H. 1974: The use of genetic relationships among cattle breeds in the formulation of rational breeding policies: An example with South Devon (South Africa) and Gelbvieh (Germany). *Anim. Blood Groups Biochem. Genet.* **5**, 21.

Kidd, K. K., Stone, W. H., Crimella, C., Carenzi, C., Casati, M., and Rognoni, G. 1980: Immunogenetic and population genetic analyses of Iberian cattle. *Anim. Blood Groups Biochem. Genet.* **11**, 21.

Kidwell, J. F., and Kempthorne, O. 1966: An experimental test for quantitative genetic theory. *Zuechter* **36**, 163.

Kimura M. 1958: On the change of population fitness by natural selection. *Heredity* **12**, 145.

Kimura, M., and Weiss, G. H. 1964: The stepping stone model of population structure and the decrease of genetic correlation with distance. *Genetics* **49**, 561.

Kincaid, C. M., and Carter, R. C. 1958: Estimates of genetic and phenotypic parameters in beef cattle. *J. Anim. Sci.* **17**, 675.

King, J. L. 1967: Continuously distributed factors affecting fitness. *Genetics* **55**, 483.

King, J. L., and Jukes, T. H. 1969: Non-darwinian selection. *Science* **164**, 788.

Klupp, R. 1979: Genetic variance for growth in rainbow trout *(Salmo gairdneri). Aquaculture* **18**, 123.

Klupp, R., Heil, G., and Pirchner, F. 1978: Interactions between strains and environment for growth traits in rainbow trout *(Salmo gairdneri). Aquaculture* **14**, 271.

Koch, R. M., and Clark, R. T. 1955: Genetic and environmental relationships among economic characters in beef cattle. I, II, III. *J. Anim. Sci.* **14**, 775, 786, 979.

Kögel, S. 1976: Genetisch-statistische Auswertungen am Material der bayerischen Braunviehpopulation (1967–1971). *Bayer. Landwirtsch. Jahrb.* **53**, 806.

Kramser, P. 1972: Biochemische Polymorphismen und Blutgruppen bei österreichischen Rinderrassen. Thesis, Veterinary School, Vienna.

Kramser, P. 1974: Selektionsversuch bei drei Schweinerassen. Abstr. *Ann. Genet. Sel. Anim.* **6**, 153.

Krizenecky, J. 1942: Die Wirkung des Lebendgewichtes der Kuh auf die Milchproduktion und ihre Eliminierung als Laktationsfaktor. *Z. Tierz. Zuechtungsbiol.* **51**, 100.

Kronacher, C. 1932: Zwillingsforschung beim Rind. *Z. Zuechtung B* **25**, 327.

Langlet, J., and Gravert, H. O. 1961: Inzucht- und Verwandtschaftsverhältnisse bei Bullen des schwarzbunten Rindes in Schleswig Holstein. *Zuechtungskunde* **33**, 373.

Larsen, B. 1972: Proteinpolymorfi og produktionsegenskaber hos kvaeg. *Aarsk. Inst. Sterilitetsforsk.* **15**, 71.

Latter, B. H. D. 1959: Genetic sampling in a random mating population of constant size and sex ratio. *Aust. J. Biol. Sci.* **12**, 500.

Latter, B. H. D., and Robertson, A. 1962: The effect of inbreeding and artificial selection on reproductive fitness. *Genet. Res.* **3**, 11.

Lauprecht, E. 1930: Über die Vererbung körperlicher Merkmale beim Rind. *Zuechtungskunde* **5**, 241.

Lauprecht, E. 1961: Production of a population with equal frequency of genes from three parental sources. *J. Anim. Sci.* **20**, 426.

Lauprecht, E., and Walter, E. 1960: Über einige Umwelteinflüsse auf die Mast- und Schlachteigenschaften des Schweines bei dänischen Mastprüfungsgruppen. *Arch. Tierz.* **3**, 3.

Lauprecht, E., Flock, D., and Hinkelmann, K. 1966: Diallele Paarungen beim Schwein. *Z. Tierz. Zuechtungsbiol.* **83**, 178.

Lauvergne, J. J., and Lefort, G. 1973: Nouvelle méthods pour analyser le comportement de la fréquence des génes récessifs à effets visibles dans les populations bovines. *C. R. Acad. Sci. Paris D* **277**, 2793.

Lederer, J. 1977: *Der direkte Bullenvergleich*, Rinderproduktion Niedersachen E. V., Verden, Federal Republic of Germany.

Lederer, J., and Averdunk, G. 1973: Vergleich des realisierten Zuchtfortschrittes nach verschiedenen Schätzmethoden beim Fleckvieh in Bayern. *Zuechtungskunde* **45**, 179.

Leithe, H. 1972: Einflüsse auf und Beziehungen zwischen Melkbarkeit und Mastitisparametern. Thesis, Tierärztl. Hochschule Wien.

Lerner, I. M. 1950: *Population Genetics and Animal Improvement*. Cambridge University Press, Cambridge.

Lerner, I. M. 1954: *Genetic Homeostasis*. Wiley, New York.

Lerner, I. M. 1958: *The Genetic Basis of Selection*. Wiley, New York.

Le Roy, H. C. 1958: Die Abstammungsbewertung. *Z. Tierz. Zuechtungsbiol.* **71**, 328.

Levene, H. 1953: Genetic equilibrium when more than one ecological niche is available. *Am. Nat.* **87**, 331.

Levene, H. 1963: Inbred genetic loads and the determination of population structure. *Proc. Natl. Acad. Sci. USA* **50**, 587.

Lewontin, R. C., and Cockerham, C. C. 1959: The goodness-of-fit test for detecting natural selection in random mating populations. *Evolution* **13**, 561.

Lewontin, R. C., and Hubby, J. L. 1966: A molecular approach to the study of genic heterozygosity in natural populations. II. *Genetics* **54**, 595.

Lewontin, R. C., and Krakauer, J. 1973: Distribution of gene frequency as a test of the theory of the selective neutrality of polymorphisms. *Genetics* **74**, 175.

Li, C. C. 1967: Genetic equilibrium under selection. *Biometrics* **23**, 397.

Li, C. C. 1976: *Population Genetics*. Boxwood Press, Pacific Grove, California.

Lindström, U. B. 1978: Selection intensity for milk yield in 1970–1977 in the Finnish Ayrshire. *J. Sci Agric. Soc. Finland* **50**, 445.

Lörtscher, H. 1937: Variationsstatistische Untersuchungen an Leistungserhebungen einer British-Friesian Herde. *Z. Zuechtung B* **39**, 257.

Lörtscher, H. 1947: Umfang und Aufbau der Herdbuchzucht. *In* H. Wenger, ed. *Das Simmentaler Fleckvieh der Schweiz*, p. 26. Schweiz. Fleckviehzuchtverband, Bern, Switzerland.

Lush, J. L. 1947: Family merit and individual merit as bases for selection. Part 1, 2. *Am. Nat.* **81,** 241, 362.

Lush, J. L. 1948: *The Genetics of Populations*. Mimeograph, Dept. Animal Science, Iowa State College Press, Ames, Iowa.

Lush, J. L., and Straus, F. S. 1942: The heritability of butterfat production in dairy cattle. *J. Dairy Sci.* **25,** 975.

Lush, J. L., Holbert, J. C., and Willham, O. S. 1936: Genetic history of the Holstein–Friesian cattle in the United States. *J. Hered.* **27,** 61.

Lush, J. L., Lamoreux, W. F. and Hazel, L. N. 1948: The heritability of resistance to death in the fowl. *Poult. Sci.* **27,** 375.

Lyon, M. F. 1954: Stage of action of the litter-size effect on absence of otoliths in mice. *Z. Indukt. Abstammungs-Vererbungsl.* **86,** 289.

Mackay, T. F. C. 1980: Genetic variance, fitness, and homeostasis in varying environment.: An experimental check on the validity of the theory. *Evolution* **34,** 1219.

Magee, W. T. 1965: Estimating response to selection. *J. Anim. Sci.* **24,** 242.

Malécot, G. 1948: *Les Mathématiques de l'Heredité*. Masson, Paris.

Manwell, C., and Baker, C. M. A. 1980: Chemical classification of cattle. 2. Phylogenetic tree and specific status of Zebu. *Anim. Blood Groups Biochem. Genet.* **11,** 151.

Mason, I. 1964: Genetic relations between milk and beef characters in dual-purpose cattle breeds. *Anim. Prod.* **6,** 31.

Mason, I. L., and Robertson, A. 1956: The progeny testing of dairy bulls at different levels of production. *J. Agric. Sci.* **47,** 367.

Mather, K. 1943 Polygenic inheritance and natural selection. *Biol. Rev.* **18,** 32.

Mayr, E. 1963: *Animal Species and Evolution*. Belknap Press, Cambridge, Massachusetts.

Mayrhofer, G. 1979: Ein Verfahren zur Berechnung der Abstammungskoeffizienten nach Malécot. *Z. Tierz. Zuechtungsbiol.* **96,** 143.

McCarthy, J. C., and Doolittle, D. P. 1977: Effect of selection for independent changes in two highly correlated body weight traits in mice. *Genet. Res.* **29,** 133.

McClintock, A. E., and Cunningham, E. P. 1974: Selection in dual purpose dairy cattle: Defining the breeding objective. *Anim. Prod.* **18,** 237.

McGloughlin, P. 1980: The relationship between heterozygosity and heterosis. *In Proceedings British Poultry Breeders Roundtable*, Edinburgh.

McKusik, V. A. 1980: The anatomy of the human genome. *J. Hered.* **71,** 370.

Merrit, E. S., and Gowe, R. S. 1960: Combining ability among breeds and strains of meat type fowl. *Can. J. Genet. Cytol.* **3,** 286.

Meyer, H. H., and Enfield, D. F. 1974: Experimental evidence on limitations of the heritability parameter. *Theor. Appl. Genet.* **45,** 268.

Mi, M. P., Chapman, A. B., and Tyler, W. J. 1965: Effects of mating system on production traits in dairy cattle. *J. Dairy Sci.* **48,** 77.

Milkman, R. D. 1967: Heterosis as a major cause of heterozygosity in nature. *Genetics* **55,** 493.

Moav, R. 1966: Specialized sire and dam lines. I. Economic evaluation of crossbreds. *Anim. Prod.* **8,** 193.

Moav, R., and Wohlfarth, G. W. 1974: Magnification through competition of genetic differences in yield capacity in carp. *Heredity* **33,** 181.

Mode, C. J., and Robinson, H. F. 1959: Pleiotropism and the genetic variance and covariance. *Biometrics* **15,** 518.

Moll, R. H., Lindsey, M. F., and Robinson, H. F. 1964: Estimates of genetic variances and level of dominance in maize. *Genetics* **49**, 411.

Monteiro, L. S., and Falconer, D. S. 1966: Compensatory growth and sexual maturity in mice. *Anim. Prod.* **8**, 179.

Morton, N. E., Crow, J. F., and Muller, H. J. 1956: An estimate of mutational damage in man from data on consanguinous marriages. *Proc. Natl. Acad. Sci. USA* **42**, 855.

Mostageer, A. 1969: The use of information on mates in estimating breeding values. *Z. Tierz. Zuechtungsbiol.* **86**, 42.

Mukai, T. 1964: The genetic structure of natural populations of *Drosophila melanogaster:* I. Spontaneous mutation rate of polygenes controlling viability. *Genetics* **50**, 1.

Mukai, T. 1979: Polygenic variation. *In* J. N. Thompson and J. M. Thoday, eds. *Quantitative Genetic Variation.* Academic Press, New York.

Neiman-Sørensen, A. 1956: Blood groups and breed structure as exemplified by three Danish breeds. *Acta Agric. Scand.* **6**, 115.

Neiman-Sørensen, A. 1958: Blood groups of cattle. Thesis, Carl Fr. Mortensen, Kopenhagen.

Neiman-Sørensen, A., and Robertson, A. 1961: The association between blood groups and several production characteristics in three Danish cattle breeds. *Acta Agric. Scand.* **11**, 163.

Niebel, E. 1974: Methodik der Zuchtplanung für die Reinzucht beim Rind bei Optimierung nach Zuchtfortschritt und Züchtungsgewinn. Thesis, Universität Hohenheim.

Niebel, E., and Fewson, D. 1976: Untersuchungen zur Zuchtplanung für die Reinzucht beim Zweinutzungsrind. II. Zuchtwahl in zwei Selektionsstufen. *Z. Tierz. Zuechtungsbiol.* **93**, 169.

Nordskog, A. W., Ghostley, F. J. 1954: Heterosis in Poultry. *Poult. Sci.* **33**, 704.

Nordskog, A. W., Festing, M., Wehrli Verghese, M. 1965: Selection for egg production and correlated response in the fowl. *Genetics* **55**, 179.

Notter, D. R., Dickerson, G. E., and Deshazer, J. A. 1976: Selection for rate and efficiency of lean gain in the rat. *Genetics* **84**, 125.

Nozawa, K., Shotake, T., and Okkura, Y. 1976: Blood protein variations within and between the East Asian and European horse populations. *Z. Tierz. Zuechtungsbiol.* **93**, 60.

Nozawa, K., Shinjo, A., and Shotake, T. 1978: Population genetics of farm animals. III. Blood protein variations in the meat goats in Okinawa islands of Japan. *Z. Tierz. Zuechtungsbiol.* **95**, 60.

Oakes, M. W. 1967: The effect of different levels of management on the assessment of differences within a variety of chickens. *Anim. Prod.* **9**, 121.

Ollivier, L. 1968: Étude du determinèsme héréditaire de l'hypertrophie musculaire du porc de Piétrain. *Ann. Zootech. (Paris)* **17**, 393.

Ollivier, L. 1974a: La regression parent–descendant dans le cas de descendance subdivisées en familles de taille inegale. *Biometrics* **30**, 59.

Ollivier, L. 1974b: Optimum replacement rates in animal breeding. *Anim. Prod.* **19**, 257.

Ollivier, L., Sellier, P. and Monin, G. 1975: Déterminisme génétique du syndrome d'hyperthermie maligne chez le porc de Piétrain. *Ann. Genet. Sel. Anim.* **7**, 159.

Olsen, M. W. 1962: Twelve year summary of selection for parthenogenesis in Beltsville small White Turkeys. *Br. Poult. Sci* **6**, 1.

Osborne, K., and Patterson, W. S. B. 1952: On the sampling variance of heritability estimates derived from variance analyses. *Proc. R. Soc. Edin.* **64**, 456.

Osman, H. E. S., and Robertson, A. 1968: The introduction of genetic material from inferior into superior strain. *Genet. Res.* **12**, 221.

Owen, J. B. 1974: Selection of dairy bulls on half sister records. *Anim. Prod.* **20**, 1.

Pani, P. K. 1974: Plumage color gene (i^+), a possible modifier of cellular susceptibility. *Theor. Appl. Genet.* **44**, 17.

Pesek, J., and Baker, R. J. 1969: Desired improvement in relation to selection indexes. *Genetics* **28**, 476.

Pirchner, F. 1959: Zusammenhang von Form von Stieren mit Form und Leistung der Töchter. *Zuechtungskunde* **31**, 218.

Pirchner, F. 1968: Models in animal breeding experiments. *In Proceedings of the International Summer School on Biomathematics and Data Processing in Animal Experiments.* National Institute of Animal Science, Copenhagen, Denmark.

Pirchner, F. 1970: Eignung verschiedener Vergleichsdurchschnitte zur Nachkommenprüfung beim Rind. *Z. Tierz. Zuechtungsbiol.* **87**, 20.

Pirchner, F., and von Krosigk, M. 1973: Genetic parameters of cross- and purebred poultry. *Br. Poult. Sci.* **14**, 193.

Pirchner, F., and Mergl, R. 1977: Overdominance as cause for heterosis in poultry. *Z. Tierz. Zuechtungsbiol.* **94**, 151.

Pirchner, F., Stüber, O., and Tschoerner, F., 1960: Vererbung von Milchbestandteilen bei Höhenvieh. *Z. Tierz. Zuechtungsbiol.* **74**, 285.

Pisani, J. F., and Kerr, W. E. 1961: Lethal equivalents in domestic animals. *Genetics* **46**, 773.

Poutous, M., and Vissac, B. 1962: Recherche théorique des conditions de rentabilité maximum de l'épreuve de descendance des taureaux d'insémination artificielle. *Ann. Zootech. (Paris)* **11**, 233.

Powell, R. L., and Freeman, A. E. 1974: Estimators of sire merit. *J. Dairy Sci.* **57**, 1228.

Puff, H. 1976: Erwarteter und realisierter genetischer Fortschritt in der Niederbayerischen Schweinezucht. II. Erwarteter Zuchtfortschritt. *Z. Tierz. Zuechtungsbiol.* **93**, 135.

Rasch, D. 1967: Wieviel Töchter benötigt man zum Inzuchttest auf rezessive unerwünschte Allele? *Arch. Tierzucht* **10**, 159.

Rasmusen, B. A., and Christian, L. L. 1976: H blood types in pigs as predictors of stress susceptibility. *Science* **191**, 947.

Rasmuson, M. 1964: Combined selection for two bristle characters in *Drosophila. Hereditas* **51**, 231.

Reeve, E. C. R. 1955: The variance of the genetic correlation coefficient. *Biometrics* **11**, 357.

Reeve, E. C. R., and Robertson, F. W. 1952: Studies in quantitative inheritance. II. Analysis of a strain of *Drosophila melanogaster* selected for long wings. *J. Genet.* **51**, 276.

Reich, T., James, J. W., and Morris, C. A. 1972: The use of multiple thresholds in determining the mode of transmission of semicontinuous traits. *Ann. Hum. Genet.* **36**, 163.

Reinhardt, F. 1982: Selection in dairy cattle. *In Proceedings of the 32nd Annual Meeting of the EAAP, Zagreb, 1981. Ann. Genet. Sel. Anim.* **14**, 93.

Rendel, J. 1956: Heritability of multiple birth in sheep. *J. Anim. Sci.* **15**, 193.

Rendel, J. 1958: Studies of cattle blood groups II. *Acta Agric. Scand.* **8**, 131.

Rendel, J. 1961: Relationship between blood groups and the fat percentage of the milk in cattle. *Nature* **189**, 408.

Rendel, J. 1967: *Canalisation and Gene Control.* Logos Press, London.

Rendel, J., and Robertson, A. 1950: Estimation of genetic gain in milk yield by selection in a closed herd of dairy cattle. *J. Genet.* **50**, 1.

Revelle, T. J., and Robison, O. W. 1973: An explanation for the low heritability of litter size in swine. *J. Anim. Sci.* **37**, 668.

Richardson, R. H., and Kojima, K. I. 1965: The kinds of genetic variability in relation to selection responses in *Drosophila* fecundity. *Genetics* **52**, 583.

Rittler, A., Moser, H., Fewson, D., and Werkmeister, F., 1968: Milchleistungsprüfung beim Rind als Grundlage fuer die Zuchtwertschaetzung, I. Z. Tierz. Zuechtungsbiol. **84**, 161.

Roberts, R. C. 1966: The limits to artificial selection for body weight in the mouse. I. The limits attained in earlier experiments. Genet. Res. **8**, 347. II. The genetic nature of the limits. Genet. Res. **8**, 361.

Roberts, R. C., and Mendell, N. R. 1975: A case of polydactily with multiple threshold in the mouse. Proc. R. Soc. Lond. B **191**, 427.

Robertson, A. 1952: The effect of inbreeding on the variation due to recessive genes. Genetics **37**, 189.

Robertson, A. 1956: The effect of selection against extreme deviants based on deviation or on homozygosis. J. Genet. **54**, 236.

Robertson, A. 1957: Optimum group size in progeny testing and family selection. Biometrics **13**, 442.

Robertson, A. 1959a: The sampling variance of the genetic correlation coefficient. Biometrics **15**, 469.

Robertson, A. 1959b: A simple method of pedigree evaluation in dairy cattle. Anim. Prod. **1**, 167.

Robertson, A. 1960: A theory of limits in artificial selection. Proc. R. Soc. B **153**, 234.

Robertson, A. 1961: Inbreeding in artificial selection programs. Genet Res. **2**, 189.

Robertson, A. 1964: The effect of non-random mating within inbred lines on the rate of inbreeding. Genet. Res. **5**, 164.

Robertson, A. 1965: The interpretation of genotypic ratios in domestic animal populations. Anim. Prod. **7**, 319.

Robertson, A. 1973: Gene frequency distributions as a test of selective neutrality. Genetics **81**, 775.

Robertson, A. 1977: The effect of selection on the estimation of genetic parameters. Z. Tierz. Zuechtungsbiol. **94**, 131.

Robertson, A., and Asker, W. A. 1951: The genetic history and breed structure of British Friesian cattle. Emp. J. Exp. Agric. **19**, 113.

Robertson, A., and Lerner, I. M. 1949: The heritability of all-or-none traits: Viability of poultry. Genetics **34**, 395.

Robertson, A., and Rendel, J. 1950: The use of progeny testing with artificial insemination in dairy cattle. J. Genet. **50**, 21.

Robertson, F. W. 1957: Studies in quantitative inheritance. XI. Genetic and environmental correlations between body size and egg production in Drosophila melanogaster. J. Genet. **55**, 428.

Rollins, W. C., and Howell, C. E. 1951: Genetic sources of variation in the gestation length of the horse. J. Anim. Sci. **10**, 797.

Rønningen, K. 1969: Studies on selection in animal breeding. I. The efficiency of two-stage selection compared with single stage selection with respect to progeny testing in animal breeding. Acta Agric. Scand. **19**, 149.

Rønningen, K. 1971: Tables for estimating the loss of efficiency when selecting according to an index based on false economic ratio between two traits. Acta Agric. Scand. **21**, 33.

Rønningen, K. 1976: The estimation of genetic parameters for all-or-none traits. Z. Tierz. Zuechtungsbiol. **93**, 226.

Rutzmoser, K. 1977: Genetische Korrelationen zwischen Milch- und Fleischleistung beim bayerischen Fleckvieh. Bayer. Landwirtsch. Jahrb. **54**, 836.

Rutzmoser, K., and Pirchner, F. 1979: Zur Schätzung genetischer Parameter der Fleischleistung stationsgeprüfter Mastbullen. II. Z. Tierz. Zuchtungsbiol. **96**, 151.

Sandler, L., and Novitzki, E. 1957: Meiotic drive as an evolutionary force. *Am. Nat.* **91**.

Sarkissian, J. K., and McDaniel, R. G. 1967: Mitochandrial polymorphism in maize. I. Putative evidence for *de novo* origin of hybrid-specific mitochondria. *Proc. Natl. Acad. Sci. USA* **57**, 1262.

Schebler, A. 1976: Zur Frage des Zuchtzieles in der Rinderhaltung aus der Sicht des Molkereiwirtschaft. *Zuechtungskunde* **48**, 79.

Scherfler, P. 1972: Blutgruppen und biochemischer Polymorphismus beim Waldviertler Blondvieh. Thesis, Veterinary School, Vienna.

Schierman, L. W., and Nordskog, A. W. 1961: Relationship of blood type to histocompatibility in chickens. *Science* **134**, 1008.

Schilling, E., Thormaehlen, D. 1976: Dichtegradientenzentrifugation von Kaninchenspermien und das Geschlechtsverhaeltnis. *Zuchthygiene* **11**, 113.

Schittmayer, F. 1971: Blutgruppen und biochemischer Polymorphismus beim Kärntner Blondvieh. Thesis, Veterinary School, Vienna.

Schlager, G., and Dickie, M. M. 1966: Spontaneous mutation rates at five coat color loci in mice. *Science* **151**, 205.

Schneeberger, M., Freeman, A. E., and Berger, P. J. 1980: Costs and risks for sire selection strategies in artificial insemination. *J. Dairy Sci.* **63**, 491.

Schneeberger, M., Freeman, A. E., and Boehlje, M. D. 1981: Estimation of a utility function from semen purchases from Holstein sires. *J. Dairy Sci.* **64**, 1713.

Schnell, F. W. 1975: Type and variety and average performance in hybrid maize. *Z. Pflanzenzuecht.* **74**, 177.

Schönmuth, G., Wilke, A., Bergner, H., Wirthgen, B., Muenchenmeyer, R. 1966: Beziehungen zwischen verschiedenen Genotypen beim Rind und einigen biochemischen und physiologischen Kennwerten. I. *Arch. Tierzucht* **9**, 141.

Schulte-Coerne, H. 1976: Untersuchungen über die Auswirkungen einer Einkreuzung von amerikanischen Brown-Swiss Tieren in das württembergische Braunvieh. Thesis, Universität Hohenheim.

Sciuchetti, A. 1935: Ein Beitrag zur genetischen Analyse der Schweizerischen Braunviehrasse. *Arch. Julius Klaus-Stift.* **10**, 85.

Searle, S. R. 1963: The efficiency of ancestor records in animal selection. *Heredity* **18**, 351.

Searle, S. R. 1965: The value of indirect selection: I. Mass selection. *Biometrics* **21**, 682.

Seebeck, R. M. 1973: Sources of variation in the fertility of a herd of zebu, British, and zebu × British cattle in Northern Australia. *J. Agric. Sci.* **81**, 253.

Selander, R. K. 1976: Genic variation in natural populations. *In* F. J. Ayala, ed., *Molecular Evolution*. Sinauer, Sunderland, Massachusetts.

Sellier, P. 1970: Hétérosis et croisement chez la porc. *Ann. Genet. Sel. Anim.* **2**, 145.

Sen, B. K., and Robertson, A. 1964: An experimental examination of methods for the simultaneous selection of two characters, using *Drosophila melanogaster*. *Genetics* **50**, 199.

Sheridan, A. K. 1980: A new explanation of egg production heterosis in crosses between White Leghorns and Australorps. *Br. Poult. Sci.* **21**, 85.

Sheridan, A. K., and Barker, J. S. F. 1974: Two-trait selection and the genetic correlation. II. *Aust. J. Biol. Sci.* **27**, 75.

Shinjo, A., Mizuma, Y., Nishida, S. 1973: The effect of X-ray irradiation on the fitness in inbred lines of Japanese quails. *Jpn. Poult. Sci.* **10**, 226.

Shrode, R. R., Stone, B. H., Rupel, I. W., Leighton, R. E. 1960: Changes in repeatability with changes in herd environment. *J. Dairy Sci.* **43**, 1343.

Simon, D. L., and Flock, D. K. 1971: Zuchtwertschätzung unter Berücksichtigung genetischer Unterschiede zwischen Vergleichswerten. I. *Zuechtungskunde* **43**, 388.

Skarman, S. 1965: Crossbreeding experiment with swine. *Lantbrukshogsk. Ann.* **31**, 3.

Skjervold, H. 1977: The effect of foetal litter size on milk yield: Cross fostering experiments with mice. Z. Tierz. Zuechtungsbiol. 94, 66.

Skjervold, H., and Fimland, E. 1975: Evidence for a possible influence of the fetus on the milk yield of the dam. Z. Tierz. Zuechtungsbiol. 92, 245.

Skjervold, H., and Langholz, J. J. 1964: Factors affecting the optimum structure of A.I. breeding in dairy cattle. Z. Tierz. Zuechtungsbiol. 80, 25.

Smith, C. 1962: Estimation of genetic change in farm livestock using field records. Anim. Prod. 5, 239.

Smith, C. 1964: The use of specialized sire and dam lines in selection for meat production. Anim. Prod. 6, 337.

Smith, C. 1967: Improvement of metric traits through specific genetic loci. Anim. Prod. 9, 349.

Smith, C. 1969: Optimum selection procedures in animal breeding. Anim. Prod. 11, 433.

Smith, C. 1974: Concordance in twins: Methods and interpretation. Am. J. Hum Genet. 26, 454.

Smith, C. 1977: Use of stored frozen semen and embryos to measure genetic trends in farm livestock. Z. Tierz. Zuechtungsbiol. 94, 119.

Smith, C., and Bampton, P. R. 1977: Inheritance of reaction to halothane anaesthesia in pigs. Genet. Res. 29, 287.

Smith, H. F. 1936: A discriminant function for plant selection. Ann. Eugen. 7, 240.

Smithies, O. 1955: Zone electrophoresis in starch gels: Group variation in the serum proteins of normal human adults. Biochem. J. 61, 629.

Sokal, R. R., and Sneath, P. H. A. 1973: Principles of Numerical Taxonomy. W. H. Freeman, San Francisco.

Soller, M., Bar-Anan, R., and Pasternack, H. 1966: Selection of dairy cattle for growth rate and milk production. Anim. Prod. 8, 109.

Sperlich, T. D. 1973: Populationsgenetik. Gustav Fisher Verlag, Stuttgart, Federal Republic of Germany.

Sprague, G. F., and Tatum, L. A. 1942: General vs. specific combining ability in single crosses of corn. J. Am. Soc. Agron. 34, 923.

Stegenga, T. 1970: Kongenitale Missbildungen beim Rind—Ursachen und Massnahmen. Zuchthygiene 5, 4.

Stephenson, A. G., Wyatt, A. J., and Nordskog, A. W. 1953: Influence of inbreeding on egg production in the domestic fowl. Poult. Sci. 32, 510.

Stormont, C. 1959: On the application of blood groups in animal breeding. In Proceedings of the Xth International Congress on Genetics, Volume 1, p. 206. University of Toronto Press, Toronto, Canada.

Stuber, C. W., and Moll, R. H. 1969: Epistasis in maize. I. F_1 Hybrids and their S_1 progeny. Crop Sci. 9, 124.

Sutherland, T. M. 1965: The correlation between feed efficiency and rate of gain, a ratio and its denominator. Biometrics 21, 739.

Suttner, K. 1980: Einfluss der Älpung von Jungrindern auf deren spätere Leistung und auf Blutparameter. Thesis, Technische Universität München.

Sved, J. A., Reed, T. A., and Bodmer, W. F. 1967: The number of balanced polymorphisms which can be maintained in a natural population. Genetics 55, 469.

Syrstad, O. 1966: Studies on dairy herd records. III. Estimation of genetic change. Acta Agric. Scand. 16, 3.

Tallis, G. M. 1959: Sampling errors of genetic correlation coefficients calculated from analysis of variance and covariance. Aust. J. Stat. 1, 35.

Tallis, G. M. 1962: A selection index for optimum genotype. Biometrics 18, 120.

Tantawy, A. D., and Reeve, E. C. R. 1956: Studies in quantitative inheritance. IX. The effects of inbreeding at different rates in *Drosophila melanogaster. Z. Indukt. Abstam-mungs-Vererbungsl.* **87**, 648.

Taylor, S. C. S. 1965: Genetic correlation during growth of twin cattle. *Anim. Prod.* **7**, 83.

Taylor, S. C. S., and Craig, J. 1967: Variation during growth of twin cattle. *Anim. Prod.* **9**, 35.

Tebb, G. 1957: Inbreeding effects contrary to the direction of selection in a poultry flock. *Poult. Sci.* **36**, 402.

Teehan, T. J. 1980: Estimation and estimates of individual heterosis, maternal heterosis and recombination loss from a crossbreeding experiment with sheep. *In Proceedings of the 30th Annual Meeting of the EAAP, Harrogate, 1979. Ann. Genet. Sel. Anim.* **12**, 116.

Thomson, G. M., and Freeman, A. E. 1970: Environmental correlations in pedigree estimates of breeding value. *J. Dairy Sci.* **53**, 1259.

Timoféeff-Ressovsky, H. 1937: Studies on the phenotypic manifestation of hereditary factors, I. *Genetics* **12**, 128.

Touchberry, R. W. 1951: Genetic correlations between five body measurements, weight, type and production in the same individual among Holstein cows. *J. Dairy Sci.* **34**, 242.

Touchberry, R. W. 1963: Heritability of milk and milk fat yield and fat percent at different levels of milk yield. *J. Dairy Sci.* **46**, 620.

Touchberry, R. W., Lush, J. L. 1950: The accuracy of linear body measurements of dairy cattle. *J. Dairy Sci.* **33**, 72.

Touchberry, R. W., Rottensten, K. and Andersen, H. 1960: A comparison of dairy sire progeny tests made at special Danish progeny test stations with tests made in farmer herds. *J. Dairy Sci.* **42**, 529.

Turner, H. N. 1959: Ratios as criteria for selection in animal or plant breeding, with particular reference to efficiency of feed conversion in sheep. *Aust. J. Agric. Res.* **10**, 565.

Utz, H. F. 1969: Mehrstufenselektion in der Pflanzenzüchtung. Thesis, Universität Hohenheim.

Vandepitte, W. M., and Hazel, L. N. 1977: The effect of errors in the economic weights on the accuracy of selection indices. *Ann. Genet. Sel. Anim.* **9**, 87.

Vangen, O. 1980: Realized genetic parameters in a two-traits selection experiment in pigs. *In* A. Robertson, ed., *Selection Experiments in Laboratory and Domestic Animals.* Commonwealth Agriculture Bureaux, Slough, U.K.

Van Vleck, L. D. 1964: Sampling the young sire in artificial insemination. *J. Dairy Sci.* **97**, 441.

Van Vleck, L. D. 1970: Misidentification and sire evaluation. *J. Dairy Sci.* **53**, 1697.

Van Vleck, L. D. 1972: Estimation of heritability of threshold characters. *J. Dairy Sci.* **55**, 218.

Van Vleck, L. D. 1978: Economic weights for direct and fetal genetic effects in choosing sires. *J. Dairy Sci.* **61**, 970.

Van Vleck, L. D., and Bradford, G. E. 1964: Heritability of milk yield at different environmental levels. *Anim. Prod.* **6**, 285.

Van Vleck, L. D., and Bradford, G. E. 1965: Comparisons of heritability estimates from daughter–dam regression and paternal half-sib correlation. *J. Dairy Sci.* **48**, 1372.

Van Vleck, L. D., and Henderson, C. R. 1961: Empirical sampling estimates of genetic correlations. *Biometrics* **17**, 359.

Van Vleck, L. D., and Henderson, C. R. 1962: Measurement of genetic trend. *J. Dairy Sci.* **44**, 1705.

Van Vleck, L. D., and Henderson, C. R. 1963: Bias in sire evaluation due to selection. *J. Dairy Sci.* **45**, 976.

Von Krosigk, G. M. 1959: Genetic influences on the composition of cow's milk. Thesis, Iowa State University, Ames, Iowa.

Von Krosigk, G. M., and Pirchner, F. 1974: Genotype by type of housing interaction in populations of White Leghorn under reciprocal recurrent selection. *Brit. Poult. Sci.* **15**, 11.

Von Krosigk, G. M., McClary, C. F., Vielitz, E., and Zander, D. V. 1972: Selection for resistance to Marek's disease and its expected effect on other important traits in White Leghorn strain crosses. *Avian Dis.* **16**, 11.

Wahlund, S. 1928: Zusammensetzung von Populationen und Korrelationserscheinungen vom Standpunkt der Vererbungslehre aus betrachtet. *Hereditas* **11**, 65.

Wallace, B. 1958: The role of heterozygosity in *Drosophila* populations. *In Proceedings of the Xth International Congress of Genetics*, Volume 1, p. 408, University of Toronto Press, Toronto, Canada.

Walter, E. 1960: Optimale Selektionsverfahren. *In* E. Schilling, ed., *Rationalisierung der viehwirtschaftlichen Erzeugung.* Max Planck Institut Mariensee, Federal Republic of Germany.

Waters, N. F., and Burmester, B. R. 1961: Mode of inheritance of resistance to Rous Sarcoma virus in chickens. *J. Natl. Cancer Inst.* **27**, 655.

Watson, J. H. 1960: Milk yield and butterfat percentage of twin and single-born cattle under experimental and field conditions. *Anim. Prod.* **2**, 67.

Weber, F. 1957: Die statistischen und genetischen Grundlagen von Körpermessungen am Rind. *Z. Tierz. Zuechtungsbiol.* **69**, 225.

Weber, F., and Lörtscher, H. 1958: Untersuchung über die Vererbung der Punktiererergebnisse beim Rind als Grundlage für eine neue, vereinfachte Punktierung. *Landwirtsch. Jahrb. Schweiz* **72**, 121.

Weinberg, W. 1908: Über den Nachweis der Vererbung beim Menschen. Jahreshefte Verien Vaterl. *Naturkunde Wuerttemberg* **64**, 368.

Weinberg, W. 1909: Über Vererbungsgesetze beim Menschen. *Z. Indukt. Abstammungs-Vererbungsl.* **1**, 277, 377, 440; **2**, 276.

Weir, B. S., Cockerham, C. C., and Reynolds, J. 1980: The effect of linkage and linkage disequilibrium on the covariances of noninbred relatives. *Heredity* **45**, 351.

Wiener, A. S., Lederer, M., and Polaves, S. H. 1930: Studies in isohem-agglutination. IV. On the chances of proving non-paternity. *J. Immunol.* **19**, 259.

Wiener, G. 1961: Population dynamics in fourteen lowland breeds of sheep. *J. Agric. Sci.* **51**, 21.

Willham, O. S. 1937: A genetic history of Hereford cattle in the United States. *J. Heredity* **28**, 283.

Willham, R. L. 1964: The covariance between relatives for characters composed of components contributed by related individuals. *Biometrics* **19**, 18.

Winters, L. M., Kiser, O. M., Jordan, P. S., Peters, W. H. 1935: A six year study of crossbreeding swine. *Minnesota Agric. Exp. Sta. Bull.* **320**.

Witt, W., and Döring, H. 1955: Typuntersuchungen am schwarzbunten Niederungsrind. *Z. Tierz. Zuechtungsbiol.* **64**, 287.

Wriedt, C., and Mohr, O. L. 1928: *Amputated,* a recessive lethal in cattle; with a discussion on the bearing of lethal factors on the principles of livestock breeding. *J. Genet.* **20**, 187.

Wright, S. 1921: Systems of mating. I. Biometric relationship between parent and offspring. *Genetics* **6**, 111.

Wright, S., and Mc Phee, H. C. 1925: An approximate method of calculating inbreeding and relationship from livestock pedigrees. *J. Agric. Res.* **31**, 377.

Yamada, Y. 1977: Evaluation of the culling variate used by breeders in actual selection. *Genetics* **86**, 1977.

Yamada, Y., Bohren, B. B., and Crittenden, L. B. 1958: Genetic analysis of a White Leghorn closed flock apparently plateaued for egg production. *Poult. Sci.* **37**, 665.

Yoder, D. M., and Lush, J. L. 1937: A genetic history of the Brown Swiss cattle in the United States. *J. Hered.* **28**, 154.

Yoo, B. H. 1980: Long-term selection for a quantitative character in large replicate populations of *Drosophila melanogaster*. *Genet. Res.* **35**, 1.

Young, L. D., Johnson, R. Y., Omtvedt, I. T., and Waters, L. E. 1976: Postweaning performance and carcass merit of purebred and two-breed cross pigs. *J. Anim. Sci.* **42**, 1124.

Young, S. S. Y. 1961: A further examination of the relative efficiency of three methods of selection for genetic gains under less restricted conditions. *Genet. Res.* **2**, 106.

Young, S. S. Y., and Weiler, H. 1961: Selection for two correlated traits by independent culling levels. *J. Genet.* **57**, 329.

Zurkowski, M., and Bouw, J. 1966: Changes in frequencies of blood group genes in breeding cattle. *Genet. Pol.* **7**, 197.

Zwiauer, D. 1975: Ausmass elektrophoretisch darstellbarer Variabilität im Rinderblut. *In Proceedings of the Annual Meeting Deut. Ges. Züchtungskunde, Bonn.*

Zwiaver, D., Menken, M., Stranzinger, G., and Dempfle, L. 1980: Auswirkung einer Zentromerfusion auf Form- und Leistungsmerkmale beim Rind. *Zuchthygiene* **15**, 97.

Index